Life of the Past

FOURTH EDITION

William I. Ausich
Ohio State University

N. Gary Lane
Indiana University

PRENTICE HALL, Upper Saddle River, New Jersey 07458

Library of Congress Cataloging-in-Publication Data

Ausich, William I. (Willaim Irl),
    Life of the past. — 4th ed. / William I. Ausich, N. Gary Lane.
       p.   cm. — (Prentice hall earth science series)
    Rev. ed. of: Life of the past / N. Gary Lane. 3rd ed. c1992
    Includes bibliographical references and index.
    ISBN 0-13-896069-0 (pbk.)
    1. Paleontolgy   I. Lane, N. Gary.   II. Lane, N. Gary. Life of
the past.   III. Title.   IV. Series.
QE711.2.L35 1999
560 — dc21   98-18770
                                              CIP

Executive Editor: Bob McConnin
Production Editor: Kim Dellas
Page Layout: Kim Dellas, Jim Sullivan
Manufacturing Manager: Trudy Pisciotti
Cover Designer: Karen Salzbach

© 1999, 1992, 1986, 1978 by Prentice-Hall, Inc.
Simon & Schuster/A Viacom Company
Upper Saddle River, NJ 07458

Printed in the United States of America

10 9 8 7 6 5 4 3 2 1

ISBN 0-13-896069-0

Prentice-Hall International (UK) Limited, *London*
Prentice-Hall of Australia Pty. Limited, *Sydney*
Prentice-Hall Canada Inc., *Toronto*
Prentice-Hall Hispanoamericana, S.A., *Mexico City*
Prentice-Hall of India Private Limited, *New Delhi*
Prentice-Hall of Japan, Inc., *Tokyo*
Simon & Schuster Asia Pte. Ltd., *Singapore*
Editora Prentice-Hall do Brasil, Ltda., *Rio de Janeiro*

Contents

Preface

Life of the Past is an introduction to the history of life on Earth–the basic principles and processes, the ecologic and paleoecologic organization, the rich history of past life forms, and the major events that shaped this history. Life of the Past is a textbook for a college or university course on the history of life on Earth and is appropriate either for general students or for beginning geology and biology students. It can also be a supplement for an historical geology course

The fourth edition places a greater emphasis on the paleoecologic progression of floras and faunas through time than previous editions. Also, more emphasis is given to identifying critical events that shaped life on Earth and to discussion of the processes responsible for this history. A new series of diagrams outlining changes in floras and faunas has been added as a guide to students so that the overall patterns of change are not lost among the details. Chapters in the fourth edition are slightly rearranged to have a clearer separation between the principles of Earth history and paleontology and the specific details of oceanic life and terrestrial life.

The fourth edition incorporates recent paleontologic research, both factual information and alternative interpretations of data. Also, a number of worldwide websites are listed at the end of Chapter 1. From the websites given, additional sites can also be located. The web is an ever-increasing source of data and information for all fields, including paleontology, and it is an important supplement to textbooks and monographs on paleontology.

Think about the world in which we live. We are surrounded by green–green plants that range from tiny woodland flowers to giant trees. They clothe the landscape. The vast majority of green plants, more than 250,000 species, reproduce by flowers and bear seeds enclosed in fruits. The largest animals today, both on land and in the oceans, are mammals–animals that have hair, are warm-blooded, suckle their young, and give live birth outside a shelled egg. In oceans, the most abundant forms of life are thousands of species of clams and snails and hordes of bony fish.

However, Earth has not always been like this. The life of our modern world has only been in existence with this general composition for approximately the last 65 million years, which is slightly more than 1 percent of Earth history. Life on Earth has changed continuously through time; no species has lived more than a few million years. This book is about the continuous succession of life that has inhabited Earth. *Life of the Past* is concerned with the history of life from the earliest records (which are more than 3 billion years old) to the present. It is concerned with the various communities of organisms that inhabited both the oceans and land. The evolutionary history of plants and animals with or without backbones, as well as smaller, mostly single-celled life forms, are included in this text.

If we consider time periods old enough, Earth had no green plants on dry land. Later, when green plants first appeared, they did not have seeds, flowers, or fruits but, instead, reproduced by spores. We will consider times when oceanic life was dominated by groups that are rare or extinct today. Clams and snails were few and far between. Fish lacked jaws or teeth and were harmless feeders. Other animals, relatives of squid and octopi, were the dominant predatory animals in the oceans. The vast majority of all species that have existed on Earth are now extinct—only a few million species are alive today.

Instead of focusing on a specific group or age of fossils, *Life of the Past* emphasizes the complexity and diversity of life through time and the evolutionary and ecological processes that brought about changes.

The text is arranged in three parts. First, general principles of Earth history and paleontology that are prerequisite for the remainder of the book are discussed. These principles include geologic time; the relationship of fossils to the rocks in which they occur; the origin of Earth and life; and principles of evolution, continental drift, and paleobiogeography.

The second segment consists of the fossil record of life in the marine environment. This section considers the earliest Precambrian fossil record, life during the Cambrian, the history of suspension feeders and deposit feeders, major groups of marine plankton and predators. The roles of the major groups of marine organisms are discussed for marine communities, as are the major evolutionary and extinction events in the marine environment.

The final part of the book concerns the history of life on land–the terrestrial environment. The origin of life on land and the history of both plants and animals are discussed. Because most students are more interested in plants and vertebrates, these chapters contain a more detailed account of specific groups than given for ancient marine life.

Like any other special discipline, whether in the sciences or the humanities, paleontology has its own vocabulary of technical terms. These words commonly capsulize important concepts, and their precise use helps to sharpen the understanding of major ideas. Trying to remember many unfamiliar words taxes the memory and may be a barrier to full appreciation of new knowledge. This book attempts to alleviate this problem by minimizing the technical terms. Short lists of important key words are given at the end of every chapter. Each term is defined in the glossary at the end of the book.

A few pertinent references are included at the end of each chapter. These references are annotated as to their level of difficulty or content. They have been updated with more recent works, although older books are included because many of these are still the best sources of information.

The purpose of this book is that it be used in the classroom, and the classroom and students are the ultimate test of its effectiveness. Please forward to the authors any suggestions or ideas for improvement and enhancement of this textbook.

Acknowledgments

The following individuals read all or part of the manuscript for the first edition: Robert L. Anstey, John C. Kraft, Ronald L. Parsley, and Walter C. Sweet.

The following individuals made numerous suggestions for improvement of the second edition: Lawrence H. Balthaser, Stanley S. Beus, John H. Ostrom, Norman M. Savage, and Robert E. Sloan. John H. Ostrom, Richard H. Miller, Norman M. Savage, and Robert E. Sloan also read a preliminary draft of the chapter on primate and human evolution that was new in the second edition.

For the third edition, the following reviewers provided valuable comments and suggestions: Lawrence H. Balthaser, Allen J. Kihm, Norman Savage, Daniel B. Blake, William Miller III, Robert J. Foster, Ronald E. Martin, and Ronald L. Parsley.

We acknowledge our former paleontology teachers, who include Robert W. Baxter, Harold K. Brooks, Raymond C. Moore, Frank Peabody, Peter P. Vaughn, and Robert W. Wilson for NGL; and Daniel B. Blake, J. Robert Dodd, Alan S. Horowitz, N. Gary Lane, Philip A. Sandberg, and Robert H. Shaver for WIA. They taught, encouraged, and enlightened us. For several years, NGL team-taught an introductory course in paleontology at U.C.L.A. with Clarence A. Hall, Jr., and J. William Schopf; to these individuals a debt of gratitude is owed for this association.

Those who aided in acquiring photographs for the first three editions are John A. Barron, Joyce R. Blueford, Derek Briggs, David Dilcher, J. Wyatt Durham, V. E. Garatt, Thomas M. Gibson, Donald E. Hattin, Francis M. Hueber, I. M. Kerhner, Rebecca Lindsay, D.B. Macurda, Jr., W.G. Melton, Jr., Carl Rexroad, Eugene Richardson, George Ringer, Raymond T. Rye II, J. William Schopf, Takeo Susuki, Mary Wade, E. Reed Wicander, Edward C. Wilson, and Robert J. Zakrzewski. For the fourth edition, Roy E. Plotnick provided additional photographs.

Those who greatly helped in the preparation of the fourth edition by reading preliminary chapter drafts of the manuscript or helped with other aspects of the manuscript include Rosemary A. Askin, The Ohio State University; Loren E. Babcock, The Ohio State University; Stig M. Bergström, The Ohio State University; David J. Bottjer, University of Southern California; David L. Dilcher, University of Florida; Kenneth A. Foland, The Ohio State University; Robert A. Gastaldo, Auburn University; Stephen R. Jacobson, The Ohio State University; Thomas W. Kammer, West Virginia University; Jeffrey McKee, The Ohio State University; Mark A.S. McMenamin, Mount Holyoke College; and Walter C. Sweet, The Ohio State University. Regina Ausich was the first to read the fourth edition of this book, and those who read it subsequently should thank her, as do we.

CHAPTER 1

Time and Fossils

INTRODUCTION

This book is about **fossils**, which are the preserved remains of former life on Earth. **Paleontology** is the scientific discipline that encompasses the study of fossils. Fossils include **body fossils**, which are any direct evidence of prehistoric life, such as shells and bones, and **trace fossils**, any indirect evidence of prehistoric life, such as footprints and burrows. Fossils are preserved in rocks of different ages, some quite young and others very ancient, and these enclosing rocks provide a time framework within which all fossils occur. We can think about fossils as occurring within a four-dimensional space-time framework—three-dimensional space of rocks on and under the surface of Earth and one-dimensional time. Thus, fossils occupy a four-dimensional time-space continuum that is a record of the history of life on Earth. As demonstrated throughout this book, this is a dynamic history both temporally and geographically. Today scientists are concerned with **global change** issues, such as increased amounts of greenhouse gases, deforestation, and rising sea levels due to glacial melting. In particular, many are concerned with understanding the global consequences of these changes on the future—how will climate, sea level, and life be affected in the future? The past is our best predictor of the future, and the fossil record is a storehouse of information on past climatic, sea level, and biological change. Many "global change experiments" have been "run" throughout Earth history, and the paleontological, physical, and chemical evidence of these events are encased in the geologic record. By thoroughly studying the many examples in our past and by determining how and why past biotic changes occurred, we can understand how organisms living on Earth today will respond to current man-induced changes to Earth's system. In this book, we will outline the grand patterns of life throughout its history on Earth and attempt to explain the responsible processes.

In order to understand fossils, one must also understand time, one subject of this chapter. What is time? This is a more difficult question to answer than you might assume, and one that philosophers have considered for centuries. A dictionary definition is as follows:

> That character and relation of all events and things with respect to which they are distinguished as simultaneous or successive, and as becoming, enduring, and passing away; usually conceived as a dimension of reality, distinguished from the spatial by the fact that the order of temporal succession is irreversible. [1]

---

[1] *Webster's New International Dictionary*, 3rd ed., s.v. "time."

Time is real, it is filled or can be recognized by a series of events, and it is irreversible. You and I can return to a specific point in space on Earth, but we can never return to a past moment of time. Time is envisaged as a more elusive element of our existence. This is even more difficult to comprehend when considering the vastness of geological time, metaphorically referred to as "deep time."[2]

Life on this planet has been a series of multitudinous events that occurred in this time dimension. Fossils are the record, preserved in rocks, of this biological history.

THE MEANING OF FOSSILS

The word *fossil* comes from the Latin *fossilium*, which means "dug from beneath the surface of the ground." For example, a mole is spoken of as having a fossorial way of life. As originally used by medieval writers, a fossil was any stone, ore, mineral, or gem from an underground source. Some of the earliest books on mineralogy are called books of fossils. During the eighteenth century, this broad meaning was gradually restricted to objects in rocks that are parts of once living organisms—bones, shells, leaves, wood, and so on.

For many years there was heated debate about the significance of fossils. Some believed that all fossils resulted from the single Noachian flood of Genesis. Others thought that fossils grew in place in the rock or had been placed there by Satan to betray humans. Fossils were found that clearly were parts of plants or animals that no longer lived on Earth. This was problematic. The question of whether some species had become extinct raised a debate concerning the perfection of organic creation. Gradually, after the Dark Ages, fossils came to be accepted as records of ancient life.

Fossils demonstrate two truths about this planet on which we live. First, many species once existed and later became extinct. Second, there has been a succession of plants and animals that evolved through time; the communities of life that have existed on Earth have changed through time, sometimes gradually and sometimes rapidly, both on land and in oceans.

THE BEGINNINGS OF PALEONTOLOGY

One of the most intriguing questions that we can ask about fossils is how early scholars came to realize that fossils are the remains of once living plants and animals that became entombed in rocks. Certainly, this idea about the true nature of fossils did not come easily or quickly into Western thought. This section deals primarily with developments in western Europe, mainly because we have the best-documented record of the progress of thought in this area. We will also briefly consider early ideas about fossils in China (Figure 1.1).

The earliest apparent mention of fossils is in writings by early Greeks, perhaps as early as the sixth century B.C. A very early discussion of fossil fishes and marine shells from the mountains of Greece is attributed to Xenophanes, but his writings on the subject have not survived and credit to him comes from a manuscript prepared 900 years later by Hippolytus.

[2]McPhee, J. *Basin and Range.*

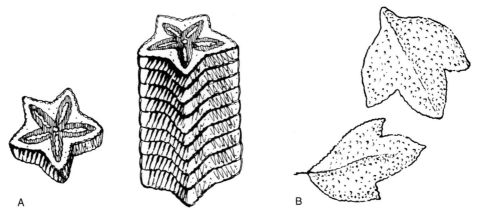

Figure 1.1   Two of the earliest illustrations of fossils. (A) Woodcut of Mesozoic pentagonal crinoid stems from a book by Conrad Gesner written in 1565. Although many early works mention fossils, this is the first known book in western Europe to provide illustrations. (B) A sketch that is interpreted as two fossil brachiopods, called swallow stones by the Chinese. The specimens may represent a Paleozoic spiriferid brachiopod. The illustration is taken from the Chinese book *Pên Tshao Kang Mu*, by Li Shih-Chen, printed in 1596, 31 years after Gesner's work.

Another early writer who discussed fossils was Theophrastus, also a Greek, who lived about 300 B.C. He set the study of fossils on an erroneous path that took many centuries to correct. Theophrastus was a student of Aristotle and believed in basic elemental virtues such as fire, water, and air. He attributed bones found in rocks on the island of Lesbos to a plastic virtue, a characteristic of the rocks that caused the bones to grow within them. Thus, he negated the idea that the bones may have belonged to creatures that were once alive quite independent of the rocks within which their remains were found.

The Aristotelian point of view held unarguable force among learned men of Europe for many centuries. The naturalistic explanations of the early Greeks gave way to mythical or magical explanations that had little relationship to reality. Scholars repeated the words of earlier writers without original thought or criticism. No one took the trouble to examine rocks and fossils firsthand. This situation persisted from about A.D. 200 to about A.D. 1400, and this period has been called the Great Interruption in Western thought. Medieval ideas were so fanciful and basically otherworldly that the very word *medieval* has come to stand for any ideas that are old-fashioned and out of date.

As Western thought emerged from these dark ages, writings concerning fossils were very uneven in content and understanding. Whereas one writer seems to be on the correct track, he is followed a hundred years later by another writer who is still parroting Theophrastus, although in somewhat modified form. Here is an example.

One of the best-written descriptions of fossils during this early time was by Georg Bauer, a German who wrote under the pseudonym of Agricola during the early 1500s. He is known primarily for his works on mining, minerals, and ores and is sometimes called the "Father of Mineralogy." Agricola also wrote quite detailed descriptions of a number of fossils, especially in one book, *De Natura Fossilium* (1546). One should not jump to the conclusion that this book really concerns fossils, because Agricola used *fossilium* in the original Latin sense of anything dug up from the earth. Fossils to him included stones, ores, crystals, gems, and any other objects obtained from the ground. He thought that some objects, such as fossil sharks' teeth, had grown within the rocks but that other fossils had been alive and had been turned into rock, or petrified, by a stone-producing juice within the rocks. However, a hundred years later the Welshman Edward Lhwyd wrote that marine fossils resulted from the eggs of sea creatures being washed onto beaches, dried, then blown by the wind until they fell into crevices in the rock where they hatched and grew into stonified copies of their living counterparts. Thus, it took several hundred years for universal agreement to be reached that fossils were indeed the remains of organisms that were once alive.

Once natural scientists accepted that fossils were the remains of organisms, fossils were traditionally explained as the remnants of plants and animals killed and buried by the great biblical flood. This concept was widely accepted for many years, and it resulted in two conclusions: first, all fossils were the same age, and second, all fossils were relatively recent in age. Biblical scholars, typified by Bishop Ussher (1664) of Ireland, counted the number of generations represented in the genealogical chapters of the Old Testament and assigned a reasonable number of years to each generation. These scholars carried this back to Adam and Eve and assumed that Earth and all of its life had been formed just a few days before humans first appeared. This resulted in a calculation that Earth was formed in the year 4004 B.C., and for many years this age was printed in the margin of the King James edition of the Old Testament. Bishop Ussher later refined this calculation and decided that Earth was formed on October 29 in 4004 B.C. at 9:30 in the morning.

By the eighteenth century, geologists and paleontologists, who began to seriously study rock strata and their contained fossils, rather quickly began to realize that such a recent age for Earth was impossible and that Earth must be very old, indeed. A relatively young age for Earth did not seem a sufficient amount of time to account for the tens of thousands of feet of rocks that are stacked upon one another. This young age did not seem to be a sufficient amount of time to account for the deformation of rocks into high mountain ranges that were later destroyed by erosion. Surely, enormous amounts of time were necessary to wear down such lofty peaks. Furthermore, fossiliferous rocks may be several thousands of feet thick in any one locality, and it is unlikely that a single flood event could have produced such a thick sequence of sediments. As a better stratigraphic framework was developed, it was found that the entire assemblage of fossils in the lower (and presumably older) rock layers was completely different from the fossils in higher (younger) rock layers. This evidence also refutes the idea that all these animals lived at the same time. The proponents of the flood as a major feature of Earth history countered the rising skepticism of scientists with many ingenious arguments, and some aspects of their points of view still linger among those who do not fully appreciate the fundamental principles of geology.

Once fossils became appreciated in a rational, modern way, a greatly accelerated rate of discoveries and advances became possible. By 1800 the French anatomist George Cuvier was able to compare the bones of extinct animals with those of living ones and to reconstruct in detail the appearance of extinct animals. About this same time the English canal engineer William Smith discovered that fossils were distinctive in each different rock layer, so that fossils could be used to identify rocks over wide areas of England. Using this relationship, in part, he also prepared one of the very first geologic maps showing the distribution of rock layers over the surface of southern England.

Smith's discoveries were quickly followed by various British and continental geologists who demonstrated that distinctive fossil assemblages could be recognized over most of western Europe. Thus, the way was paved for unraveling the sequence of fossils on all the continents and for establishing a worldwide time scale for reconstructing Earth history.

Much of the initial work on fossils was undertaken by individuals who had little formal training in geology. Such courses were not part of university curricula during this period. Merchants, doctors, and noblemen all took part in this pioneering work. However, as more and more fossils were gathered, the first natural history museums in Europe were created during the eighteenth century. Here, collections of fossils from many different areas could be assembled and studied. Gradually, universities began to include geology and paleontology in their offerings, as the economic need for geological materials became more and more prevalent.

In the United States much pioneer work on fossils was undertaken by a series of state-organized agencies, normally called state geological surveys, which began during the 1830s. The impetus for creation of these organizations was also mainly economic—to find valuable coal or ore deposits—but much work on fossils was undertaken because of their value in matching rocks of the same age in different areas. Almost the entire North American continent was unknown territory as far as fossils were concerned, and many important discoveries of new life forms were made. Even our third president, Thomas Jefferson, was intrigued with fossils. Jefferson was interested in many aspects of natural history and was sent a collection of large bones of extinct animals, including mastodons, from Big Bone Lick, Kentucky, on the Ohio River between Cincinnati and Louisville. Jefferson thought that these elephantine animals might still be alive within the interior of the continent, and when he sent out the Lewis and Clark expedition in 1804, he cautioned them to keep an eye out for such animals.

Beginning in 1850, federally sponsored expeditions to the West resulted in many important fossil finds. Most spectacular among these were the world-famous dinosaur beds and fossil mammal localities discovered in several western states and Canadian provinces. This aspect of paleontology—exploration and discovery—is still an important and valuable part of the science. Many parts of the world's continents have not been adequately explored for fossils even to this day. Only superficial knowledge is available for fossil assemblages in many parts of Central and South America. Alaska was poorly known until the oil exploration of recent years; and many parts of eastern Asia, especially China, still have not been adequately surveyed. Important new fossil discoveries continue to be made and many of the more important and popular ones are regularly reported in newspapers and news magazines. Beyond the excitement of exploration and discovery, it is essential to continue to expand our knowledge of fossils throughout the world because today scientists are asking global questions about climate, evolution, extinction, sea level, and so on. The resolution of our answers to such global concerns can only be improved by a well-resolved global fossil data base.

GEOLOGIC TIME

Time is perhaps most comprehensible in terms of how we measure this dimension. Years, months, days, minutes, and seconds are all familiar concepts to us. We deal with them so often that we are comfortable talking and thinking about them. But what about millions, billions of years in geologic time? We have a vague uneasiness trying to grasp the enormous span of time expressed by such numbers because such lengths of time are far beyond our own experience. This is "deep time." We can grasp the idea of centuries in terms of life spans, generations, and years; even a few centuries or a few thousand years do not really make us shake our heads—but billions of years? One way is to study the events that have taken place during these millennia. By relating the succession of events to a scale of time measured in a familiar unit, such as years, we can gradually become accustomed to the idea of what we generally call the **geologic time scale**.

Geologic time differs from other types of time only with regard to its immense span or duration. There is no essential difference between this kind of time and any other, except duration and perhaps the way in which we measure or estimate the duration of geologic time.

Geologic time is unique because it fills a special interval of time and has a unique duration. The historian is only concerned with time as far back as written historical records are preserved—a few thousand years at the most. The archeologist or anthropologist is concerned with time only as far back as the records of humans (bones and artifacts) are preserved. For many years these scientists dealt only with the last 1 million years of Earth history, but recently humanlike fossils have been found in rocks as old as 4.2 million years, tripling the duration of the framework within which the study of hominids is conducted. Astronomers, on the other hand, work with time durations that are much longer than those commonly considered in geological time. The origin and evolution of the universe involved lengths of time that are probably 10 to 20 billion years. We do not know exactly how old the universe is, but it is clearly much older than our own planet. Geologists and paleontologists are generally concerned with the interval of time that is intermediate between the very long times that concern astronomers and the few millions of years within which the anthropologist works.

As we shall see, Earth is judged to have come into being approximately 4.5 billion years ago, although no rocks that old have ever been found. It is unlikely that we will ever find a rock on Earth's surface that was present during the initial formation of the planet. There have been too many changes in Earth's crust for such a rock to have a chance of survival. However, rocks have been found that can be accurately and confidently dated as 3.96 billion years old, so we know that Earth must be at least that old. The fossil record, the subject of this book, begins in rocks nearly the age of these oldest rocks. Fossil remains of very primitive organisms have been found in rocks that are 3.5 billion years old. Again, considering the changes that have taken place on Earth, it seems unlikely that evidence of life will be found in rocks that are very much older than this, and we are probably quite lucky to have discovered any information about this very early life. Thus, as far as a time scale for fossils is concerned, we begin approximately 3.5 billion years ago.

How close to the present day do we come with our time scale? We said at the start of this chapter that fossils are the preserved remains of past, or ancient life. How old must a shell or a bone be in order to be judged a fossil? Unfortunately for the student in search of neat, concise answers, we cannot provide one to this question. A humorous, but wrong, answer is that if a dead organism no longer has any odor of decay (i.e., it doesn't smell), it is a fossil. Fossils are defined as the remains of prehistoric life, so fossils must be at least a few thousand years old. The remains of an animal that died and was buried and preserved during historical times is not a fossil but is considered the remains of a modern organism. However, because historical time differs from place to place, a precise age cannot be given to set the upper limit of prehistory and the beginning of the study of fossils.

HOW DO WE TELL GEOLOGIC TIME?

There are really two different answers to this question. We can measure the **absolute ages** of units of rocks in years by analyzing naturally occurring radioactive elements that are present in minute quantities in certain rocks and minerals. This method of radioactive age dating is comparatively new, having been started in the early 1900s and expanded and refined since then. Long before this method was developed, geologists and paleontologists had worked out ways to determine the **relative ages** of different rocks (i.e., Rock A is older or younger than Rock B). Furthermore, they observed several phenomena that could be used to attempt estimates of the age of Earth and of some rocks that occur in Earth's crust. We will trace the historical developments in the quest to tell geologic time, beginning with early attempts to determine the relative ages of rocks and fossils.

The conceptual breakthroughs for understanding relative geologic time were made by **Nicolas Steno**. Steno was a Danish physician in the service of the Grand Duke of Tuscany in northern Italy. He traveled throughout northern Italy and sought to understand his observations on various rock types, strata, and stratigraphic relationships. Although the concepts Steno developed seem quite intuitive to us, in 1669 when Steno proposed these ideas, they were revolutionary concepts that provided a very different perspective for interpreting sedimentary strata. First, Steno proposed that sedimentary rocks were usually deposited in horizontal layers, which he termed the **principle of original horizontality** (Figure 1.2). These horizontal layers form a succession of rock types in any given area. The rock layers are stacked like a deck of cards, one on top of the other. Scientists have realized for several centuries that the rock layer at the bottom of the "deck" is the oldest, first-formed layer, and that the layers above are progressively younger. This is called the **principle of superposition**, which simply states that in an undisturbed sequence of layered rocks, an older layer is always below a younger layer. Steno's third principle, the **principle of lateral continuity**, states that rock units continue laterally for great distances, even globally. Geologists now recognize that this third principle is not always true; all rock units pinch out laterally, or they are wedge shaped. The scale of this wedge-shaped geometry varies from small channels a meter or less in width to some rock units that cover nearly an entire continent.

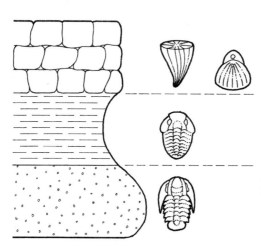

Figure 1.2   A vertical sequence of sedimentary rocks with rock types indicted by conventional symbols. A sandstone (stipple pattern) is the oldest (lowest) rock layer; a shale (horizontal lines) is intermediate in age; and a limestone (block pattern) is the youngest. Fossils from each layer are shown at the right. Superposition indicates that the sandstone and shale with trilobites are of a different, older age than the limestone with horn corals and brachiopods.

These principles were understood and applied in western Europe during the 1600s and 1700s before any studies of rocks were undertaken in North America. Geologists gradually came to realize two things: one, certain distinctive rocks were confined to specific parts of the rock sequence, and two, different layers of rocks commonly contained unique suites of fossils. These early studies had an economic motive, such as coal production. European geologists studying coal-bearing rocks determined that beds of coal could be located above or below a certain sequence of rocks that also contained shales, sandstones, and a few thin limestone beds. This sequence of coal-bearing rocks became known as the **Carboniferous**, after the carbon of which the coals are composed. Other rocks in England and France included thick beds of chalk, and these beds were eventually called the **Cretaceous**, coined from the Latin word for chalk. These early geologists realized that beds of chalk did not occur above or below a certain rock interval and that coal beds were always below (older than) those that contained chalk. The entire sequence of sedimentary rock strata was gradually reconstructed by geologists literally walking on the outcroppings of rock across western Europe, principally Great Britain, France, and Germany. In doing so, they also learned that certain kinds of fossils were confined to specific rock horizons, and this observation was used to piece together the relative ages of scattered rock outcroppings over increasingly larger areas.

Not all the sedimentary rocks so studied were still in their original horizontal positions. In some areas the rocks had been folded, broken, or displaced during mountain building (Figure 1.3). Some layers were distorted until they were completely upside down. Fossils came to be an increasingly useful tool for unraveling these complex disturbances of Earth's crust. In other areas it was determined that thick sequences of rocks were missing; older rocks with distinctive fossils were found immediately below much younger rocks. Elsewhere, these layers could be separated by several hundreds or thousands of feet of fossil-bearing sedimentary beds. As these complex relationships were unraveled, a **relative time scale** was gradually developed. It is relative because it can only differentiate whether any object or event is older or younger than another; no years of duration can be attached. By the 1860s, fossil-bearing rocks in western Europe were divided into three great **eras** of geologic time: the

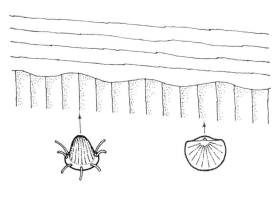

Figure 1.3   The lowest rock layers are standing vertically. They were originally horizontal sandstones that were tilted later and eroded during mountain building. Because they are vertical, it is difficult to tell which layers are younger and which are older. The two kinds of fossil brachiopods shown were found in the layers indicated. We know from study of rocks that are still horizontal elsewhere that the spiny brachiopode on the right is younger than the nonspinose one on the left. Therefore, the top of the section of vertical rock layers is to the right. The younger, slightly tilted rocks above the vertical beds are clearly much younger than the vertical layers, having been deposited on top of them. The surface between the two sets of rock layers records an interval of time when mountain building and erosion took place and no rock layers were laid down. There is a gap in the fossil record at this site, and the surface separating the two sets of rocks is called an unconformity.

Paleozoic, Mesozoic, and **Cenozoic** (Table 1.1). These names mean, respectively, "ancient life," "middle life," and "young or recent life." These three eras are grouped into the Phanerozoic Eon. The Paleozoic was further divided into six **periods** of geologic time. The oldest three, **Cambrian, Ordovician**, and **Silurian**, were all named for rocks recognized in Wales. Cambria is the ancient Roman name for Wales, and the Ordovices and Silures were two of the early Celtic tribes of this area. The **Devonian** is next youngest and is named for Devonshire, England. The **Carboniferous** derived its name from the coal its strata encompasses, as already mentioned. In North America, the Carboniferous is not used, but instead this time is divided into two periods, the **Mississippian** and **Pennsylvanian**, after rocks of the upper Mississippi River valley and those of western Pennsylvania. In North America the coal-bearing strata of this time are Pennsylvanian. The final and youngest period of the Paleozoic is the **Permian**, which was named for the town of Perm, just west of the Ural Mountains in Russia.

The Mesozoic Era is divided into three periods beginning with the **Triassic**, so named for rocks in Germany where rocks of this age occur in a three-fold (tri-fold) sequence based on color: the Red, White, and Brown Trias. The **Jurassic** is named from the Jura Mountains of northeastern France and adjacent Germany and Switzerland, and the chalk-bearing **Cretaceous** forms the top of the Mesozoic sequence. The Cenozoic Era is divided into three periods, the **Paleogene, Neogene**, and **Quaternary**. (In older texts, the Paleogene and Neogene were combined into the Tertiary.) The Cenozoic is further divided into **epochs**, so that the Paleogene Period is composed of the **Paleocene, Eocene**, and **Oligocene** Epochs; the Neogene Period is composed of the **Miocene** and **Pliocene** Epochs, and the Quaternary is divided into the **Pleistocene** and **Holocene**. These Cenozoic epochs were originally defined based on the percentage of extinct species of molluscs contained within the rocks of each. Thus, for example, 96

Table 1.1 THE GEOLOGIC TIME SCALE (AGES IN MILLIONS OF YEARS)

Eons	Eras of Time	Periods of Time	Epochs of Time	Age (millions of years)	Duration (millions of years)	Major Biological Events
PHANEROZOIC	CENOZOIC	Quaternary	Holocene	0.01	1.6	Extinction of large land mammals in northern hemisphere and South America. Evolution of human beings. Rapid shifts in marine and terrestrial communities in response to four major glaciations.
			Pleistocene	1.6		
		Neogene	Pliocene	5	21	Extensive radiation of flowering plants. Extensive radiation and evolution of mammals. Co-evolution of insects and flowering plants. Dominance of gastropods and pelecypods in the oceans.
			Miocene	23		
		Paleogene	Oligocene	35	42	
			Eocene	56		
			Paleocene	65		
	MESOZOIC	Cretaceous		146	81	First flowering plants. Extinction of terrestrial, marine, and aerial reptiles. Extinction of ammonoid cephalopods. Radiation of primitive mammals.
		Jurassic		208	62	Gymnosperms (cycads, conifers, ginkgos), ammonoid cephalopods, and dinosaurs dominant. Radiation of marine reptiles; first birds; flying reptiles.
		Triassic		245	37	Depauperate marine faunas; dominance of ammonoid cephalopods and mammal-like reptiles. Origin of mammals and dinosaurs.

Era	Period	Duration (millions of years)	Age (millions of years before present)	Distinctive features
PALEOZOIC	Permian	45		Extinction of trilobites, blastoids, many other marine invertebrates. Dominance of mammal-like reptiles. Decline of amphibians. Evolution of fusulinids.
			290	
	Pennsylvanian*	33		Origin of reptiles. Evolution of fusulinids. Algal-sponge reefs and banks. Extensive coal-swamp forests. Many primitive insects.
			323	
	Mississippian*	39		Echinoderms and bryozoans dominant in the oceans. Amphibians on land. First appearance of coal-swamp forests.
			362	
	Devonian	46		Extinction of many marine groups. Oldest land vertebrates. Many corals, br achiopods, and echinoderms. Extensive radiation of land plants and fishes.
			408	
	Silurian	31		Oldest land life: land plants, scorpions, and wingless insects.
			439	
	Ordovician	71		First diverse marine communities. Dominance of brachiopods, bryozoans, corals, graptolites, and nautiloid cephalopods.
			510	
	Cambrian	33		First vertebrates (jawless fishes). First metazoans with skeletons. Dominance of trilobites. Marine faunas of low diversity. No known land life.
			543	
PRECAMBRIAN	Proterozoic	4060	540–2500	Origin of life, prokaryotes, eukaryotes, and metazoa.
	Archean		2500–4000	
	Priscoan		4000–4600	

*Together, these periods are known as the Carboniferous outside North America.

percent of clams and snails in Eocene strata were thought to be extinct, whereas only 10 percent were extinct during the Pliocene. This percentage subdivision of the Cenozoic was abandoned when it was discovered that this history was based on the local extinctions of western Europe that were not necessarily global in extent. However, the names, with revised definitions, are still in use. The last 2 million years are the Quaternary, which is divided into the Pleistocene and Holocene. We are living during the Holocene, or recent times, but it is unclear whether the Holocene is really a separate episode of Earth history or simply another interglacial interval of the Pleistocene. The Pleistocene is commonly referred to as the Ice Age, because this is when four major advances and retreats of continental glaciers punctuated North American history (five episodes occurred in Europe). The cold periods with glaciers are termed glacial intervals, and the warmer periods are interglacial intervals. This glacial epoch did much to shape the landscape of northern North America and Europe, and we are not certain whether glaciers will return again soon (soon in terms of geologic time).

Rocks older than the Paleozoic were once thought to have been deposited prior to the emergence of life, and this ancient period of Earth history was called the "Azoic." We now know that this is not true, and this very long interval of Earth history is called the **Precambrian** Eon. The Precambrian is divided into the **Archean, Proterozoic**, and **Priscoan** Eras.

This brief review of the historical development of the time scale should make the lists of names more meaningful than they may appear at first glance. It took scores of geologists walking over the ground and puzzling over rocks for several lifetimes to fit together the jigsaw puzzle of rocks at Earth's surface into a coherent and correct time sequence. They did it with no clear concept of the true ages of these rocks calibrated in years before the present.

HOW OLD IS EARTH?

While geologists were untangling the rock record and reaching a decision that Cambrian rocks were the oldest fossil-bearing strata (we now recognize much older fossils), other scientists were attempting to solve the riddle of the age of Earth. They were trying to find some type of natural, irreversible clock that would be reliable for measuring the age of Earth in years. Our common household clocks and calendars are based on periodic, regularly recurring events: the rotation of the Earth on its axis and the rotation of Earth around the sun. But no such regularly recurring events were known that would reach far enough back in time to be useful for measuring the age of Earth or events in Earth history. Instead, scientists attempted to find a one-way, irreversible series of events that could be calibrated in years (Figure 1.4). One such attempt was made by the English physicist Lord Kelvin. He assumed that Earth had originally been a molten mass that had cooled and solidified to its present state. He was able to measure the heat flow of rocks at Earth's surface as approximately 40 calories per year per square centimeter, and he also knew that the temperature within Earth increases with depth (measurable in deep mines and wells), approximately 2 degrees centigrade per 100 meters (this value varies from place to place). By measuring the heat conductivity of rocks with knowledge of the temperatures at which com-

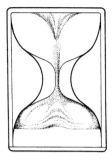

Hourglass clock:
> rate of cooling of Earth's surface;
> rate of salt accumulation in the
> oceans;
> rate of sediment accumulation in
> the oceans.

Early attempts to estimate the age of Earth were based on reversible processes, here called hourglass clocks, rather than on irreversible processes.

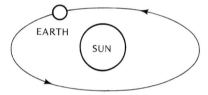

Rotational clocks:
> revolution of Earth
> about the sun;
> rotation of Earth
> on its axis.

Clocks for measuring time in ordinary terms are based on regularly repeating phenomena that recur at definite intervals of time.

Figure 1.4   Different kinds of clocks. Those used to measure the age of Earth, above; those used for ordinary time, below.

monly occurring rocks melt, in 1897 he estimated that Earth's surface would have been unable to support life approximately 20 to 40 million years ago, which was his estimate for the age of Earth. Kelvin's dating method was ingenious, but we now know that his estimate was far too low. The error that undermined his calculations (of which he could not have been aware) was that much of the heat that is generated within Earth is not residual heat remaining from the original molten state of Earth. Rather, it is generated by decay of radioactive elements within the deep layers of Earth. This heat flow is and has been relatively constant over very long periods of time. So despite the fact that heat flow has been a naturally occurring process throughout Earth history, it is not an irreversible process; it cannot be used as our geologic clock.

A second similar attempt was made by the Irish geologist, John Joly, in 1899. Joly reasoned that Earth's ocean waters were fresh when first formed and that oceanic salt content was derived as dissolved substances were carried into the oceans by rivers. Therefore, the age of Earth could be calculated by estimating the total flow of river waters into the oceans, the salt content of this fresh water, and the total salt content of the oceans. He estimated that it took 90 to 100 million years for the oceans to acquire their present saltiness. Like Kelvin, Joly's estimate was much too low. Again, the error was because of a lack of information that the process under consideration was a reversible process. We now know that many cubic miles of salt have been precipitated from the oceans during Earth history and buried deeply within the crust. Thus, much of the salt brought to the oceans was later removed through precipitation, causing a serious underestimate of the age of Earth.

A third attempt to estimate the age of Earth was made by several geologists who utilized known rates of sediment accumulation. They reasoned that if the thickest, most complete sections of strata were summed for all periods of geologic time (30 to 100 kilometers) and if it is assumed that sediment accumulated at a rate of approximately 1 centimeter per year, it would be possible to estimate the age of Earth at least from the beginning of the Cambrian. This simple calculation again yielded an estimate that was far too short because there is no place on Earth where sedimentation has been complete, because sedimentation rates vary greatly, and because many of these sediments were eroded after deposition. Thus, again, estimates using this method are certain to be too short.

Despite several attempts to determine the age of Earth, a naturally occurring, irreversible clock had not been discovered. However, near the close of the nineteenth century, this changed when **radioactivity** was first discovered by the French physicist Antoine Henri Becquerel in 1896. Much of the pioneering work on radioactivity was completed by Madame Currie and her husband, Pierre Currie. They discovered that natural decay of certain unstable, or radioactive, elements occurred at a constant rate. Furthermore, others recognized that this process of radioactive decay was irreversible and could be applied as a method to determine the age of the rocks containing these decaying elements. The first date of a rock, using radioactive decay, was calculated in 1905 by Sir Ernest Rutherford, a British physicist working in Canada. As analytical methods were refined, additional absolute rock dates were made until finally, in 1930, the first true geological time scale in millions of years was published by Sir Arthur Holmes of Scotland. Geochronologists continue to refine these dates as better techniques and instruments are developed.

All the early dates were determined using the radioactive element uranium. Uranium occurs naturally in rocks only in radioactive form. In this state, the nucleus of the atom is unstable. It contains excess energy that is released quite independent of surrounding atoms, temperature, or pressure. As this energy is released, the uranium nucleus is altered, and the radioactive **parent element** is said to decay to a **daughter element**. Decay continues until, ultimately, a daughter element that is stable results. In the case of uranium, the nuclei go through a complex series of changes until the final stable daughter is in the form of the element lead.

We refer to the different forms of each element as an **isotope**. Isotopes are defined by the composition of the nucleus. For example, lead is, by definition, the element that has 82 **protons** (positively charged particles) in the nucleus. Two lead isotopes are Pb^{206} and Pb^{207}. The nucleus of each of these contains 82 protons, but the number of **neutrons** (neutrally charged particles) differs; there are 124 and 125, respectively. The sum of the protons and neutrons, 206 and 207, is referred to as the **mass number**.

Radioactive uranium also occurs as different isotopes with different mass numbers. Two isotopes of uranium that occur in nature are U^{238} and U^{235} (Table 1.2). Each of these isotopes decays to a different isotope of lead, Pb^{206} and Pb^{207}, respectively.

Table 1.2 NATURALLY OCCURRING RADIOACTIVE ELEMENTS USED IN DATING ROCKS.

Parent Isotope	Daughter Isotope	Half-life in Years
Uranium 235	Lead 207	704×10^6 (1 million)
Uranium 238	Lead 206	$4,470 \times 10^6$
Thorium 232	Lead 208	$14,010 \times 10^6$
Potassium 40	Argon 40	$1,250 \times 10^6$
Rubidium 87	Strontium 87	$48,800 \times 10^6$
Carbon 14	Nitrogen 14	5,730

Ordinary lead, Pb^{204}, is not the stable product of radioactive decay. This decay of uranium is known to occur at a constant rate. If, for example, a mineral started with a given quantity of radioactive U^{238}, and that uranium decayed to produce a known amount of Pb^{206}, one could readily calculate the length of time for that amount of U^{238} to produce that quantity of Pb^{206}. It is this rate of radioactive decay that is a natural, irreversible clock that is used for absolute age-dating in years. The rate of decay is expressed in what is called a half-life. A **half-life** is the amount of time needed for one half of the atoms of the radioactive element to decay. The other atoms are still radioactive, and within a second half-life, one half of those atoms will also decay. So with this type of probabilistic decay, after two half-lives only one quarter of the original atoms remain. Some radioactive elements have extremely short half-lives—one millionth of a second; whereas others have very long half-lives, such as U^{238}, with a half-life of 4.47 billion years (Table 1.2). Small amounts of naturally occurring uranium are found in rocks and dated with this technique. With study of radioactive decay, it became immediately obvious that these dates made the age of Earth very much older than any of the earlier guesstimates.

Because all the initial dates obtained from radioactive materials used lead-uranium, the number of available dates grew slowly because uranium minerals are rare. Later, other minerals such as zircon, which contain only a few parts per million of uranium, were used as new and more delicate instruments were designed to measure radioactivity. The number of dated rock samples is now very large, especially because other naturally occurring parent-daughter combinations were discovered, such as potassium-argon and rubidium-strontium (Table 1.2).

The most widely cited radioactive date reported to the general public is C^{14}. The most common form of carbon is C^{12}, but C^{14}, which is radioactive, occurs in small quantities and decays to the stable daughter N^{14}. The half-life of this radioactive combination is only 5,730 years; but because carbon is incorporated into living tissue, this radioactive dating technique is extremely useful for dating very young geologic events. Bone, wood, or ashes that are 50,000 years old or younger can be dated using C^{14} dating. Because of the short half-life, any older materials have lost too much C^{14}.

The result of all this age-dating is that geologists are now able to tie the previously constructed relative geologic time scale of eras and periods to an absolute time scale in years before the present. In order to do this accurately, an igneous rock layer that has been dated radioactively must be related to fossil-bearing beds just above or below. As a result, geologists now know that the boundary between the Cenozoic and Mesozoic eras occurred approximately 65 mya (million years ago) or Ma (*mega annum*). The Paleozoic-Mesozoic boundary occurred approximately 245 mya, and the base of the Cambrian (the Precambrian-Paleozoic boundary) occurred approximately 543 mya (Table 1.1). The periods of geologic time range in age from a 78-million-year duration for the Cretaceous Period to approximately 30 million years for the Silurian and Cambrian Periods. The oldest rock dated on Earth is approximately 3.96 billion years old. The Earth is thought to have been formed 4.5 billion years ago (bya), so the Precambrian is more than three quarters of known Earth history.

KEY TERMS

absolute age	Jurassic	Pleistocene
Archean	mass number	Pliocene
body fossil	Mesozoic	Precambrian
Cambrian	Miocene	principle of lateral continuity
Carboniferous	Mississippian	principle of original
Cenozoic	Neogene	horizontality
Cretaceous	neutron	principle of superposition
daughter element	Oligocene	Priscoan
Devonian	Ordovician	Proterozoic
Eocene	Paleocene	proton
epoch	Paleogene	Quaternary
era	paleontology	radioactivity
fossil	Paleozoic	relative age
geologic time scale	parent element	relative time scale
global change	Pennsylvanian	Silurian
half-life	period	Steno, Nicholas
Holocene	Permian	trace fossil
isotope	Phanerozoic	Triassic

WEB RESOURCES

At the end of every chapter, a short annotated bibliography is given where additional data, illustrations, and explanations can be found. In many cases the listed references are primary sources for the information in that chapter. A relatively new and increasing source for paleontologic and geologic resources is the worldwide web. The following sources are recommended, and additional sites can be located on those listed.

Field Museum of Natural History
<http://www.fmnh.org/>

The Field Museum of Natural History is in Chicago, Illinois. The museum has comprehensive paleontological displays, and the website highlights museum displays and work of Field Museum scientists.

Florida Museum of Natural History
<http://www.flmnh.ufl.edu/>

The Florida Museum of Natural History is in Gainesville, Florida. Their website contains a variety of types of information on vertebrate paleontology, paleobotany, and invertebrate paleontology.

National Museum of Natural History, Smithsonian Institution
<http://www.mnh.si.edu/>

The Department of Paleobiology at the National Museum of Natural History includes invertebrate paleontologists, paleobotanists, and vertebrate paleontologists. Their website reflects this diversity and includes, among many things, information on dinosaurs, the Burgess Shale, and shark teeth.

Paleobotanical Section of the Botanical Society
<http://www.botany.org/>

Paleobotanists form a section of the Botanical Society. This website gives information about this society.

Paleontological Research Institution
<http://www.englib.cornell.edu/pri/pri1.html>

At Ithaca, New York, the Paleontological Research Institution is a research and educational paleontological organization. Among other things, the Paleontological Research Institution website allows you to browse through virtual museum drawers and to take virtual field trips.

Paleontological Society
<http://www.uic.edu/orgs/paleo/homepage.html>

The Paleotological Society webpage describes this, the largest paleontological society in North America, and provides links to other paleontological resources around the world.

Royal Tyrell Museum of Palaeontology
<http://www.tyrell.magtech.ab.ca/>

The Royal Tyrell Museum of Paleontology is in Drumheller, Alberta, Canada. This museum has excellent dinosaur exhibits. Their website has many things, but focuses on these dinosaurs, including a virtual exhibit of their displays.

Tree of Life
<http://www.phylogeny.arizona.edu/tree/phylogeny.html>

The Tree of Life is a comprehensive information source about the morphology, diversity, and phylogeny of life on Earth. Webpages are available for nearly every organism group from fungi to dinosaurs.

United States Geological Survey
<http://www.usgs.gov/>

The United States Geological Survey is the primary geological agency in the United States government. This website has a wide variety of geologic and paleontologic information available.

University of California at Berkeley Museum of Paleontology
<http://www.ucmp.berkeley.edu/>

The University of California at Berkeley Museum of Paleontology is in Berkeley, California. This extensive site concentrates on all life, geologic time, and evolutionary thought. Their pages on life include information on ecology, fossil record, life history, morphology, and systematics.

READINGS

All modern introductory textbooks in physical geology and in historical geology include chapters on the geologic time scale and on radioactive age-dating. Most cover the details of radioactive decay in more detail than in a general paleontology text such as this book. The reader is urged to consult one of these textbooks for additional information.

Berry, W.B.N. 1968. *Growth of a Prehistoric Time Scale Based on Organic Evolution*. Freeman. 158 pages. Primarily historical in treatment, this book traces the growth and development of the relative time scale in considerable detail. Radioactive dating is not covered.

Edwards, W.N. 1967. *The Early History of Palaeontology*. The British Museum (Natural History) Publication 658. 58 pages. Brief discussion of the historical ideas on fossils leading to their correct understanding as evidence of ancient life.

Eicher, D.I. 1968. *Geologic Time*. Foundations of Earth Science Series. Prentice-Hall. 150 pages. This book includes discussion of the historical development of the time scale, geologic time in relation to stratigraphy, the study of rock strata, and radioactive dating.

Harland, W.B., and others. 1990. *A Geologic Time Scale*. Cambridge University Press. 131 pages. A comprehensive reference book that details all the modern geologic time scales presently in use.

McPhee, J. 1980. *Basin and Range*. Farrar, Straus, and Giroux. 216 pages. A very well written book in which the author accompanies geologists as they do their science.

Needham, J. 1954. *Science and Civilization in China*, Volume 3. Cambridge University Press. 620 pages. The authoritative study on science in ancient China. Volume 3 contains information on Chinese studies of fossils.

Toulmin, S., and Goodfield, J. 1965. *The Discovery of Time*. The Science Library. Harper Torchbooks. 279 pages. Written by two historians of science, this book examines the gradual recognition of time as an entity and covers early geological and biological discoveries that expanded the concept of time. Very well written and an important book in the history of science.

CHAPTER 2

The Organization of Life

INTRODUCTION

Life on this planet is today, and has been during the past, incredibly diverse. Between 1.5 and 5 million living species have been described, and more species are being named every day. It is impossible for any person to keep up with this enormous body of literature and to understand all species in any detail. If we add to this large number of living organisms the number of named extinct fossil species, estimated at 225,000, it is just that much more overwhelming. In order to make sense of this complexity, ordering principles for arrangement of the seeming chaos of named species is needed. Ordering into a few major categories will reduce the complexity to a level that can be more readily grasped. This is one of the purposes of the classification of life.

Some categories of life are well known and easily understood by everyone—plants and animals, insects, reptiles, ferns, flowering plants, and bacteria—although many may be hard-pressed to precisely define each of them. What is the basis for such classifications? The answer is **evolution**. Each unit of classification, large or small, is judged to contain organisms that are more closely related to each other in an evolutionary, ancestor-descendant sense than to organisms contained in other different but equal units of classification. For example, two kingdoms of organisms that are commonly recognized are the plant and animal kingdoms. All organisms included within the plant kingdom are more closely allied to each other than to any organisms within the animal kingdom.

In order to make the evolutionary relationships among organisms explicit, an ordered ranking of categories is used; small categories denote small series of quite closely related organisms, and increasingly higher categories include more and more remotely related organisms. Such a scheme is called a **hierarchy**. **Domains** of life are the largest, most comprehensive, and broadest units of classification. The standard scheme for animals is as follows:

Domain
    Kingdom
        Phylum (or Division)
            Class
                Order
                    Family
                        Genus
                          Species

The **species** is the fundamental unit in this hierarchy. Individuals of a species are reproductively isolated from all other species in nature, producing a series of interbreeding populations. Individuals of one species share anatomical traits, making them recognizable as the same. The sharing of genetic material exclusively among members of a species produces the unique identity of a species; however, most species are identified based on their anatomy (Figure 2.1). All fossil species are recognized from their anatomy. The fundamental anatomical and ecological reality of species can be confirmed from comparison of the scientific treatment of species and the treatment by indigenous peoples. Biologists have found that the most primitive hunter-gatherer peoples differentiate the animals in their environment into the same basic species as biologists—although, of course, with different names and not in a hierarchy.

All species that are more closely related to one another than they are to other species are grouped together in a **genus**. The genera (plural of genus) in turn are grouped into a series of families, and so on, into increasingly comprehensive categories up to **phylum** (division is used for plants), **kingdom**, and domain. All individuals grouped into the same category should share a single common ancestor. As an example, consider the common domestic cat. The scientific name of the species is *Felis domesticus*. The first part of this name is the genus, and the genus *Felis* includes other cat species such as the bobcat, lynx, and mountain lion. This genus is included in the **family** Felidae, which contains other cat genera (e.g., *Smilodon*, the extinct saber-toothed cat). This family, along with others such as dogs, is encompassed within the **order** Carnivora, the flesh-eating or carnivorous mammals. This order and many others are grouped into the **class** Mammalia. Mammals are all animals that have hair and suckle their young. Mammals, along with fish, amphibians, reptiles, and birds, are grouped together into the phylum Chordata, which is animals that have a spinal cord running along their backs. This phylum and others are classified together into the kingdom Animalia, consisting of life forms that are typically large, multicellular, and unable to manufacture their own food. In turn, Animalia is in the domain Eucarya.

In an attempt to develop an evolutionarily based classification at the highest level, **domains** have been introduced as the highest classification of life on Earth. The three domains are listed below with contained traditional kingdoms:

Domain Bacteria
Domain Archaea
Domain Eucarya
    Kingdom Protista
    Kingdom Fungi
    Kingdom Animalia
    Kingdom Plantae

A sixth, quite questionable kingdom has recently been proposed to encompass many of the highly enigmatic fossils of the late Precambrian Ediacaran fauna. This new kingdom is named the Vendobionta, but this conclusion is not universally accepted (see Chapter 8).

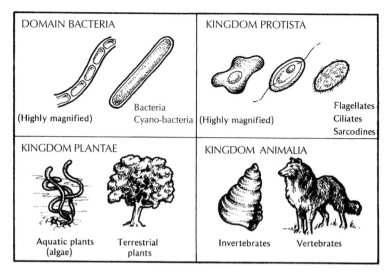

Figure 2.1   Major divisions of life that are important as fossils. The Bacteria are prokaryotes; the protists, plants, and animals are eukaryotes.

DOMAINS BACTERIA AND ARCHAEA

Members of the domains **Bacteria** and **Archaea** are the simplest and most poorly organized kinds of life on Earth. Together, these include the **prokaryotes** (Monera and Prokaryotae of previous classifications) and will probably be composed of several kingdoms and more than 20 phyla when these domains are more thoroughly studied. In this book we will refer to the fossil prokaryotes as bacteria and **cyanobacteria** (formerly blue-green algae) as distinguishable fossil representatives of the domain Bacteria. Prokaryotes differ from all other life in that they lack a well-differentiated, complex nucleus within the cell. The DNA in prokaryotes is not separated from the rest of the cell contents by a nuclear membrane (Figure 2.2). The principal distinguishing characteristic of prokaryotes from **eukaryotes** is that their metabolic biochemistries are vastly different. Furthermore, within the prokaryotes there are wide biochemical differences, reflected in part by the different ways that they feed.

Domains are distinguished at the molecular level. Members of each domain have a fundamentally distinct molecular organization, such as the composition of RNA and many other aspects. Both the Bacteria and the Archaea are prokaryotes, and the Eucarya are eukaryotes. There are no known Archaea fossils, but living forms include methanogens, sulfur-dependent forms, and several others.

In general, prokaryotes use complex molecules for food production. As such these organisms are called **autotrophs** (self-feeding), in contrast to **heterotrophs** (different-feeding), which are organisms that cannot manufacture their own food and instead capture organic material by predation, scavenging, or some other means. In order to combine materials into food, energy is required. One energy source is the sun, and organisms that use the sun would be called **photoautotrophs** because they use light (photo-) as a source of energy. Plants and some protists are also photoautotrophs.

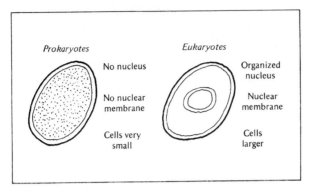

Figure 2.2 The basic dichotomy of life. Prokaryotes are the smallest and simplest forms of life. They lack an organized nucleus, DNA is not organized into a nucleus, and the cells are quite small. All other life is eukaryotic—the nucleus is surrounded by a membrane and cell size is larger.

Other prokaryotes are **chemoautotrophs** because they produce their own food through chemical reactions other than photosynthesis, such as $SO_4^= \rightarrow H_2S$, to release energy in the food synthesis processes.

KINGDOM PROTISTA

The kingdom **Protista** consists of more than 10 phyla. Protists are eukaryotes that (with one exception) lack chloroplasts for autotrophy. They are composed of a variety of microorganisms and most are single-celled (Figure 2.3). Protists are confined to aquatic habitats, either marine, freshwater, or within other organisms; some are photosynthetic and others consumers. There are three methods of locomotion among the protists: flagella, cilia, and pseudopods. Flagella are a single, long, whiplike structure

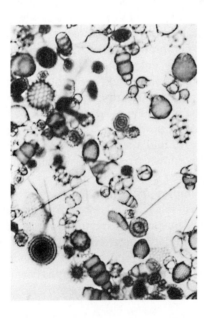

Figure 2.3 A strew slide of fossil radiolarians, a common form of protist preserved as fossils. These specimens from the southwestern Pacific are Miocene in age and highly magnified. (Courtesy of Joyce R. Blueford, U.S. Geological Survey.)

that beats rapidly for locomotion. Cilia are short structures that cover the organism and beat for movement. Pseudopods are extensions of the cell wall; this type of slow movement is utilized by amoebas.

Protists that are common in the fossil record are foraminifera, dinoflagellates, coccolithophorids, and diatoms.

KINGDOM FUNGI

Fungi are eukaryotes that reproduce with spores and that completely lack both cilia or flagella. Fungi typically undergo sexual reproduction; they are not photosynthetic but consume food through absorption. Fungi do have a fossil record that consists of trace fossils that record fungal infestation on plants and animals. However, fungi are not common elements of the fossil record and will not be treated further.

KINGDOM ANIMALIA

Kingdom **Animalia** is very diverse, including both sponges and man. Thus, there is a wide range in the complexity of organization of organisms contained within this kingdom. Animalia differ from other organisms by being multicellular, by consuming food, and by engaging in sexual reproduction through a sperm and an egg. In order to explain how the major animal groups are related to one another, it is necessary to introduce the basic body organizational schemes for animals. First is the level of organization of the cells that form the body (Figure 2.4). The cells may be largely independent of each other and either not much differentiated or highly specialized. This is called the **cellular grade of organization**. All protistans have this cellular grade of organization, and the sponges (phylum **Porifera**) are also at this cellular level. Next complex is the **tissue grade of organization** in which groups of cells are organized into tissues that have specific functions, such as muscular or nervous tissue. The phylum **Cnidaria** (jellyfish, corals, and sea anemones) is at this level. Different kinds of tissues may be arranged into organs, and these organs may form large, complex organ systems, such as the digestive, nervous, and excretory systems of higher animals. All animals above the Cnidaria have the **organ grade of organization**.

A second characteristic that is used to arrange the animal phyla is symmetry. If an animal body can be divided into two equal halves by a plane in only one way, this design is called **bilateral symmetry**. Your body and that of all other vertebrate animals has bilateral symmetry. If numerous planes can divide the body into two equal halves, an animal has either **radial** or **spherical symmetry**, depending upon the three-dimensional nature of the symmetry. Finally, if no plane of symmetry exists, the animal is considered to be asymmetrical. Although exceptions exist, the symmetry of an animal commonly reflects its fundamental ecology. A sessile (attached or very slow moving) organism must place a high premium on defensive strategies. It must be prepared to defend itself from either the environment or from predators approaching from any direction. Hence, it is advantageous for a sessile organism to have a radial or spherical symmetry in order to protect itself in all directions. Alternatively, a mobile organism

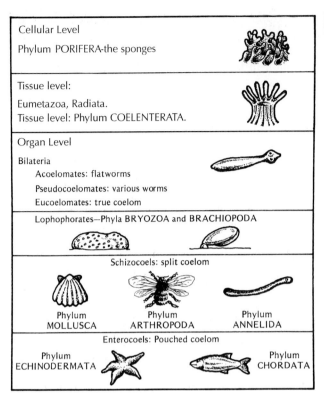

Figure 2.4 Levels of organization in the animal kingdom.

that is moving and meeting the environment "head on" needs to concentrate its sensory apparati where it meets the environment. Thus, a mobile organism typically has a head, and bilateral symmetry typically follows.

A third characteristic among animals is the number of primordial, dermal cell layers that are present in the early larval stages. These **dermal layers** develop during growth into the various tissues and organs of animals. The simplest animals lack differentiated dermal layers. More advanced animals may have an **ectoderm** (outer skin) and an **endoderm** (inner skin), and even more advanced animals have a third, middle layer, the **mesoderm** (middle skin).

A fourth important characteristic has to do with the presence or absence of an internal fluid-filled cavity in the body, called a **coelom**, and how that coelom is formed. The coelom forms from the mesoderm in advanced animals (coelomates) and is lacking in less complex animals (acoelomates). A pseudocoelomate is an organism with a coelom that is only partly formed from the mesoderm. In these animals, the coelom is also partly lined with ectoderm and endoderm.

Finally, segmentation is important. This is the division of an animal body, from head to tail, into a series of sections with repeated body parts.

These fundamental features of animals are used to formulate a classification of animals between the kingdom and phylum levels (Table 2.1). For the sake of simplicity,

Table 2.1   **MEMBERS OF THE ANIMAL KINGDOM, TO THE PHYLUM LEVEL, THAT ARE DISCUSSED IN THIS BOOK.**

CELLULAR GRADE OF ORGANIZATION, Symmetry radial, spherical or absent
> **Phylum Porifera:** Sponges
> **Phylum Archaeocyathida:** Sponge-like organisms confined to Cambrian Period

TISSUE GRADE OF ORGANIZATION, Symmetry radial
> **Phylum Cnidaria:** Corals, jellyfish, sea anemones

ORGAN GRADE OF ORGANIZATION, Symmetry principally bilateral
> **Acoelomates:** No coelomic body cavity.
> Includes many worm groups, such as flatworms and tapeworms, which are not important in the fossil record.

> **Pseudocoelomates:** Coelomic body cavity partly lined with mesoderm.
> Includes groups such as rotifers and some worms that are not important in the fossil record.

> **Eucoelomates:** Coelomic body cavity lined completely with mesoderm.
> **Lophophorates:** Possess tentacular food-gathering lophophore.
>> **Phylum Brachiopoda:** Solitary; skeleton in two valves.
>> **Phylum Bryozoa:** Colonial; most with skeleton, many forms.
> **Schizocoels:** Coelom originates from a split in the mesoderm.
>> **Phylum Mollusca:** Unsegmented except for one small primitive group (Monoplacophora). May lose bilateral symmetry; mostly with calcareous exoskeleton secreted by mantle. Includes Gastropoda, Bivalvia, Cephalopoda, and other lesser classes.
>> **Phylum Annelida:** Segmented worms. Body elongate; lacking appendages; jaws chitinous and may be preserved as fossils; no other hard parts.
>> **Phylum Arthropoda:** The joint-leg animals; conspicuously segmented. Includes crabs, shrimp, barnacles, spiders, insects, and trilobites. Insects are the most diverse group of organisms living today.
> **Enterocoels:** Coelom originates from outpockets of embryo gut.
>> **Phylum Echinodermata:** Spiny-skinned animals. Conspicuous five-sided symmetry that is most commonly a bilateral derivative. Also contain a water vascular system. Living members include asteroids (starfish), echinoids (sea urchins, sand dollars), holothurians (sea cucumbers), and crinoids. Some extinct members include blastoids, diploporans, helicoplacoids, and rhombiferans.
>> **Phylum Hemichordata:** Morphologically diverse group of living and extinct groups that possess gills and a dorsal nerve cord but lack a notochord. Graptolites are important fossil hemichordates.
>> **Phylum Chordata:** Possess gills and a dorsal nerve cord, and a notochord. Living chordates are fish, amphibians, reptiles, birds, and mammals. The extinct conodonts have most recently been assigned to the Chordata.

we will avoid giving names to the ranks of the various categories, but this subdivision is a useful way to picture the basic ways in which animal phyla are similar and dissimilar.

The animal kingdom is divided into more than phyla, including the sponges at a cellular grade of organization (Figure 2.5), the Radiata, and the Bilateria. The Radiata consist primarily of the phylum Cnidaria (corals, jellyfish, and sea anemones), which

Figure 2.5   A large slab on which are preserved the siliceous skeletons of many Devonian glass sponges, *Dichtyophyton*, from New York. The actual spicules have been dissolved or replaced by pyrite (iron sulfide). The slab is about 1 m in height (Courtesy of Field Museum of Natural History, Chicago.)

have radial symmetry and a tissue grade of organization. An extinct phylum, the Archaeocyathida probably had either cellular or tissue grade organization. The Bilateria includes virtually all other animals; they have bilateral symmetry with a distinguishable head and tail. The Bilateria are also sometimes called the eumetazoa (true metazoa), where **metazoa** are the multicellular animals. The radially symmetrical echinoderms are exceptions to this symmetry rule, and many of them, like Cnidaria, are sessile animals fixed to the sea floor.

The bilateral animals are divided into three groups based on the presence, absence, or degree of development of the coelom. The acoelomate animals such as the flatworms and tapeworms are not important as fossils. Similarly, the pseudocoelomates include several worm phyla and are not common as fossils. The eucoelomates make up the remainder of animals and represent most of animal life preserved in the fossil record. Eucoelomates are divided into three major groups, the lophophorates, protostomes, and deuterostomes.

The **lophophorates** consist of three phyla, and two are common as fossils. These are judged to be closely related because all have the conspicuous, distinctive structure called the lophophore. This is a ring of tentacles associated with the mouth. The tentacles, which serve primiarily for food gathering, are covered with cilia and mucus. Two lophophorate phyla, Brachiopoda and Bryozoa, are very important in the fossil record. Representatives of these two groups are quite different in appearance. The phylum **Brachiopoda** are relatively large, have two external valves or shells that enclose the animal, and consist of solitary individuals. The brachiopods and clams are the only two larger animals having two valves or shells that are common as fossils. The phylum **Bryozoa** (commonly called moss animals) are all colonial, and each individual of a colony is microscopic in size. The common skeleton of the colonial bryozoan is composed of calcite and is readily preservable as a fossil.

The **protostomes** (first mouth) consist of three major phyla: the Mollusca, Annelida, and Arthropoda. In early larval stages, the coelom is formed by the splitting of the solid mass of mesodermal cells to form the internal coelomic cavity (Figure 2.6). In protostomes the initial larval opening in embryos becomes the mouth. Superficially, the adults of these three phyla do not look alike at all. The phylum **Mollusca**, except for one primitive group, have unsegmented bodies, whereas the annelid and arthropod bodies are conspicuously segmented. Molluscs are very important fossils, especially three groups, class Bivalvia (clams), class Gastropoda (snails), and class Cephalopoda (including squid, octopi, chambered nautilus, etc.). All molluscs typically have a hard calcite or aragonite shell, so they are commonly preserved. The **Annelida** (segmented worms) do not have a hard exoskeleton and, consequently, are not common fossils. Annelid legs are not jointed, which distinguishes them from the arthropods. The phylum **Arthropoda** is the most diverse group of animals alive today, mainly because it includes hoards of insects. Although crabs, lobsters, shrimp, and other arthropods are very common today, they are relatively rare as fossils (Figure 2.7). Insects are also rare as fossils because they do not have a mineralized exoskeleton and are not marine. The most common fossil arthropods are trilobites and ostracodes. Trilobites have been extinct since the close of the Paleozoic. Ostracodes are also common as fossils, but they are easy to overlook because they are typically very small. Ostracodes have a calcareous external shell that is composed of two valves. In this respect ostracodes are similar to brachiopods and bivalves, but they are much smaller. Eurypterids are another group of extinct arthropods that were important during the Paleozoic.

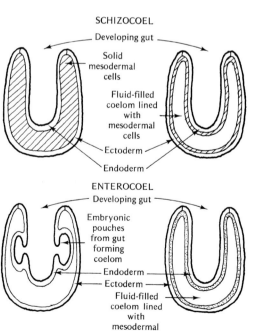

Figure 2.6   The two different modes of forming the coelom in the eucoelomates. In schizocoels, proliferating mesodermal cells divide to form a fluid-filled cavity, the coelom. In enterocoels, pouches from the side of the gut expand to form the coelom.

Figure 2.7 A fossil arthropod, *Eryon*, from Jurassic rocks of Germany. It is related to modern lobsters and crayfish. Note the segmentation of the body and division of the legs into segments. The specimen is 9 cm long. (Courtesy of Field Museum of Natural History, Chicago.)

The final group of bilateral animals is the **deuterostomes** (second mouth). In contrast to the protostomes, pockets from the gut of deuterostomes enlarge to eventually become the mesodermally lined coelom, and the initial larval opening in embryos of deuterostomes becomes the anus. The mouth is formed secondarily. Three phyla of deuterostomes are especially common as fossils: Echinodermata, Hemichordata, and Chordata. Animals in the phylum **Echinodermata** are sometimes called "spiny-skinned animals." They are exclusively marine and most have calcareous spines or nodes on the skeleton. The echinoids (e.g., sand dollars, sea urchins), starfish, and holothurians are the most common living echinoderms. However, during the Paleozoic, the phylum was much more diverse. The Echinodermata is divided into more than 20 classes. Of these, only five are still living, and all but one of the other classes are confined to rocks of the Lower and Middle Paleozoic. Crinoids are still living and were very important during the Paleozoic and early Mesozoic. Extinct echinoderm classes mentioned in this book include blastoids, diploporans, helicoplacoids, and rhombiferans.

The phylum **Hemichordata** include the living, simple, wormlike forms called acorn worms, and the colonial pterobranchs. These are thought to be related to an extinct group of colonial animals (the graptolites) that flourished during the Lower and Middle Paleozoic.

The final animal phylum is the phylum **Chordata**. Chordates are, of course, very important animals today and have been important during the geologic past. Vertebrates are chordates characterized by having a bony spinal column. Fish are the most primitive and earliest known vertebrates. They are followed in evolution by the most primitive land tetrapods (four-legged), the amphibians, which include toads, frogs, and salamanders. These in turn gave rise to the reptiles, snakes, lizards, turtles, and alligators. From the reptiles, birds and mammals arose. Both birds and mammals are warm-blooded vertebrates. All these groups have long and important fossil records, and the tracing of their evolutionary development through time is one of the most important stories to be told in paleontology.

KINGDOM PLANTAE

The most important characteristics of the kingdom **Plantae** is that its members are eukaryotic organisms that do, by and large, manufacture their own food. They accomplish this through photosynthesis, the production of reasonably complex organic molecules—sugars and carbohydrates—from simple inorganic starting materials—water, carbon dioxide, and traces of minerals in solution. Plants use these complex organic molecules for food. In order to combine these materials into food, energy is required, and plants that use sunlight as an energy source are called **autotrophs** (self-feeding), and more specifically **photoautotrophs**. Plants are mostly multicellular but many are unicellular.

Plants can be subdivided into major groups based on the presence or absence of **vascular tissue** and habitat. Vascular means circulatory. The blood system or arteries and veins in humans is their vascular system. Plants with vascular tissue are treated here as the true land plants and are called either **tracheophytes** or vascular plants (Table 2.2). Bryophytes (mosses and liverworts) are the second group of plants. They live on land but do not have a vascular system. Bryophytes are very primitive, generally small, and confined to moist habitats. Algae, another group of plants, live in water and lack vascular tissue (Figure 2.8).

ALGAE

Algae live in both marine and fresh water. They are most common close to the water surface, where they receive ample sunlight. They cannot live in the deep, dark parts of the ocean. Many algae are floaters and have oil, fat, or gas inclusions that make them about the same density as water. The major groups of algae are traditionally based on

---

Table 2.2 MEMBERS OF THE PLANT KINGDOM, TO THE DIVISION LEVEL, THAT ARE DISCUSSED IN THIS BOOK.

ALGAE: Photosynthetic aquatic plants that lack vascular tissue.
  Division Chlorophyta: Green algae
  Division Rhodophyta: Red algae
  Division Bacillariophyta: Diatoms

BRYOPHYTES: Photosynthetic land plants that lack vascular tissue; not important as fossils. Mosses and liverworts.

TRACHEOPHYTES: Photosynthetic land plants with vascular tissue.
  Division Rhyniophyta: Spore-bearing plants without true leaves or roots.
  Division Lycophyta: Spore-bearing plants with true leaves or roots and with simple leaves continuously along stem in whorls and sporangia in cones.
  Division Sphenophyta: Spore-bearing plants with true leaves or roots and with simple leaves from stem at nodes and sporangia in cones.
  Division Pteridophyta: Spore-bearing plants with true leaves or roots and with complex leaves and sporangia on bottom of leaves.
  Gymnosperms: General term for most seed-bearing plants that lack flowers.
  Angiosperms: Flowering plants.

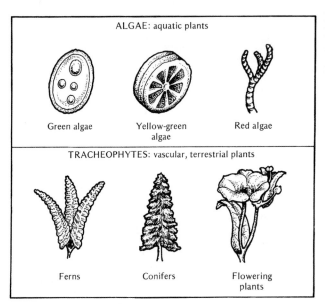

Figure 2.8 Common kinds of plants preserved as fossils. Major groups of land plants include spore-bearing ferns as well as the seed-bearing gymnosperms (conifers) and flowering plants.

the color of the algae, which is due to photosynthetic pigments that generally absorb different wavelengths of light. The green algae (phylum Chlorophyta) are mainly uni-celled or simple filaments of cells. They generally do not have skeletal parts and, thus, are not especially common as fossils. Green algae are important in two ways. First, they are thought to have been the ancestors of the land plants. One reason is that chloro-phyll is the dominant pigment in both groups. Second, green algae have been identified in late Precambrian rocks of Australia, making them the oldest known eukaryotic organisms (see Chapter 8).

The diatoms (phylum Bacillariophyta) is a group that is very conspicuous as fos-sils. Diatoms have a skeleton composed of opaline silica ($SiO_2 \cdot nH_2O$) that consists of two parts that fit together like a pillbox. These are fossils in Mesozoic and younger rocks. It is unclear whether they came into existence during the Mesozoic, whether they were present but lacked a skeleton during the Paleozoic, or whether they were present in older rocks and the microscopic skeletons simply were not preserved. They are so abundant in some Miocene rocks in California that they compose much of the rock, which has economic value in cleansing powders, toothpaste, and other products.

Red algae (phylum Rhodophyta) are important as fossils. Many of these have a complex calcareous skeleton, reflecting their complex internal structure, and their remains have been identified in many fossil reefs. They are generally not unicellular but, rather, multicellular, with whole organisms marble- to fist-sized.

VASCULAR PLANTS

Vascular plants had aquatic ancestors and made the successful transition from liv-ing in water to living on dry land—one of the most severe and difficult habitat changes possible. All these plants are characterized by specialized vascular tissues. Vascular tissues are primitively in the center of the main plant stem but, secondarily, may be situated near the surface of the body. The central core of vascular tissue provides for uptake from the soil and distribution of water and dissolved materials throughout the plant. This tissue is called xylem. Other vascular tissue, called phloem, transports manufactured food from photosynthetic areas to other parts of the plant.

The vascular plants are very diverse. They include small, very primitive plants that did not develop true leaves or true roots, as well as the most advanced, youngest, and most conspicuous of vascular plants—those with flowers. A very great majority of the plants with which you are familiar and that you see all around you are flowering plants, or angiosperms. Other vascular plants include ferns, conifers, and several other smaller groups that are much reduced in variety today but that were important parts of the landscape of earlier Earth history.

KEY TERMS

algae	dermal layer	metazoa
Animalia	deuterostomes	Mollusca
Annelida	domain	order
Archaea	Echinodermata	organ grade of organization
Arthropoda	ectoderm	photoautotrophs
autotrophs	endoderm	phylum (pl., phyla)
Bacteria	Eucarya	Plantae
bilateral symmetry	eukaryotes	Porifera
Brachiopoda	evolution	prokaryotes
Bryozoa	family	Protista
cellular grade of	Fungi	protostomes
organization	genus (pl., genera)	radial symmetry
chemoautotrophs	Hemichordata	species
Chordata	heterotrophs	spherical symmetry
class	hierarchy	tissue grade of organization
Cnidaria	kingdom	tracheophytes
coelom	lophophorates	vascular plants
cyanobacteria	mesoderm	vascular tissue

READINGS

You may consult any modern college-level textbook in biology, botany, or zoology for a discussion of the organization of life. These texts will also include one or more classifications of life. Do not expect these texts to present identical classifications; each will differ from the others in some details.

Blackwelder, R.E. 1963. *Classification of the Animal Kingdom*. Southern Illinois University Press. 80 pages. A detailed listing of the animal kingdom without illustration or explanation. One of many different classifications of animals.

Bold, A.C. 1960. *The Plant Kingdom*. Prentice-Hall. 106 pages. An excellent short discussion of the major features of monera, algae, and higher plants.

Lane, N.G. 1990. "Census of Past and Present Life." Journal of Geological Evolution 38:119–122. A summary of the numbers of known living and fossil genera or species of life.

Margulis, L., and Schwartz, K.V. 1988. *Five Kingdoms: An Illustrated Guide to the Phyla of Life on Earth*, 2nd ed. Freeman. 378 pages. Descriptions and illustrations of 89 phyla of life.

Parker, S.H., ed. 1992. *Synopsis and Classification of Living Organisms*. McGraw-Hill. 2 vols., 1,232 pages. A detailed compilation of all known living organisms. Tabulations of numbers of species can be extracted from the text.

C H A P T E R   3

Rocks and Fossils

INTRODUCTION

We have discovered how fossils fit into a time framework and how they may be of widely different ages. Fossils are also studied within the context of a physical framework provided by the rocks within which they are found. This chapter is devoted to a brief discussion of the relationship between rocks and fossils.

FOSSILIZATION

How and why does any living thing, after it dies, become a fossil? The odds against any specific individual becoming a fossil are very great indeed. To become a fossil, an organism must ordinarily have some part its body that does not easily decay. Such parts are called **hard parts** or a **skeleton**. In vertebrate animals the parts most readily preserved are the bones and teeth. Invertebrate animals may have an external shell that is composed of preservable minerals. The most common invertebrate shells are composed of the minerals **calcite** or **aragonite** (both are calcium carbonate, $CaCO_3$). Other invertebrate skeletons are composed of silica (SiO_2), which is the same as quartz; however the silica of skeletons has water added to it, forming opaline silica that is not nearly as stable as quartz. Another type of skeleton is a **chitinophosphatic** shell. This consists of microscopic layers of chitin, a complex molecule, and calcium phosphate, the mineral apatite (CaP_4). Arthropod exoskeletons are formed by a combination of layers of chitin, protein, and a waxy material; and the protein layer may be calcified. In plants the parts most commonly preserved contain the organic material cellulose, which is quite resistant to decay. The woody parts of plants and the leaves are the most readily preserved, along with spore and pollen grains that are composed of a cellulose layer surrounded by a waxy layer.

The great majority of fossils consist of only the preserved hard parts. However, extra-special preservation conditions can occur that are suitable for the preservation of soft tissues. When these conditions occur, typically, many organisms are preserved. Such a circumstance is called a soft-bodied fauna or flora, and we will discuss some of these because they add greatly to our knowledge of the history of life on Earth.

Just having hard parts is not enough to ensure **fossilization**. All hard parts are subject to decay by physical, chemical, or biological processes. If a dead shell lies on the sea floor, it will be bored by algae and sponges. A shell may be washed onto a beach and broken by waves (Figure 3.1). Sunlight and bacteria will eventually crumble a bone exposed on dry land. We know that most wood and leaves in a forest rot and decay. In order for

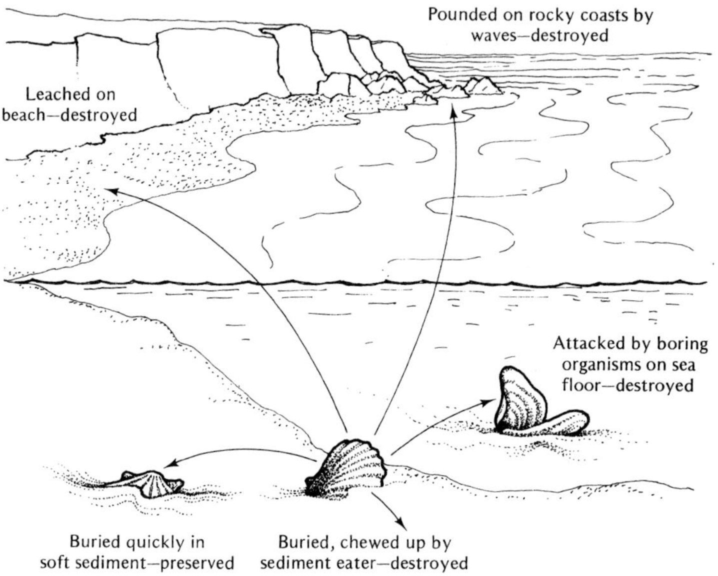

Figure 3.1   The fates of hard parts.

any hard parts to be preserved as fossils, they must become buried. Sand, silt, or mud may be washed over the skeleton, sealing it from water and air. Burial reduces bacteria, the principal agents of decay, and helps prevent destruction of the skeleton. The more rapid the burial after the death of an animal or plant, the greater the likelihood the specimen will eventually become a fossil. Indeed, most very well-preserved fossils were probably buried alive! Thus, rapid burial is a prerequisite to fossilization.

Sediments accumulate in areas that are called **basins of deposition**, that is, places where sand or mud are deposited. Such sites are the most likely places for skeletons to be buried. Any plant or animal that lives in or near a basin of deposition has a greater chance of being preserved as a fossil than an organism that lives far from such a site. Thus, the leaves of a tree growing along a lakeshore are more likely to be preserved than the leaves of a tree living on a hillside where erosion takes place.

If burial by sediments is so important for fossilization, the environment in which any plant or animal lives has much to do with the probability of preservation. Plants and animals that live on dry land, rather than in water, are less likely to be preserved. Sediment deposition occurs over large areas in the oceans, where less erosion takes place than it does on dry land. Thus, marine plants and animals have a better chance of being fossilized than terrestrial organisms. A clam that lives in a burrow in the mud has already buried itself and is a prime candidate for fossilization. Our fossil record of birds and flying insects is quite poor, partly because these animals commonly do not live close to sites of deposition.

PRESERVATION

Once a skeleton has been buried and removed from the zone of decay, it still has a long way to go before becoming a fossil. Sediment must accumulate more and more thickly over the skeleton, protecting it from burrowing animals and from erosion. Eventually, the sediment in which the organism is entombed becomes a **sedimentary rock** through the process of lithification. Lithification is a function of both compaction and cementation of loose sediment grains. In carbonate rocks (limestones), cementation occurs near the surface, but other rocks may require deeper burial before cementation can occur. Deeper burial places a greater weight on the sediment, which squeezes water from between the grains of sand and mud; this is compaction. Eventually, cementation also occurs. The end result is a hard, firmly bound sedimentary rock, so called because it was formed of what were initially loose particles of sediment.

There are many different kinds of sedimentary rocks, based on the original materials of which they were composed and the conditions of burial. In general, two major types of sedimentary rocks are recognized: **siliciclastic** and **non-siliciclastic**. Siliciclastic sedimentary rocks consist of those rocks formed by the transportation of discrete particles to a depositional site where they become buried. The particles may be moved or transported by water, ice, or wind. Different kinds of siliciclastic rocks are recognized on the basis of the sizes of the sedimentary particles. Thus, the rock **sandstone** is composed predominantly of sand-sized particles (Figure 3.2), **siltstone** is made of silt-sized particles, and **mudstone** or **shale** is made of clay-sized particles or mud. One of the most common sedimentary rocks is shale; it is composed of clay and can be split into thin layers. The coarsest siliciclastic rocks, composed of gravel, are called **conglomerates**.

Non-siliciclastic sedimentary rocks are those formed of particles that have not been transported long distances and have been chemically precipitated, either organically or inorganically. Rock salt is an example of a chemically precipitated non-siliciclastic rock. The most abundant non-siliciclastic rocks and the ones of greatest interest to paleontologists are **limestones**. These are formed of calcium carbonate and are typically composed of fossil fragments; thus, they were organically precipitated. Whereas organically precipitated shells are the most conspicuous feature of many limestones, most are formed by a combination of organic and inorganic precipitation, and some are completely derived through inorganic precipitation. Only a few limestones are completely unfossiliferous. Most limestones are marine, but some are freshwater (from lakes and rivers) in origin. There are many different kinds of limestones, too numerous

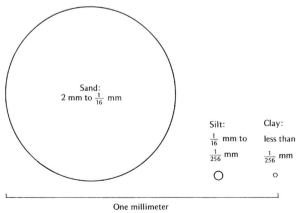

Figure 3.2   Relative sizes of common sedimentary particles, highly magnified. Sand grains range from 2 mm to 1/16 mm, silt particles are 1/16 to 1/256 mm in diameter, and clay particles are smaller than 1/256 mm.

Sand:
2 mm to $\frac{1}{16}$ mm

Silt:
$\frac{1}{16}$ mm to
$\frac{1}{256}$ mm

Clay:
less than
$\frac{1}{256}$ mm

One millimeter

to expand on here, based on differences in texture and on the components that make up the rock.

Two other main types of rocks exist: **igneous** and **metamorphic rocks**. The former consists of rocks that were once molten, such as lava from a volcano. Obviously, such rocks rarely have any fossils preserved in them, although some terrestrial plant and animal fossils are known to have been preserved by being buried under a lava flow. Metamorphic rocks are those that form by alteration of either igneous or sedimentary rocks due to high temperatures and pressures. If a metamorphic rock is formed from a sedimentary rock, such as marble from limestone or slate from shale, and if the rock is not too severely altered, fossils may still be preserved. Fossils in metamorphic rocks are relatively rare and are commonly quite poorly preserved; however, they are extremely important for understanding the origin of the sedimentary rock.

As sedimentary rocks are formed, they undergo conspicuous chemical and physical changes. Consequently, it is not surprising that the skeletons of organisms preserved in these rocks also undergo many changes from their original state. If a rock is quite young, say 1 to 3 million years old, contained fossils may not have been altered too much, if at all. Such young fossils may closely resemble shells that one would pick up on a beach, except that any coloration would probably be leached or faded. On the other hand, many (but not all) fossils that are older than a few million years have evidence of alteration in one or several ways. Therefore, preservation has two basic categories: **original preservation** and **altered preservation**. Some of the most unusual and spectacularly preserved fossils are those that remain in an unaltered condition.

A **mummy** is a specimen of a once-living organism in which some of the soft tissues are preserved, commonly muscles, skin, and tendons. Both artificial and natural mummies are known, the former produced especially by the early Egyptians, who used sodium salts and tars to preserve humans and many other animals, including snakes, bulls, cats, and beetles. All internal organs and the brain were removed during the initial stages of mummification and preserved separately. Artificial mummies, of course, are not fossils.

Natural mummies occur through some unusual happenstance of nature. The most common natural mummies are frozen mammoths preserved in the permafrost regions of Alaska and Siberia (Figures 3.3 and 3.4), but others include mastodons, horses, bison, woolly rhinoceroses, musk oxen, moose, and more. At least some of these mummified animals seem to have been suddenly entombed when they became stuck in soft, thawed sediment during a summer or when they fell into an ice crevasse. Stomach contents are preserved in some mummies, and it has been possible to do albumin and collagen typing on some of the preserved tissues. These tests confirm that the mammoth is much more closely related to the living Asian and African elephants than to the extinct mastodon.

For several centuries, the tundra of Siberia has yielded tons of fossil ivory in the form of fossil mammoth tusks. There has been an ages-long trade in this ivory from Siberia to China, where tusks, generally known as dragon bones, are ground up and used in Chinese medicines. In Alaska there are reports that fossil mammoth hair was so common in Pleistocene gravels that it clogged the dredges of placer gold miners.

In addition to frozen mummies, desiccated mummies are also known. All the water has been dried out of the flesh of these mummies, thus rendering the material

Figure 3.3   The carcass of an adult male mammoth found in 1900 on the bank of the Berezovka River in eastern Siberia. The individual was approximately 35 years old, with a height of about 9 ft. The geologic age is approximately 44,000 years. The mount is a dermoplastic reproduction of the original. (Courtesy of V.E. Garutt, Zoological Institute, St. Petersburg, Russia.)

Figure 3.4   The mummified carcass of a male baby mammoth found in 1977 in Magadan Province in eastern Siberia. The specimen was found in a sandpit at a depth of 6 ft in permafrost; it is emaciated, about 8 months in age, and slightly more than 3 ft tall. The geologic age is about 40,000 years, based on radiocarbon dating. (Courtesy of V.E. Garutt, Zoological Institute, St. Petersburg, Russia.)

unsuitable for bacterial action. Some of the most spectacular desiccated mummies are of extinct ground sloths preserved in caves or other cavities in the southwestern United States. The hair, skin, and dried tissues of one of the animals was found in a fumarole cave in a volcanic crater in New Mexico (Figure 3.5). In some instances, the mummies occur in caves that also contain piles of sloth dung up to one meter high, and this waste material has yielded much information about the plant diet of these extinct animals (Figure 3.6).

Even natural human mummies are known. A very old human burial in Egypt has been discovered that contained a mummified person that had not been treated with mummification procedures (Figure 3.7). In 1992, a natural mummy of a human was discovered melting from a glacier in the Italian Alps. This Stone Age man, approximately 5,300 years old, is the oldest intact human ever discovered.

The oldest known mummy is a hadrosaur (duck-billed) dinosaur from the Late Cretaceous of Wyoming. This animal was suddenly suffocated and preserved beneath a dense cloud of hot volcanic ash. This specimen gives us an unparalleled view of what dinosaur skin was like and also includes valuable information on muscles and tendons in a bipedal dinosaur (Figure 3.8).

What are some of the things that can happen to a fossil as it is buried deeper in the earth and sediments gradually lithify into rocks (Figure 3.9)? The most common type of alteration is that in which hard parts have all microscopic pore spaces filled by minerals precipitated from interstitial waters. This process is called **permineralization**, and makes the hard parts heavier and denser than they were originally. The open cell

Figure 3.5   Skeleton of an extinct ground sloth, *Nothotheriops*, from a fumerole at Aden Crater, New Mexico. Small particles of skin and hair are attached to the extremities of the legs. (Courtesy of Paul S. Martin, University of Arizona, and Peabody Museum, Yale University.)

Figure 3.6   View of the interior of Rampart Cave in northwestern Arizona showing piles of fossil ground sloth dung on the floor of the cave. The dung has been radiocarbon dated at 11,000 years. (Courtesy of Paul S. Martin, University of Arizona.)

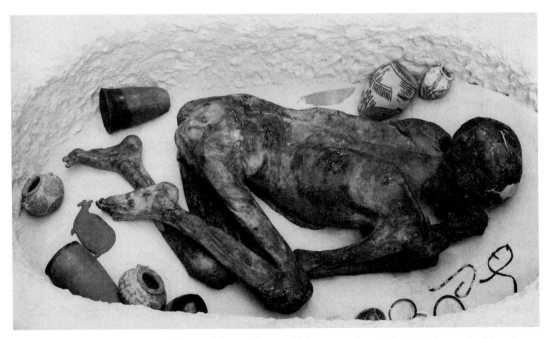

Figure 3.7   A natural mummy that is several thousand years old, from a predynastic burial in the sands of Egypt. (Courtesy of Trustees of the British Museum.)

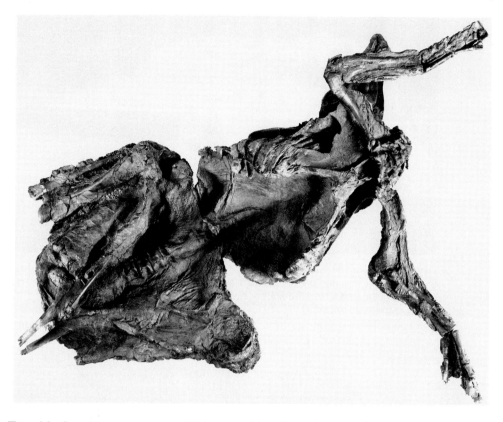

Figure 3.8   One of the rare mummies of dinosaurs and one of the oldest mummies known. This mummy includes the skeleton with preserved skin, muscles, and tendons of the ornithischian dinosaur, *Edmontosaurus annectens*, of Cretaceous age from the Kirtland Formation in San Juan County, New Mexico. (Neg. no. 35607 courtesy of the Department of Library Services, American Museum of Natural History.)

spaces in bones and wood become filled to form petrified wood or bone (Figure 3.10). Even the external shells of invertebrate animals (e.g., clam shells), which seem to the naked eye to be solid, have microscopic pores that are filled with mineral matter during fossilization. Virtually every fossil that is more than a few million years old and that looks like it is unaltered has some permineralization. A simple test for permineralization is to compare the weight of a fossil with that of a modern shell of similar size and shape; if permineralized, the fossil will be heavier.

Much more drastic things can happen to a skeleton than simply having its pores filled. The pressure from being buried under a thickness of rock or chemical changes in the interstitial waters can cause the original shell material to undergo **recrystallization**. Virtually all shells and bones have a detailed microscopic structure dependent on the way in which the animal grew the crystalline material of the skeleton. These microstructures are commonly visible in permineralized shells, but this has become obliterated where the fossil material was recrystallized.

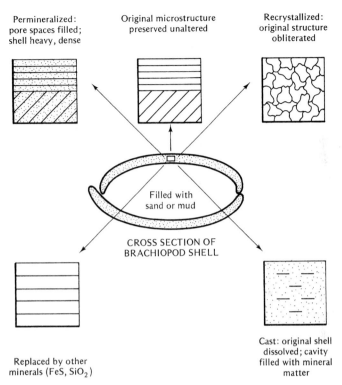

Figure 3.9   Different modes of fossil preservation of shell material.
Small section of a brachiopod shell is enlarged to diagrammatically show
the various ways in which hard parts may be preserved.

The fossil may also undergo **replacement** (Figure 3.11). In this process, the original material is dissolved by chemical action and other minerals are substituted so that a replica of the original shell is formed of foreign material. By far the most common replacing mineral is silica, which usually replaces skeletons composed of the mineral calcite. The process takes place on a one-to-one volumetric basis; that is, a given volume of original material is replaced by the same volume of silica. Thus, the size and shape of the fossil is not disturbed. In those instances involving limestones, the silicified fossils may be selectively etched from the rock using dilute acids. Exquisitely detailed fossils, otherwise impossible to chip from the rock, may be obtained in this way. The next most common replacing material is **pyrite**, iron sulfide (FeS_2) or fool's gold. Fossils replaced by pyrite are especially attractive as they have a shiny golden color. Pyrite is generally formed in organic-rich shales, with iron and sulfur resulting from the process of organic matter decay.

Another condition of preservation is the formation of molds and casts. Let us suppose that you surround a shell with potter's clay and then fire it in an oven, simulating the rock-forming processes of lithification. Then, you break open the clay and there is an impression of the shell on the inside of the clay. Such an impression is called a **mold** (Figure 3.12). Now, let us imagine that you somehow dissolved away the shell

Figure 3.10   Broken log of petrified wood from the Petrified Forest National Park, Arizona, approximately 67 cm (2 ft) across. The open, woody cells of the plant tissue have been filled with silica precipitated from percolating water in the sediments. Growth rings are well preserved.

Figure 3.11   A brachiopod shell of Permian age from the Glass Mountains of West Texas. The original calcareous shell was selectively replaced by silica in the rock. Dissolving the limestone matrix with acid released the shell with its long, delicate spines. The specimen is enlarged. (Courtesy of National Museum of Natural History.)

Figure 3.12   Molds of a large starfish, *Devonaster*; a small ophiuroid echinoderm; several brachiopods; and other fossils. The original shell material has been removed by solution, leaving behind only an imprint of the fossil in the rock. The specimens are from Devonian rocks of New York. The slab is approximately 25 cm (10 in) across. (Courtesy of National Museum of Natural History.)

that you placed in the clay. The molds of the exterior of the shell are still in place, and you now fill the vacant space with plaster of paris. You have formed a **cast** of the shell. Both casts and molds occur in nature (Figure 3.13). Shells or bone may be dissolved after burial, and the vacant spaces may or may not become filled later with foreign material, this is one method in which fossils are replaced. Molds and casts are especially common in coarse-grained rocks, such as sandstones, because waters can easily percolate through these rocks to dissolve shells.

There are many other ways in which fossils may be preserved (Figure 3.14). Insects are known from fossil amber, which is the hardened resin of certain conifer trees (Figure 3.15). Black shales, rich in organic matter and deposited on a sea floor depleted in oxygen, may have preserved remnants of an entire soft animal, because decay is retarded in such a setting. This type of preservation is called **carbonization**. In carbonization, all the light, more mobile organic molecules are driven off during compaction and diagenesis, but carbon molecules are very stable, remaining behind as a thin carbon film on the rock (Figure 3.16).

These hard and soft parts of ancient life are **fossils**, defined in Chapter 1 as any preserved remains of ancient life. Indications of ancient life other than shells, bones, and body parts also exist. These are called **trace fossils**, which are defined as any *indirect* evidence of ancient life. Many types of trace fossils exist, including, among others, footprints of ancient animals (Figure 3.17) in mud, burrows and trails of marine worms and crustaceans, or ancient excrement (Figure 3.18). Even certain chemicals may be considered trace fossils. When chlorophyll from green plants decomposes, it breaks down into organic molecules. Two of these, pristane and phytane, are very stable over geologic time. Pristane and phytane molecules have been recovered from rocks more than 2 billion years old, and these provide indirect evidence for the early existence of

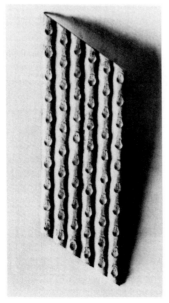

Figure 3.13   Natural cast of the stem of an extinct, primitive tree (lycopod, *Sigillaria*) from the Pennsylvanian of Illinois. The distinctive scars on the cast are places where leaves attached when the plant was alive. (Courtesy of Field Museum of Natural History, Chicago.)

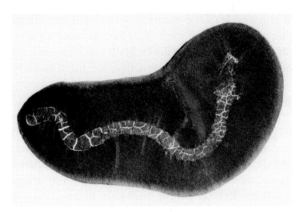

Figure 3.14   Preserved impression of the soft parts of a holothurian (an echinoderm that lacks hard parts except for microscopic spicules) of Pennsylvanian age from Illinois. The specimen, about 10 cm long, is preserved in an ironstone nodule that presumably formed soon after the animal died, preventing decay. (Courtesy of Field Museum of Natural History, Chicago.)

Figure 3.15   A caddis fly preserved in amber, the hardened resin of certain kinds of coniferous gymnosperms. The specimen is much enlarged and is Oligocene in age from East Prussia. The wing venation, hairs, bristles, and eye facets are all preserved. (Courtesy of National Museum of Natural History.)

Figure 3.16   Carbon film on an extinct arthropod belonging to a group called the eurypterids. These were especially common during Silurian time. The specimen is about 25 cm long. (Courtesy of Field Museum of Natural History, Chicago.)

Figure 3.17 Footprints of two fossil vertebrates, much reduced, of Permian age from Arizona. Can you tell the direction in which each animal was moving and which one walked over the area first? (Courtesy of National Museum of Natural History.)

organisms capable of photosynthesis. These and other chemical trace fossils that are diagnostic of former life are called **chemical fossils**.

In some respects, given the odds against preservation, it is amazing that so many different kinds of plants and animals are preserved in the fossil record. On the other hand, it is true that the record of life is not perfect. Given the vicissitudes of burial and fossil alteration, the record is quite good; occasionally truly remarkable fossils are preserved, such as fossil mummies, insects in amber, or body outlines in black shales. Another reason for incomplete knowledge of the fossil record is that paleontologists have not uniformly explored Earth for fossils. We have only recently begun to obtain much information about fossils and sedimentary rocks from the ocean basins. A deep-sea drilling project, first called JOIDES (Joint Oceanographic Institutions for Deep Earth Sampling), now called ODP (Ocean Drilling Program), has only been operational since 1963. We know much more about the rocks and fossils exposed on land because they are much more accessible. However, even on land, we know a lot more

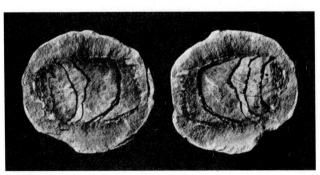

Figure 3.18 Two halves of a fish coprolite that have been preserved in a concretion of Pennsylvanian age from Illinois. A coprolite is the preserved feces of an animal; they are common fossils at some localities. (Courtesy of Field Museum of Natural History, Chicago.)

about the fossils of western Europe and North America than we know about those from many other parts of the world. This is mainly because most paleontologists have been trained in these two areas. The paleontology of parts of South America, Africa, Asia, and Antarctica is still in an exploratory state. Finally, our knowledge of fossils collected on land areas is largely limited to those rocks that are exposed on the surface today. Many rock layers have been removed by erosion, and their contained fossils destroyed. Other rocks are completely buried under younger rocks and may be accessible only in mines or by taking cores of rocks from oil-well drilling. Despite all these difficulties, paleontologists have put together a remarkably complete record of life on Earth. We are missing many smaller details, and we continue to argue about the causes of phenomena exhibited by fossils. However, the history of life is certainly well documented in its broad outlines.

ENVIRONMENTS OF DEPOSITION

Life today is very diverse, and all plants and animals are restricted to certain habitats and environments. On the broadest scale, for example, there are very few animals that can live both in water and on dry land. We expect that the same situation applied to ancient life—that fossil plants and animals were restricted to specific environments. The rocks in which we find specific fossils may or may not provide a record of the environment in which the ancient organisms once lived. Fossils may provide such a clue if an organism died and was buried at or near the site where it lived. However, we know that in some cases hard parts can be transported far from their place of origin. We find, for existence, that logs of ancient tree trunks were rafted down rivers, carried into the sea, and eventually sank far from the terrestrial environment where they grew. Waves and currents may also transport empty shells from one site to another in the ocean. Paleontologists must always be aware of the potential for fossil transportation, and many criteria are used to assess this possible problem. However, despite the potential for transportation, the vast majority of macroscopic fossils probably have high fidelity for preservation within the environment where they lived—either at their living site or, more commonly, moved from the living site but still preserved within the basic environment where they lived. This is verified by study of modern sediments, in which shells of marine organisms are preserved in the sediment of the environment in which they lived. Even for many terrestrial vertebrate organisms that were swept from their living sites during floods, bones are simply deposited in streams within the same valley where the organisms lived. Microscopic organic remains, such as oceanic plankton or spores and pollen from terrestrial plants, have a much higher transportation potential.

Not only are the remains of life different from place to place in rocks, but the physical aspects of the environment also differ. The floor of the ocean today is covered by very different kinds of sediment from place to place. A coastal bay may have sandy beaches and bars, the center of the bay may be floored by mud, and rocky shorelines may have coarse gravel seaward. All these sediments may be buried and changed into sedimentary rocks, and yet they will all be of the same age and will be in close proximity to one another. If we look at the animals living on or in a sandbar, a muddy bottom, or a gravel bed in a single bay, we would find that each environment is occupied by a different suite of animals. Both the physical and biological characteristics of sedimentary rocks provide clues to the depositional environment of any layer of fossiliferous sedimentary rock. We commonly have very specific knowledge of what the sea floor was like at a given place and time because fossils reflect and record the nature of that sea floor. However, we may have to make inferences or do especially detailed studies if we hope to answer questions about the temperature, salinity, or water depth of an ancient setting. Reconstruction of ancient environments is part of sedimentology, and the study of the interaction of ancient organisms with their environment is called **paleoecology**. In order to be successful, such research must take into account all data that can be obtained by study of both the rocks and the fossils that they contain. Studying one without the other may lead to inconclusive results or misinterpretations.

THE TIME-SPACE PROBLEM IN PALEONTOLOGY

It is very important in paleontology to be able to demonstrate whether or not different rocks and fossils in different areas are of the same age or of different ages. If a paleontologist studies only a single outcrop of a rock or a single vertical section of a rock, the problem is relatively simple—superposition alone determines the age relationships. However, if rocks and fossils are studied over some geographic area (one square mile or several states), the problem arises as to whether or not rocks and fossils in different areas are contemporaneous or perhaps of different ages. Questions about time relations over an area are referred to as the time-space problem, and they cannot be answered by the study of fossils alone. Several aspects of sedimentary rocks must be considered.

STRATIGRAPHY

The study of stratified (layered or bedded) rocks is called **stratigraphy**. A stratigrapher attempts to decipher the sources, distribution, limits, and environments of deposition of sedimentary rocks. In any stratigraphic study, the first step is to divide a sequence into working units that can be measured and that can have their physical characteristics and fossil content described. Imagine a large hill with rocks exposed from top to bottom that consists of beds of shale, sandstone, conglomerate, and limestone. Each rock type must be delimited, and the sequence of rock types must be worked out. Depending on its thickness, each rock type occupies a certain area on the hill and has a definite interval of altitude on the hill, for example, from 200 to 250 meters above sea level.

Now let us suppose that there are several other hills in the immediate area, each of which also has a rock sequence exposed. The stratigrapher wants to match these rocks from one hill to another. How is this accomplished? The technique that is used is called **geologic mapping**. A geologic map shows the distribution of different rock types over an area (Figure 3.19), thus differing from a road map that shows only towns and roads. The sequence of rocks on the hill is divided into basic mapping units called **formations**. They are based on physical characteristics of the rocks. Each for-

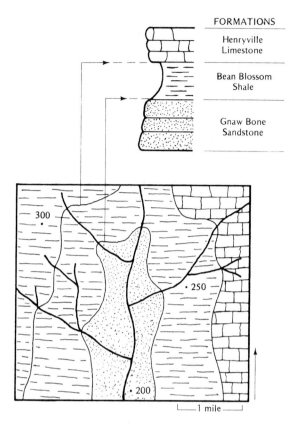

FORMATIONS

Henryville
Limestone

Bean Blossom
Shale

Gnaw Bone
Sandstone

300

· 250

· 200

1 mile

Figure 3.19 Use of formations to construct a geologic map. At the top is a section of rocks with three mappable rock units (formations). The map below shows the distribution of these three rock types over an area. Bold lines are streams; light lines indicate the positions of the boundaries between formations. Numbers give heights above sea level. Note that the oldest formation, the sandstone, occupies the valley floor and that the youngest formation, the limestone, occupies the highest ground.

mation has definite upper and lower boundaries at which changes to other formations occur. Every bed of rock must be contained within a single formation. There can be no gaps or overlaps. Formations are given geographic names, such as St. Louis Limestone, Chattanooga Shale, or Columbus Limestone, after towns, streams, mountains, or other features. Although not strictly part of the rock description, the fossil content of all formations must also be described.

As an example, let us suppose that at the base of our hill, we can see 20 meters of exposed sandstone; directly above this is 10 meters of limestone, followed by 15 meters of shale. The transition from one rock type to another is quite abrupt so that the change from sandstone to limestone forms a visible line on the side of the hill. That line marks the boundary between two different formations, and it can be accurately marked on a map. The line between the limestone and the shale can also be delimited. Once the positions of these boundaries have been established on one hill, the different rock types can be traced to an adjacent hill with the lines indicating the boundaries between formations carried along from one hill to another. In this way, a geologic map showing the geographic distribution of each formation can be constructed. If the mapping is done over a large enough area, the map may then show the geographic limits of different formations. Some units, such as the limestone mentioned, may thin and disappear, so that the shale is directly above the sandstone. Other new formations may also appear in the rock section.

Geologic mapping relies on application of Steno's three general principles for stratigraphic interpretation of sedimentary rocks: **principle of original horizontality**, **principle of lateral continuity**, and **principle of superposition** (see Chapter 1).

At this point, you may well ask, what does all of this have to do with fossils? The process of matching sequences of rocks from one area to another is called correlation. In making a geological map, a stratigrapher correlates or matches the shale formation and the limestone formation across an area. This is done largely on the basis of the physical characteristics of the rocks, and it is called **lithostratigraphic correlation**. The stratigrapher may then ponder, "I know that the shale unit is the same rock unit across this area, but is the shale the same age over the entire area? Perhaps while the mud was being deposited to the west, limestone was being deposited to the east."

There may be more than one approach to answer this question. First, a clear way to determine the temporal relationships across an area is to identify a short-duration, unique event. Strata that record short-duration occurrences in geologic time are commonly called **event beds**. For example, during the eruption of a volcano, volcanic ash is dispersed across a broad region and deposited in a relatively short time. Thus, ash would appear as a single sedimentary layer. Through the processes of lithification and diagenesis, a volcanic ash bed is transformed into a bed of clay that is called **bentonite**; this bed would represent a virtually instantaneous, in geological time, time marker across a large area. Event beds could also be formed by a large storm.

Unfortunately, event beds are uncommon, and all but the most catastrophic event will not be preserved in rocks globally. The second, and standard, way to determine time relationships among rocks is through the study of fossils. Every animal and plant species that ever lived on Earth lived during a unique interval of geologic time, beginning when the organism originated and ending when it became extinct (or today if still living). Some species lived for relatively short periods of geologic time, whereas others lived for long periods. By carefully collecting fossils through the rock sequence, a paleontologist can pinpoint the level at which any species first appears and where it disappears. By plotting these vertical (time) ranges through rocks, a pattern of species duration through time can be established, and the pattern can be compared with similar patterns in adjacent rock sequences. This endeavor is called **biostratigraphy**, the study of the temporal ranges of fossils and their use in correlating rocks in time (Figure 3.20). The correlation of fossil occurrences from one area to another is called **biostratigraphic correlation**. An interval of rock, whether it is a sandstone, limestone, or shale, that contains a distinctive assemblage of fossils is called a fossil zone. One or more of the species may be confined to that interval, or the interval may be characterized by a distinct suite of concurrent species.

In order for a fossil to qualify as a good **zonal fossil**, it generally should have a reasonably short geological time span—a few million years. The fossil brachiopod *Lingula* is known from the Ordovician period to the Holocene, a span of 500 million years. Obviously, *Lingula* is not a good zonal fossil. In addition to a short duration, a zonal fossil should, ideally, be reasonably common. Furthermore, it is helpful if it is distinctive so that it is readily recognized, and it should be a plant or animal with a

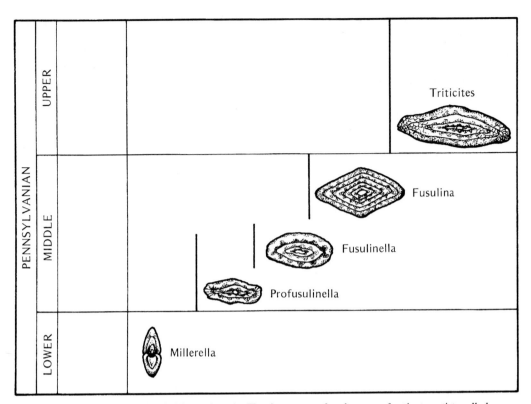

Figure 3.20   Use of fossils in biostratigraphy. The time range of each genus of extinct protists called fusulinids is shown, with diagrammatic cross sections of each type. The fossils mark a sequence of fossil zones in rocks of Pennsylvanian age that is used in biostratigraphy.

wide geographic distribution, i.e., not restricted to a specific environment. A fossil brachiopod may have lived only on soft, muddy bottoms and, thus, may only be present in shales and mudstones. Therefore, this brachiopod may have a restricted distribution and may not be a good zonal fossil. Alternatively, a fish may have lived above all types of bottoms—sandy, silty or gravelly, and when it died become entombed in any of these, resulting in a wide geographic distribution. Generally, those plants and animals that lived in the water, swimming or floating, make better zonal fossils in the oceans than animals that lived on the bottom and were controlled by bottom conditions. On land, spores and pollen that travel through the air are excellent zonal fossils.

Fossil zones are arranged to completely divide rock sequences. Thus, Jurassic rocks are divided worldwide into 74 zones and 144 subzones, based largely on fossil ammonoid cephalopods. Because the duration of this time period is 64 million years, we can discriminate zones into intervals of time less than 1 million years by using these fossils. With subzones, time intervals can be determined, on average, by less than .5 million years. The system of using fossils for correlation is founded on the **principle of faunal and floral succession**. This was first recognized in the nineteenth century by **William Smith**, a British surveyor and canal engineer, who documented the orderly succession of fossils among different strata in England. Succession is the product of organic evolution and occurs as species replace one another through time; older plants and animals become extinct and new ones evolve. Because of the sequential changes in life through time, we can use fossil succession to characterize and date rock strata. We can do this not only on a local basis, but we can also correlate rocks that are widely separated on different continents.

Biostratigraphy is a very important task of the paleontologist. In order to make environmental analyses of rocks and fossils, it is necessary to be quite certain that the rocks being studied are of the same age, and biostratigraphy provides this means.

FACIES

As briefly mentioned, sediments that are being deposited today are not done so uniformly everywhere. On some ocean bottoms, sand is being deposited, whereas on others limestones are accumulating. On land, rivers may deposit sand on the beach, and during floods, fine muds are deposited over the flood plain. All these different kinds of sediment are contemporaneous; that is, they are being deposited at the same time. This has always been the case on Earth, so the rock record for any given time is very diverse.

The principle that we have just described is part of a broader unifying geologic principle called **uniformitarianism**, which is a fancy word that means that the same processes that are operating on Earth today operated during the geologic past. This is commonly summarized in the phrase, "the present is the key to the past." There have certainly been differences in rates of processes through geologic time, as well as differences in the emphases of certain processes, but the basic physical, chemical, and biological processes operating through time are considered to have been the same. For example, there have been times during the past when there were immense coastal swamps, the plant debris of which formed extensive coal beds in the rock record. Such coastal swamps are much restricted today, but the basic process of coal formation is judged to be the same.

In an ancient bay, mud could have been deposited in the center of the bay, and this mud could give way to sandy muds and, finally, to sands along the shore. When changed into rocks, the muds and sands become shales and sandstones that intergraded laterally with each other. Such changes, when they occur within rocks are called a **facies** change. Facies are a lateral change in the character of rocks due to a lateral change that occurred during the deposition of sediment (Figure 3.21). Not only would the physical aspects of the sedimentary rocks have changed, but we would expect the fossil content of the rocks to also have changed. We would not expect to find many of the same animals living on a muddy bottom and a contemporaneous sandy bottom, because animals are adapted to different environments. Such lateral changes in the occurrence of fossils are called **biofacies** changes.

Facies changes in rocks and contained fossils are one of the main clues that paleontologists use to solve the time-space problem. We know that the face of Earth is changing constantly. Rivers shift their courses, lakes fill up with sediment and are destroyed, the sizes of shallow marine bays shrink or expand, and sea level rises and falls. These kinds of changes are what cause different rock types to be deposited in the same area through time. At times during the past, North America was almost 75 percent covered by shallow oceanic seas; these seas retreated, only to flood parts of the continent again. During a time of flooding, offshore muds were deposited on top of nearshore sands. During a retreat of the sea, the opposite may have occurred;

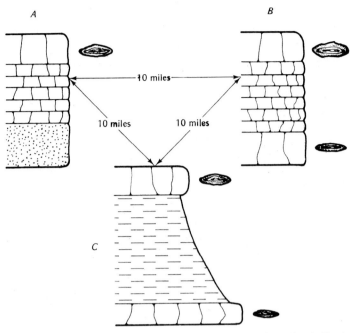

Figure 3.21   Facies and correlation. Three stratigraphic sections of rock 10 mi from each other are shown. The limestone at the top contains one species of *Fusulina*. The lower limestone contains another species of *Fusulina* at sections B and C. The thin-bedded limestone in the middle of the section at A and B is a facies of the thicker shale at section C. The basal sandstone at A may or may not be a facies of the lower limestone at B and C (see Figure 1.2 for standard key for rock types).

nearshore sands may have been laid down on top of offshore muds. This would produce a rock sequence of sandstone overlaid by shale, in turn overlaid by sandstone (Figure 3.22). This results in a dynamic stratigraphic record. In a single stratigraphic section, the lower sandstone is older than the shale, and the shale is older than the upper sandstone. However, on a regional basis, some shale is younger than the sandstone below and younger than the sandstone above, but not all of it. Some shale would be of the same age as the sandstones, and these would be facies of one another. Sorting out age differences versus facies differences is one of the primary tasks of paleontologists using biostratigraphy and fossil zones. It should be obvious why fossils that were not strongly controlled by the environment would be most helpful in determining age relationships. Alternatively, fossils that were strongly influenced by environment when they were alive would be the most helpful tools in characterizing environments that were present during the past. Such fossils, called **facies fossils** because they are strictly controlled by the environment, are of special interest in paleoecological interpretations.

Another goal of paleontology is to reconstruct the past as accurately and in as much detail as possible. A paleoecological study may reveal where and for how long particular environments were present over an area. It may be possible to construct **paleogeographic maps** showing the distribution of ancient environments (Figure 3.23).

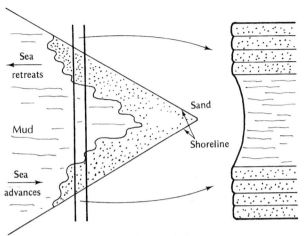

Figure 3.22   Idealized diagram of an advance and retreat of the sea on the left, with sand being deposited close to the shoreline and mud offshore, with the rate of sediment accumulation constant. The rocks that result in one area (shown by the double lines) from this transgression and regression are shown on the right. Shale is in the middle of the section, indicating maximum advance of the sea, with nearshore sandstones above and below.

Such maps might show ancient shorelines where land met the sea, areas of coastal sand dunes or offshore sand bars, river channels, deltas, islands, ancient coral reefs, and so on. The plants and animals that lived on, in, and above these environments can be analyzed in terms of communities, associations of organisms that once lived together. Communities and environments can be traced forward and backward through time, revealing changes that occurred, environmental shifts, and addition or deletion of organisms to communities during the course of evolution. Such comprehensive analyses must be based on careful study of both the physical and biological aspects of rocks. Neither can be neglected. The age relationships of the rocks must be evaluated and time-space problems solved; otherwise, rocks that are of different ages may be judged erroneously to be contemporaneous facies of each other.

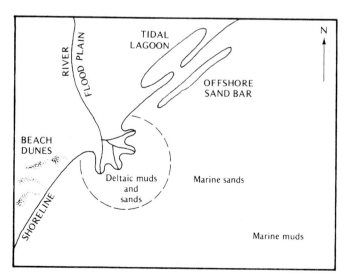

Figure 3.23   A simple paleogeographic map showing reconstruction of ancient environments over an area. Such a map depicts ancient, contemporaneous facies and is reconstructed using all available rock and fossil information.

KEY TERMS

altered preservation
aragonite
basin of deposition
bentonite
biofacies
biostratigraphic correlation
biostratigraphy
calcite
carbonization
cast
chemical fossils
chitinophosphatic
conglomerate
event beds
facies
facies fossil
formations
fossil

fossilization
geologic mapping
hard parts
igneous rock
limestone
lithostratigraphic correlation
metamorphic rock
mold
mudstone
mummy
non-siliciclastic
original preservation
paleoecology
paleogeographic map
permineralization
principle of faunal and floral
    succession
principle of lateral continuity

principle of original
    horizontality
principle of superposition
pyrite
recrystallization
replacement
sandstone
sedimentary rock
shale
silica
siliciclastic
siltstone
skeleton
Smith, William
stratigraphy
trace fossil
uniformitarianism
zonal fossil

READINGS

Introductory college-level textbooks on physical and historical geology include discussions of sedimentary rocks, environments of deposition, facies, and correlation. The latter books also discuss the uses of fossils in correlation, biostratigraphy, and preservation. These books are designed for introductory geology courses and provide much more detail than is possible to cover in a paleontology course.

Boggs, S., Jr. 1985. Principles of Sedimentology and Stratigraphy, 2nd ed. Prentice-Hall. 774 pages. The emphasis in this book is on broad-scale interpretations of sedimentary rocks with an emphasis on facies and environments of deposition.

Feldman, R.M., and others. 1989. *Paleotechniques*. Paleontological Society Special Publication 4. 358 pages. This is the only up-to-date book in English dealing with all the various laboratory techniques used for working with fossils.

Lockley, M.G., and Rice, A. 1990. *Volcanism and Fossil Biotas*. Geological Society of America Special Paper 244. 125 pages. A fascinating account of sudden burial of organic remains by volcanic ash; includes an interesting chapter on plant remains preserved at Pompeii.

Rixon, A.E. 1976. *Fossil Animal Remains: Their Preparation and Conservation*. Athlone Press. 304 pages. A useful handbook of techniques for working with invertebrate and vertebrate fossils.

Stanley, S.M. 1993. *Exploring Earth and Life Through Time*. W.H. Freeman. 538 pages. A comprehensive historical geology textbook.

Origins of Earth, Oceans, Atmosphere, and Life

INTRODUCTION

Why consider the origin of Earth in a book about fossils? The answer is that the origin and early evolution of life are thought to be closely tied to the composition of the early atmosphere and oceans. These prime features of Earth have not always been in existence, and they were not the same during early Earth history as they are today. Obviously, the planet Earth formed before the oceans and the atmosphere existed; therefore, the processes of Earth's formation were important parts of these interrelated events.

ORIGIN OF THE EARTH

In Chapter 1, on geologic time, it was established that Earth is indeed very ancient, older than 4.5 billion years. We know that Earth is a **planet** with one moon in a solar system that includes one star, the sun, and planets orbiting around the sun at various distances. Our solar system is a tiny part of an immense nebula of the stars represented by the familiar Milky Way Galaxy, and this galaxy is only a small fraction of the large number of nebulae and galaxies that make up the universe. The origin of Earth cannot be understood in isolation, because any theory that explains Earth's origin must also explain the remainder of the solar system. It is highly improbable that Earth formed alone and uniquely, independent of its nearest neighbors. Any explanation of Earth's or the solar system's origins must take into account and satisfactorily explain a series of important observations made by astronomers:

1. The solar system is part of a nebula.
2. The sun represents most of the mass of the solar system (2×10^{30} kilograms).
3. The planets have most (98 percent) of the angular momentum of the solar system. The angular momentum is mass times velocity times distance. Because most of the mass is concentrated in the sun, the velocities of movement of the planets are as great as are their distances from the sun.
4. Earth almost lacks noble (heavy, inert) gases, such as xenon, neon, and krypton, at least in the abundances that they are present in the sun.
5. Earth is layered with a thin outer **crust**, only a few kilometers thick; a thick **mantle** composed of very heavy rock under tremendous heat and pressure; and a liquid or molten outer **core** and a solid inner core, both composed of a material that is similar to an iron-nickel silicate.

6. Each planet has a different density, and most of them are denser than the sun. Earth has a density of 5.5 grams per cubic centimeter; the Sun and Jupiter have densities of 1.4. The differences in density mean that each planet has a somewhat different chemical composition and that each planet may have formed at a somewhat different temperature. The differences may also imply that the planets formed at somewhat different times and that there may have been a temperature gradient from the sun to the outer limits of the solar system.

Two contrasting ideas concerning the origin of the solar system have vied for scientists' acceptance for the past 400 years. The two ideas can be expressed with the question, are the planets the daughters or sisters of the sun? The older of the two ideas was originally proposed by the French philosopher and mathematician René Descartes and has the planets spinning off into space from a large, hot proto-sun (Figure 4.1). We now know that in order to accomplish this feat, a near collision of the sun with another star would have been necessary in order to pull out the material for the planets. A serious objection to this hypothesis is based on the distribution of mass and angular momentum in the solar system. If the planets are indeed daughters of the sun, the sun should still be spinning rapidly enough for it to have a much greater share of the angular momentum and somewhat less of the total mass of the solar system than it does.

The second hypothesis, that the sun and planets are sisters, states that the solar system began as an enormous **nebular cloud** of gases and dust spinning as a spiral arm of the Milky Way nebula. The cloud would have been very large, 30 or 40 light years across, with a mass two to ten times less than that of the present solar system. There would have been only a few atoms of matter per cubic centimeter. As this cloud spun slowly, the force of gravity and magnetic attraction would have caused particles to coalesce and the cloud to shrink. As dust accumulated into larger particles, they would have been collected, adhering to each other. This coagulation process would have con-

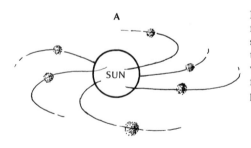

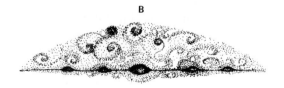

Figure 4.1   Alternate hypotheses for the formation of the solar system. (A) Planets formed from the sun. (B) A swirling dust cloud with local aggregations of matter formed a proto-sun and proto-planets.

tinued until a large part of the mass of the nebular cloud was concentrated near the center, with collisions of materials producing heat. This proto-sun would have gradually warmed until it reached a temperature at which spontaneous nuclear reactions could take place, mainly involving hydrogen. At this point the sun would have become very much hotter, and the heat would have driven off vast streams of lighter atoms into space, creating what are called **solar winds**. This wind would have stripped the frozen inert noble gases and light gases such as hydrogen from the developing proto-planets. Thus, the solar winds would have caused much of Earth's original mass to be dissipated into space, especially huge quantities of light elements, such as hydrogen and helium. Under these circumstances, it is estimated that the present mass of Earth is only about 1/1,200 of its original mass.

The dust cloud hypothesis, outlined above and in Figure 4.1B, is generally accepted by astronomers. Earth may have initially accreted in a cool condition. As Earth became larger, heat released by radioactive decay of elements within the rocks became trapped and could not escape to the surface. Thus, Earth gradually warmed and became hot to the point where melting or, perhaps, partial melting may have occurred. In a molten or semi- molten state, heavier elements and compounds slowly sank to the center of Earth, forming a dense, heavy core, thought to be composed of iron-nickel silicate minerals that were still partly in a molten state. Lighter materials "floated" to the top, forming a thin, light crust. Materials of intermediate density, those forming the bulk of Earth, formed the mantle. This differentiating phase of Earth will be termed the embryonic Earth (Table 4.1). As Earth gradually cooled, the crust solidified. There was no atmosphere, all the lighter elements having been blown away by the solar winds. There was no water on the surface, the heat of formation having vaporized any free water that may have been present. We will call this **initial Earth**. Earth gradually cooled, but as it did, it continued to receive an important source of heat. Radioactive elements trapped within Earth continually gave off heat as they decayed. This provided a major source of internal heat that is still important.

ORIGIN OF THE OCEANS AND PRIMITIVE ATMOSPHERE

Initial Earth was solid, but it lacked oceans and an atmosphere. As known, there would have been only a few processes operating on initial Earth, such as meteorite impacts and volcanic eruptions. These processes, especially volcanism, can account for the next changes on Earth.

Table 4.1   CONDITIONS OF THE LITHOSPHERE, ATMOSPHERE, AND HYDROSPHERE DURING VARIOUS STAGES OF THE DEVELOPMENT OF EARTH.

Earth Stages	Lithosphere	Atmosphere	Hydrosphere
Embryonic Earth	Molten or hot, differentiating	Original light elements	None
Initial Earth	Solid, hot and cold	None	None
Primitive Earth	Solid, more moderate temperatures	H_2O, CO_2, N, NH_3, CH_4, SO_2	Present
Present Earth	Solid	Free O_2	Present

One by-product of volcanic activity is **volcanic outgassing**. Various gases are emitted accompanying eruptions, and emission of gases continues long after the eruptive phase is completed, as fumaroles, geysers, etc. Gases emitted from the interior of Earth include water vapor (H_2O), carbon dioxide (CO_2), nitrogen (N), sulfur dioxide (SO_2), methane (CH_4), and ammonia (NH_3). Through this volcanic outgassing, an atmosphere gradually accumulated, and through condensation of the water vapor, water gradually accumulated to form oceans and lakes. Thus, over time, the conditions changed from those of initial Earth to those that we term **primitive Earth** (Table 4.1) Primitive Earth was solid with more moderate temperatures than initial Earth; oceans existed, as did an atmosphere. This atmosphere had no free oxygen. It was composed exclusively from volcanic outgassing, so it was composed of water vapor (H_2O), carbon dioxide (CO_2), nitrogen (N), sulfur dioxide (SO_2), methane (CH_4), and ammonia (NH_3). These atmospheric conditions would be poisonous to us and to nearly all life on Earth. However, the elemental composition of this primitive atmosphere was highly significant. It was primarily composed of hydrogen (H), carbon (C), oxygen (O), and nitrogen (N), and these elements are the fundamental components of the organic molecules that form living organisms.

Many of the same physical processes operating on Earth today would have occurred on primitive Earth, including the hydrologic cycle of evaporation and rain, tides, a climatic gradient from the tropics to the poles, and others.

ORIGIN OF LIFE AND THE PRESENT EARTH

Our understanding of the next stages in the development of Earth, which included the origination of life, occurred from a series of provocative experiments by Stanley Miller in 1953 (Figure 4.2). He put methane, ammonia, hydrogen, and water vapor into a closed flask at ordinary room temperature and pressure. Note that these four materials, or at least the elements that compose them, are thought to have been present in the atmosphere of the primitive Earth. No free oxygen was used. He then subjected this mixture to continuous electrical discharges for one week. The liquid phase of the mixture gradually changed color, turning a light yellow. When Miller analyzed the resultant liquid, he found that he had synthesized several **amino acids**. These are the building blocks of life, especially of large protein molecules. Miller had simulated possible conditions in the primitive atmosphere and oceans on the assumption that storms and lightning may have been frequent occurrences and that amino acids may have formed close to the air-water interface. Other experimenters tried somewhat different combinations of starting materials and different sources of energy and also synthesized amino acids. It is now known that amino acids can be formed experimentally in this way by using not only electricity but heat, ultraviolet light, sunlight, and radioactivity. Such experiments give us insight into the very first events on Earth that ultimately would lead to life.

Amino acids are constructed of small molecules that have from ten to twenty atoms each. The initial steps in forming the amino acids seem to have involved several very simple but highly reactive organic molecules such as formaldehyde (CH_2O), acetylene (H_2C_2), and hydrogen cyanide (HCN) (Figure 4.3). These have

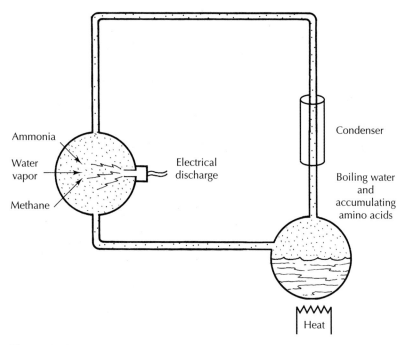

Figure 4.2   Stanley Miller's experiment that produced amino acids, the building blocks of life. Three gases, water vapor, methane, and ammonia, were exposed to repeated electrical discharge. This resulted in the gradual accumulation of amino acids and other simple organic molecules in the water.

also been detected in the experiments, and they would have rapidly combined to produce amino acids.

This process of forming organic molecules, such as amino acids, from nonbiological materials is called **abiotic synthesis**. The term means that these materials were synthesized without any living organism being present. Today, amino acids are formed naturally only by living organisms, but initially, before life existed, these building blocks were synthesized from materials by nonliving energy sources. Abiotic synthesis does not occur today because organic molecules are consumed by many organisms, especially bacteria, before complex organization can occur. As amino acids and other simple organic molecules gradually accumulated in the oceans or in fresh water (we are not certain which), some of them combined into larger and larger molecules, so that proteins or proteinlike giant molecules were eventually formed. These then combined into still larger structures that may have been cell-like. Continuation of experiments like Miller's have attempted to simulate primitive Earth conditions and have yielded larger proteins and even membrane-bound structures. One intriguing experiment was performed by Sidney Fox, a biochemist, who was able to link amino acids together by boiling them in seawater, thus simulating conditions near an undersea volcano. In the boiled mixture, he found tiny microspheres surrounded by a double-walled membrane, some of which had divided. For living systems to develop, amino

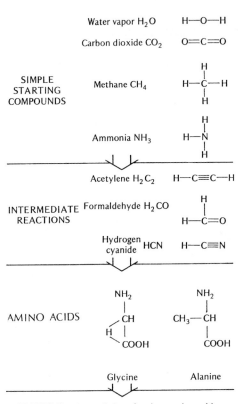

Figure 4.3   A potential pathway for the origin of complex organic molecules. Start with a few simple inorganic compounds such as water, carbon dioxide, methane, and ammonia. These may be synthesized into highly unstable, short-lived intermediate compounds that in turn may combine into amino acids, the building blocks of proteins and life.

acids must be assembled into long chains to produce protein and nucleic acid molecules. How did these very large molecules become assembled? One suggestion is that tiny crystals of clays served as templates to which the amino acids adhered in definite sequences.

In order for such cells to qualify as living entities, they would have to be self-replicating, that is, have the ability to reproduce, and they would have to utilize some energy source in order to maintain a metabolism. A replicating, metabolizing entity has not been created in the laboratory. Exactly how these latter steps of abiotic synthesis occurred is still a matter of speculation.

All of this synthesis took place in a reducing environment. There was no free oxygen that could be metabolized by primitive organisms as a source of energy. In fact, free oxygen would have been a poison to these first organisms. Also, there was no ozone layer in the atmosphere to shield Earth's surface from ultraviolet radiation that would have been lethal to some of this first-formed life. Probably the surface waters of the oceans provided the first ultraviolet shield. However, about 10 meters of water is needed to provide such a shield, so life surely did not originate right at the surface.

What would these first kinds of life have been like? They would have been very small (1 to 2 micrometers, μm) and simple (composed of a single cell). They would have been able to survive without oxygen; free oxygen would have been a deadly poison for them. In order for the organisms to have obtained nourishment, they would have been what are known as **anaerobic heterotrophs** (anaerobic because they lived in an oxygen-free environment; and heterotrophs because they used outside sources for food—they consumed food). Organisms that are able to manufacture their own food (make complex organic molecules from simple compounds) are called **autotrophs**. Green plants are a common example.

The waters of Earth at this stage probably contained large numbers of organic molecules of various sorts, formed in the ways that we have just described. These first organisms would have utilized the molecules in what has been called a primordial, pre-biotic, or organic soup. The very earliest kinds of life would have fed from the organic molecules of this "organic soup."

From these earliest anaerobic heterotrophs, how did life as we know it develop on Earth? Let's speculate, given the conditions that existed during the early Archean, on how this may have come about. As anaerobic heterotrophs expanded, their food source would have become depleted rather rapidly resulting in the Earth's first food shortage. If an organism could survive without external food, it would be at a great advantage in such a situation. Thus, it is believed that **anaerobic autotrophs** would be the second major type of life to have evolved on Earth. These microbes could take sim-ple compounds that were available in large quantities and combine them into more complex organic molecules that could serve as an internal energy source.

In order to do this, there had to be an external source of energy because the chemical reactions in which simple materials such as carbon dioxide and water are combined into a sugar require energy. One potential source of energy is that produced when certain molecules are split. For instance, energy is released when hydrogen sul-fide (H_2S) is split, and the sulfur is changed to elemental sulfur (S). These are called sulfur-reducing bacteria, and the chemical formula for their metabolism is:

$$CO_2 + H_2S \rightarrow (CH_2O)_n + S$$

Once energy-converting pigments evolved, the obvious source of energy—sun-light—could be used. This is done in green plants by the pigment chlorophyll and in bacteria and some algae by other pigments. In photosynthesis, two of the most abun-dant compounds available on Earth are used as a source of food:

$$CO_2 + H_2O \rightarrow (CH_2O)_n + H_2O + O_2$$

This is probably the single most important chemical reaction that has ever taken place on Earth. Note that both carbon dioxide and water were and are cur-rently very abundant and easy for organisms to acquire. Water and free oxygen are released as by-products of the reaction. So, this not only provided a means for early life forms to be autotrophs, but it also provided, for the first time, a source of free oxygen for Earth. The oxygen was released into a reducing environment and, initially and for some time, it would have combined with other elements to form compounds such as FeO and Fe_2O_3.

As anaerobic autotrophs flourished and produced more and more oxygen, the various reduced elements all would have become oxidized, and eventually free oxygen would have begun to accumulate in the atmosphere. As previously mentioned, free oxygen would have been a poison to organisms acclimated to an anaerobic environment. Perhaps many of these organisms became extinct and others became restricted to Earth's anaerobic habitats, but still others were able to adapt to the newly developing aerobic atmosphere. Thus, the third major type of life to appear should have been **aerobic autotrophs**. The aerobic atmosphere with the gaseous composition of the atmosphere in equilibrium with the gases in the surface waters of the ocean is like our present condition on Earth. The final major type of life to appear would have been **aerobic heterotrophs**.

KEY TERMS

abiotic synthesis	core	nebular cloud
aerobic	crust	planet
amino acids	heterotrophs	primitive Earth
anaerobic	initial Earth	solar winds
autotrophs	mantle	volcanic outgassing

READINGS

Most introductory textbooks in physical geology include chapters on the origin and early history of Earth. There is considerable variation in terms of detail and comprehensiveness. Some of these texts are listed below.

Bernal, J.D. 1976. *The Origin of Life*. World Press. This book explains the possible origin of large molecules on clay templates.

Fox, S.W., ed. 1988. *The Emergence of Life: Darwinian Evolution from the Inside*. Basic Books. 208 pages. Papers from a recent symposium on abiotic origin and synthesis of life.

Origins: Can Man Survive? (videotape). 1987. BBC, Group W Television. 30 minutes. Covers the origin of the universe, cosmology, human evolution, and artificial intelligence.

Schopf, J.W., ed. 1983. *Earth's Earliest Biosphere: Its Origin and Evolution*. Princeton University Press. 543 pages. An important book with chapters by several experts in the field. Includes information on Earth's oldest known fossils from Australia.

Organic Evolution and Extinction

INTRODUCTION

Evolution is one of the most provocative theories ever conceived. It has entered into debates not only in science but also in religion, politics, and education. The idea that species can be transformed from one to another, rather than being immutable, fixed packages of life, has changed our perspective of the world around us. For centuries one of the prime dogmas of natural history was that all life was created on Earth solely for the benefit and use of humankind. As scientists learned more and more about the fossil record, they came to realize that there were species of plants and animals that had lived during the past that no longer existed—they had become extinct. Even the fact of extinction took a long time to become established; some persistently argued that these supposedly extinct species would eventually be discovered living in still unexplored parts of Earth.

During medieval time, all life was viewed as forming a "Great Chain of Being." Each species was a link in this chain. The lowest forms of life were algae, then it moved up through higher plants, then lower forms of animals, and it culminated in the final link, human beings. Links closest to each other were most closely related, but each link was immutable and unchanging. This idea was popular for many years and still is evidenced in "missing links," a term used for animals that are supposedly or actually intermediate in relationship between two major groups of life.

This view of a static world was thoroughly upset by the evolutionary theory presented by Charles Darwin. In his monumental work *On the Origin of Species by Means of Natural Selection or the Preservation of Favoured Races in the Struggle for Life*, published in 1859, he presented many detailed examples demonstrating that species had arisen from each other and that evolution had occurred.

As envisioned by Charles Darwin, evolution by natural selection is a very logical deduction from the observations that he made and that we can all make. First, individuals within a species are not all identical; characteristics are varied (look around a classroom, look at the horses at a horse show, etc.). Second, this variation is inheritable, which is to say that characteristics are passed from parents to offspring (if two Scottish terriers have offspring, is the offspring a poodle?). Third, this variation affords different abilities to individuals of a population. Fourth, through selective breeding (artificial selection) animal and plant breeders can physically and behaviorally alter a species. This is a known fact of modern horticulture and animal husbandry. Finally, if man can artificially direct this change, it can also happen in nature, and this is evolution by **natural selection**, as described by Darwin. Natural selection occurs when the various forces of survival and reproduction in nature favor at some time one individual versus another. Through time the behavior and appearance of a species can be changed and eventually a new species can evolve.

Evolution provides the underlying foundation for all paleontology. The basic processes of evolution are essential to understanding the history of life. There have been many books written on evolution, yet, here, we must condense our discussion into a single chapter. There are three main areas of evolutionary theory and practice that concern us in the study of fossils. First is the fact that the sequence of occurrence of fossils in the rocks of different ages and their transition from one form to another through time provide the material evidence for the evolution of life that has, indeed, taken place on Earth. This aspect of evolution will not be discussed in this chapter because most of the remainder of this book is devoted to discussion of the documented record of evolution provided by fossils. The other two important facets of organic evolution are (1) the genetic basis of evolution—the ways in which individual plants and animals are constructed that allow for change to occur and for characteristics to be passed from one generation to the next, and (2) the mechanism of evolution—the ways in which species of organisms, composed of many individuals and populations, may change through time into a different species.

THE GENETIC BASIS OF EVOLUTION

We know that the offspring of any plant or animal inherits many of its characteristics from the parents. If the offspring is the result of sexual reproduction, it exhibits a blend of characteristics from each parent, being neither exactly like either the male or the female parent, but combining characteristics of each. Thus, offspring in a species with sexual reproduction have features that are uniquely their own. The science of how inheritance takes place is called **genetics**.

The transfer of heritable characters from parent to offspring is effected by the nuclei of the sex cells, the sperm and the egg. Each nucleus contains protein strands of nucleic material called **chromosomes** that have very large molecules of deoxyribonucleic acid, abbreviated **DNA**. These molecules are the transmitters of genetic information from the parent to the offspring; they ensure that the offspring will develop most of the same characters that are possessed by the two parents. Each parent contributes the same number of chromosomes to the first-formed cell of the offspring from which all other cells of the young develop, thus ensuring that each parent will have equal representation in the genetic materials of the new generation.

In a eukaryote, the DNA is located in the nucleus of the cell. A DNA molecule is composed of two helically spiral strands, each composed of a linear chain of sugar and phosphate molecules (Figure 5.1). The two spiral strands are held together by crossbars composed of four different kinds of nucleotide bases—adenine, thymine, cytosine, and guanine. The bases are arranged in very definite sequences along the helix, two bases composing each of the cross-links in the spiral, and it is this arrangement of bases that provides the genetic code of the chromosome that is passed from one generation to the next. Because of the chemical structure of the bases, adenine can only form a crossbar with thymine, and cytosine can only form a crossbar with guanine. When the

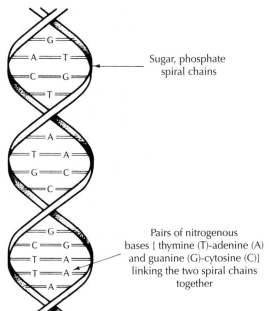

Sugar, phosphate
spiral chains

Figure 5.1   The molecular structure of DNA. Unraveling the spiral helix and splitting the pairs of cross-linking bases provides a pattern for duplication of the structure. A unit length of the DNA molecule that includes about 1,500 pairs of bases constitutes a gene, a single hereditary unit.

Pairs of nitrogenous
bases [thymine (T)-adenine (A)
and guanine (G)-cytosine (C)]
linking the two spiral chains
together

helix unravels to form a single strand, the strand contains all the information needed to build another complete, identical new strand to make a new double helix in the off-spring. If the parent strand has adenine at a specific site, the new crossbar can be completed only with thymine. The sequence provides a code for the construction of proteins and enzymes of the new individual, ensuring that the features of the parents will be passed to the offspring.

The synthesis of new molecules, proteins, and enzymes is carried out by another kind of large, complex molecule called ribonucleic acid, or **RNA**. This molecule is shorter than DNA, single stranded, and linear rather than helically spiral. It chemically copies single DNA strands, using the compatible opposite member of each of the two pairs of nitrogenous bases. That is, where the DNA strand has cytosine, the RNA molecule will insert a guanine.

Subunits of the chromosome are called **genes**. A gene includes the encoded information of a detectable heritable character. We now think each gene consists of as many as 1,500 base pairs.

If the matching of these bases is so exact, one may very well ask, how do any possible changes ever occur in the DNA molecules? Why are not all the genetic materials of all organisms exactly the same? Changes in the seqences of nitrogenous bases are called **mutations**. Mutations can occur in several ways. One or more pairs of bases may be deleted from the structure, two or more pairs may be inverted in their order, or one or more pairs of bases may flip-flop from one strand to the other. If only one or a few of the bases are involved in change, the mutation is a minor one and commonly cannot be detected. If genes are composed of 1,500 base pairs, a change in only one or two bases may have very little affect on the gene and the organism. We do know that large changes, mutations that affect a conspicuous part of the chromosome, are almost always lethal or severely deleterious to the organism. Over time, the gradual buildup of many small mutations, none of which in isolation was lethal or disadvantageous, may result in genetic changes that are advantageous to the organism. It is the gradual accretion of these many, small mutations that are considered the underlying new information on which evolution acts, but the mutations themselves do not cause evolution.

The genetic code, the sequence of nucleotide bases, of an individual is called the **genotype**, and the physical expression of the genotype is called the **phenotype**. If there are sufficient changes in the the genotype, through mutations that are neither lethal nor disadvantageous, a change in the outward appearance or behavior, the phenotype, will result. This process results in phenotypic variation. All populations are variable in one or many aspects of the individuals comprising the population. Evolution operates at the population level. It does not take place on or in a single individual, much less the genes of that individual. Thus, mutations do not result in evolution; they provide the phenotypic variation exhibited by the population.

THE MECHANISM OF EVOLUTION

Although each natural population is made of a collection of interbreeding organisms of a single species, it is not the genetic makeup of each individual that is important in evolution but, rather, the genetic composition of the entire population. This fact, that evolution takes place within populations of plants and animals, was an important, new aspect of Darwin's theory of evolution by natural selection.

VARIATION

The first aspect to be considered is **variation**. In sexually reproducing species, the genetic composition of every individual is unique. Every individual is different, or expressed in population terms, there is variation among individuals of every population. Every species is composed of one to many local interbreeding populations, and each local population will have differing types and degrees of variation. Some of this variation is the result of interaction between individuals and their surrounding environment and is due to factors such as food supply, living habitat, temperature, rainfall, or sunlight. This type of variation is not heritable and is not passed from parent to offspring, so it is of little interest here. However, some of the observed variation is genetic in origin and is due to somewhat different genotypes of the individuals in the population. As far as it is known, there is no natural population of organisms in which the genotypes and phenotypes of all individuals are identical in sexually reproducing organisms. Even among asexually reproducing organisms, mutations lead to differences among individuals of large populations. All populations include some genetic variability among their members, even though phenotypic variation may be slight or absent in rare cases. Even differences in sex are examples of genetic variation in populations.

As mentioned above, mutations provide the variations that are of most interest in the study of evolution. Nonlethal mutations in which offspring are not rendered sterile apparently occur in all natural populations but at a slow rate, producing changes in phenotype that typically become noticeable only after several generations, as the mutation is passed through more of the population.

Even if genetic variability produced by mutations can occur in viable offspring, this does not necessarily mean that the mutation confers any degree of advantage on those individuals who carry the mutation. Any mutation may be advantageous, disadvantageous, or neutral as far as the fitness or survival of the individual carrying it is concerned. This brings us to the next aspect of the evolutionary mechanism—natural selection.

NATURAL SELECTION

Given that most or all populations display genetic variability, not all individuals are equally likely to cope successfully with their environment. Especially important in this respect is success at reproduction. Those plants and animals that are successful breeders and produce viable offspring pass their genetic characteristics to the next generation. An extreme view of reproduction argued by some scientists is that the physical body of an organism is simply a vehicle designed by genes to transmit the genes to the next generation. This view is supported by examples in which organisms die after successful completion of reproduction. Annual flowering plants die after setting seed. Male black widow spiders are eaten by females after mating.

Any interaction between the individual and the environment or other organisms that tends to prevent or diminish reproductive success is lethal or disadvantageous to the genes of that organism. The plant or animal may die before it reaches breeding age, it may be unsuccessful at attracting a mate, or it may be unable to establish a breeding territory. Other disadvantageous aspects may relate to feeding, migration, predation, and so on. Conversely, one or several mutations may result in individuals that are more successful at breeding than are other members of their population. They may be able to establish a breeding territory, better utilize food resources, or resist predation. These individuals will thrive and pass their genetic characteristics to an increasingly larger proportion of the succeeding generation in the population. Mutations that confer advantages spread and become fixed in more and more individuals. This process, which favors individuals whose genetic and phenotypic characteristics favor survival and reproduction and also eliminates individuals who are less fit to cope with the environment and who have diminished breeding capabilities, is called **natural selection**. During this process, advantageous characteristics are selected for, whereas disadvantageous characteristics are selected against. Over a sufficiently long period of time, typically many generations, this process may result in a shift of the population genotype. Evolution by natural selection works because genetic change occurs within organisms at the same time that the physical and biological conditions of the environment are also changing. Therefore, a dynamic always exists. The genetic variation within the population of a species permits changes in the characteristics of a species through natural selection as organisms are constantly adjusting for survival in a changing world. Advantageous mutations can be quickly incorporated into a species' genotype, but this does not mean that all disadvantageous mutations are necessarily eliminated from the population. They may survive in low numbers, especially as a recessive genetic character. Neutral mutations may persist and spread but generally do so at slower rates than advantageous mutations.

Natural selection results in a slow to rapid, continuing shift in the genetic makeup of a population. Given enough time, this change could result in a younger population having a sufficiently different pool of genes than earlier ancestral populations, and then a new species exists. Because individuals from past generations are long since dead, there is no direct way to test for the genetic compatibility of the two populations to see whether viable offspring could be produced from interbreeding. If viable offspring could be produced, we would still consider these populations members of the same species. However, if they could not produce viable offspring, we would say that the evolution of a new species had occurred. Change in genetic makeup through time must have occurred in fossils, but we cannot test this hypothesis directly. Instead, we must infer genetic changes by studying phenotypic or morphologic changes in preserved fossils. If the observed changes in the physical aspects of fossils are sufficiently great, a paleontologist may infer that a new species developed directly from an ancestral species. The younger and older populations should be uniformly distinguishable.

Gradual change through time of one species population to another phenotypically distinct species population is called **phyletic evolution**. Phylogeny (hence *phyletic*) is the study of ancestor-descendant relationships among organisms. Notice that phyletic evolution, as we have described it, does not result in any multiplication of species. Older species A gradually evolves into younger species B as there is a gradual shift through time in certain morphologic characters used to define species. This process is sufficiently slow so that it cannot be detected among living populations.

ISOLATION

We have discussed three important aspects of the evolutionary process—variation, natural selection, and inheritance. The fourth and critical factor is **isolation**. Almost every species consists of a series of local populations of interbreeding animals. These populations may be in contact with each other at the edges of each local geographic range. As long as there is contact and migration among local populations, the entire species will maintain a genetic identity. Certainly there will be variability among local populations and differing types or intensities of natural selection. However, these differences will not be sufficiently strong to prevent an individual from one local population from interbreeding successfully with an individual from another local population. There will be a certain coherence to the gene pool of the entire species.

Let us consider a hypothetical small population of a species that somehow becomes established in an area in which it is effectively isolated from all other local populations of the species (Figure 5.2). The migrants, the founders of the new isolated unit, will have only a small sample of the total genetic variability of the species. Furthermore, the environmental conditions of the new area will not be identical to those of the area where the remainder of the populations live. Phenotypic or genotypic

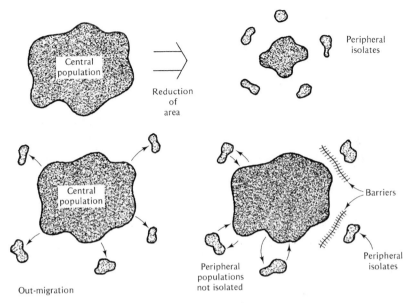

Figure 5.2   Hypothetical formation of peripheral isolated populations. Above: area occupied by a species shrinks in size, leaving behind isolated populations on perimeter of original large area. Below: peripheral populations are established, then later isolated by formation of barriers.

variables that are rare or absent in the parent population may be advantageous under new and different circumstances, and the natural selection process may also be different. In such a circumstance, the isolated population may begin to shift from the parent population both in genotype and in phenotype. Now, with time and changes in climate or habitat, it is possible for migration to occur between the members of the old species and members of the isolated population. If the changes that have taken place in the isolated population are sufficiently great, its members will not reproduce successfully with members of the old species, and a new species will have been created. In order for sufficient genetic and physical changes to take place in an isolated population, it must be genetically isolated from the parent population for a sufficient period of time for these changes to accumulate. These changes can take place relatively rapidly if the founder population is atypical genetically, and if the new area that it invades is sufficiently different from the old one.

Every species has a certain ability for dispersal, either as adults or during the early larval stages. Yet, very few species have worldwide distribution. Dispersal is limited by ecological factors—differences in moisture or rainfall, temperature, or food sources. Most species ultimately confront some kind of geographic barrier, such as a river, a mountain range, or an ocean that limits dispersal. Obviously, there are very many kinds of ecological and geographic barriers that prevent ubiquitous species. The founding of new, isolated populations outside the normal range of a species results when a few migrant individuals are able to surmount barriers and establish themselves outside the normal range of the species. Isolation could also occur if a climatic or other habitat shift severed the connection of a marginal population from the parent population (Figure 5.2).

MODES OF SPECIATION

For many years **phyletic evolution** was considered the primary evolutionary process; however, a debate during recent years concerns its overall importance to the history of life on Earth. Some paleontologists argue that phyletic evolution has recurred repeatedly and is an important mode of evolution causing both genetic change through time and splitting of lineages to increase the number of species. Others believe that an alternative evolutionary mode explains the history of life. This is commonly called **punctuated equilibrium** among paleontologists, and it is a derivative of what biologists call **allopatric speciation**. With this evolutionary mode, a fossil lineage persists for a long time with no significant morphologic changes (period of **stasis**). New species arise rapidly when a small population becomes geographically and reproductively isolated from the larger population of the species and becomes genetically distinct. Speciation with punctuated equilibrium results in an increase in the number of species. Both evolutionary modes must occur, and perhaps in different settings both can be important.

EXTINCTION

The focus of most discussions of the development of life on Earth is the evolutionary process. However, **extinction** is equally important. Earth does not have an infinite carrying capacity for species, so at some point existing species must disappear prior to the evolution of new species. More commonly, extinctions occur due to changing habitat conditions, and new species arise to fill the vacated niches.

Extinction of species has numerous causes, but ultimately it results from the reduction and eventual disappearance of all individuals of a species. The population death rate exceeds the birth rate because the species' environment has changed in some critical way and the species can neither cope with the change nor adapt to it through natural selection. These environmental changes may be physical changes, such as those in climate or biological changes. Examples of biological changes are the introduction of a new, more effective predator or a disease. The high rate of global species extinctions occurring today is largely biologically induced; they are being caused in various ways by *Homo sapiens*.

In a parallel manner to species, genera, families, and other higher categories can also become extinct. In the case of a genus, the critical juncture is reached when, for the species of that genus, the species' extinction rate exceeds the rate of speciation. Because Earth's environments are constantly changing and because species are also continually undergoing natural selection and evolution, extinction is also occurring constantly on a global scale. This general, ongoing extinction is referred to as **background extinction**. Biotic crises occur when the rates of extinction exceed the background rate. A very high extinction rate over a short interval of time is called mass extinction, which is discussed further below.

PATTERNS OF EVOLUTION AND EXTINCTION

Examination of the fossil record indicates that evolution and extinction have not been simply a random assortment of new species appearing here and there at different times and places in different ages of rocks. Instead, clusters of species may be closely related and grouped into genera. Different genera in the same or related families may, over the course of time, show remarkably similar or different trends in morphologic change that can be interpreted in terms of different or similar ways of coping with the environment. These changes can represent adaptation to the environment in various lineages and, when traced through time, can yield a variety of patterns that the course of evolution has followed. In this section, we will discuss some of these patterns and what they mean.

DIVERGENCE, CONVERGENCE, AND PARALLEL EVOLUTION

Small scale evolutionary patterns involve the morphology of organisms in the same or closely related lineages. These relationships can be illustrated on a simple diagram where time is the vertical axis and morphology is the horizontal axis (Figure 5.3). On such a diagram, a line traces the relative morphology through time, and extinction is shown by the termination of a line. **Divergent evolution** represents the evolution of a new species, which we call speciation. As currently understood, a subset of the original, parent population typically assumes a different morphology, probably rapidly, and diverges from the parent population.

In addition to divergence patterns illustrating different rates of morphologic change, other recognizable patterns are parallel and convergent evolution. Suppose

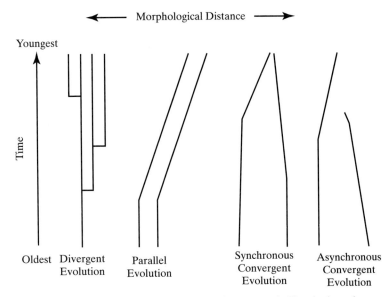

Figure 5.3 Graphical representation of evolutionary trends. Time is along the vertical axis and morphologic change is along the horizontal axis.

that two different lineages or stocks are evolving through time, and a sequence of morphologic changes appear in each lineage that duplicates similar changes in the other lineage. Such an evolutionary pattern is called **parallel evolution**. At first consideration this seems impossible, but in fact it has happened many times. It occurs due to the fact that similar organisms are adapting to similar environmental pressures, and natural selection produces similar solutions. A prime example is among the shelled cephalopods, the ammonoids. When these animals first appeared during the Devonian, the shell was divided into chambers by very simple, shelly partitions called septa. During their evolution, the septa became progressively more sinuous along the edges, more and more complex (Figure 5.4). This increase in complexity occurred in several different and distinct families and superfamilies during the Paleozoic and also during the Mesozoic.

In **convergent evolution** two or more stocks not only develop a sequence of the same structures, as in parallel evolution, but these structures come to look very similar, again in response to morphologic solutions to similar problems. The convergence may be so close that the hard parts of the animals eventually bear such a striking resemblance that their distant relationship can be revealed only through careful study. If structures evolve from a similar ancestral feature, whether they look similar and are convergent or dissimilar, these structures are called **homologous**. There are many, many examples of convergent homologous structures among fossil invertebrates, especially among brachiopods, ammonoids, and crinoids. The homologues may occur in the same time interval and be found together, in which case they are termed synchronous homologues. Alternatively, the homologous structures may be of quite different ages. For example, a few Jurassic ammonoids are virtually indistinguishable from unrelated Cretaceous ammonoids. These would be called heterochronous homologues.

Figure 5.4   Increase in complexity of ammonoid septa, from simplest at the bottom to most complex at the top. All except one septum has been omitted for clarity. These changes occurred in several different evolving lineages of ammonoids.

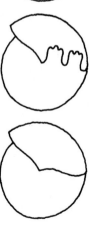

One of the classic examples of convergent evolution is the completely independent evolution of wings in those groups that have successfully invaded the aerial environment with powered flight. Insects, birds, reptiles (pterosaurs), and mammals (bats) all became successful aerial creatures through evolution of the wing. Yet, the details of wing structure in each of these groups are conspicuously different (Figure 5.5). Which are homologous? Those of birds, pterosaurs, and bats; even though different in detail, all evolved from the front limbs of vertebrates, thus they are considered homologous structures. Alternatively, the wing of an insect evolved in quite a different manner, so insect wings are not homologous. However, because they serve the same function but have different origins, insect wings are called **analogous** structures.

While speaking about the evolutionary aspects of fossils, careful distinctions must be made regarding whether the concern is with an entire population, with a complete individual, or with a specific structure borne by an individual. This is especially true when using such descriptive pairs of words as *primitive* and *advanced* or *generalized* and *specialized*. Sometimes these words are applied to whole organisms and sometimes particular morphologic features. In addition, *primitive* and *advanced* both have time connotations and suggest a "better" morphology. This is commonly ill-advised terminology, because any morphology that evolved was adaptive.

ADAPTIVE RADIATION AND MASS EXTINCTION

Large-scale evolutionary trends also exist, where the speciation and species extinctions are multiplied to a great extent or where higher-level taxa, such as genera, undergo periods of intense origination or extinction. Short time intervals during which a lineage

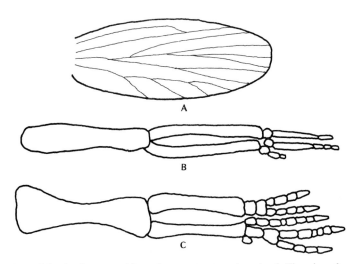

Figure 5.5   Analogous and homologous structures in animals. The wing of an insect (A) and the bones in the wing of a bird (B) are analogous in that both serve the same function, flying, but they have different morphologic and evolutionary origins. Alternatively, the bones in the wing of a bird (B) and the bones in the forelimb of a reptile (C) are homologous, because they have the same origins, even though the functions may differ.

undergoes high levels of speciation and origination of higher taxa are called **adaptive radiation** (Figure 5.6). Examples of adaptive radiation discussed in succeeding chapters include the diversification of skeletonized marine life at the base of the Cambrian, angiosperm diversification during the Lower Cretaceous, and the radiation of mammals at the beginning of the Cenozoic. Adaptive radiations typically result from multiplication of species as one or more lineages adapt to a new environmental situation. There may have been a drastic shift in the environment or a major new zone of adaptation may have become available to a species. In some way innumerable niches open to be exploited by a rapidly developing series of descendant species.

The reverse of adaptive radiation is **mass extinction**, which is a short interval of very high rates of extinction (Figure 5.6). Mass extinctions have punctuated Earth history, and the basic divisions of the geologic time scale commonly record mass extinctions. Mass extinctions are caused by a relatively rapid deterioration of the environment that results in a biotic crisis. Mass extinctions may affect a single group of organisms or organisms from many ecosystems.

Processes responsible for mass extinctions are a subject of considerable current research. Drastic, rapid deterioration of the environment can be both physical and biological. Present ideas emphasize climatic change (either cooling or warming) meteorite impacts, habitat reduction, oceanic anoxia, and evolution or invasion of a new fauna.

Through the Phanerozoic, more than ten intervals of mass extinction have been identified. The largest, most devastating mass extinction occurred at the Permian-Triassic (Paleozoic-Mesozoic boundary), which we will call the end-Permian extinctions. At this extinction, it is estimated that 84 percent of the genera on Earth became extinct. The second most devastating mass extinction was the end-Ordovician extinction; and the end-Cretaceous extinction, which is the most celebrated because during this time all dinosaurs became extinct, was the third most severe. Those mass extinctions listed in (Figure 5.7) will be discussed in later chapters.

A recent, tantalizing hypothesis states that since the beginning of the Mesozoic, mass extinctions have occurred on a cyclical basis, every 26 million years. Earth-bound

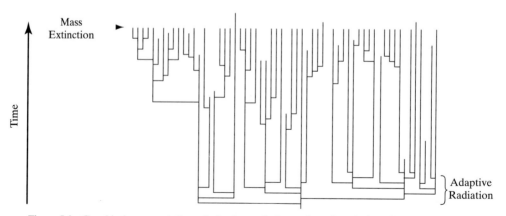

Figure 5.6   Graphical representation of adaptive radiation, a short time during which many new taxa originate, and mass extinction, a short time when many taxa become extinct.

processes with a 26 million year periodicity are not known, and several of these extinctions (see discussion of dinosaur extinction, Chapter 15) coincide with abnormally high concentrations of the element iridium, which is argued to have an extraterrestrial source. This work has sparked much research and debate concerning the roles of extraterrestrial versus Earth-bound processes and mass extinctions. Both types of processes are capable of causing widespread extinction, and detailed research is necessary to determine the role of potential processes for each mass extinction. Current ideas on various mass extinctions are discussed in later chapters.

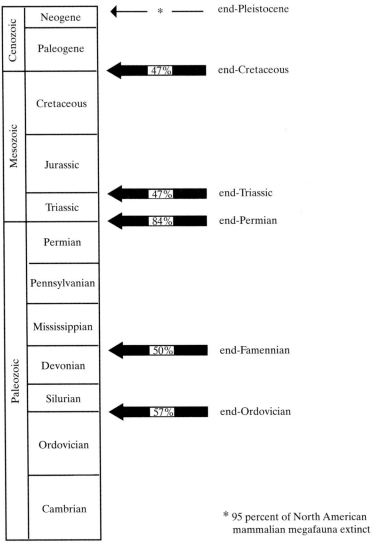

Figure 5.7   Intervals of mass extinction discussed in this book. Estimated percent generic extinction is given for each, except for the end-Pleistocene.

EVOLUTION VERSUS CREATIONISM

In recent years there has been considerable renewed controversy concerning the Darwinian theory of evolution and what is called scientific **creationism**. The latter idea assumes that the story of creation in the Bible is literally and scientifically true and, thus, deserves equal treatment in science classes, especially at the primary and secondary levels.

Not all Biblical literalists accept that creation occurred in six twenty-four hour days because a day can be interpreted as much longer in duration. The creationist's claim for scientific equality of their views has been rejected by federal courts in Arkansas and Louisiana, which have ruled that creationism is religious in nature, not scientific. Therefore, it should not be taught in science classes. This section examines some of the more important claims and counterclaims of Darwinian scientists and creationists.

Any scientific theory must be testable concerning its truth or falsity. New evidence may result in modification or rejection of a theory. In this sense, creationism is not scientific because it cannot be falsified or modified with new data. One must either accept the premise of the literal truth of the Bible or reject that premise.

The enormous span of time for life history on Earth causes problems for creationists who interpret the six days of creation in Genesis as sequential twenty-four hour days. They believe that Earth is only a few thousand years old. Efforts have been made to discredit the reliability of absolute radioactive age-dating, but these efforts have been rejected by scientists as lacking any substantial evidence. Other dating methods, such as the length of time needed for light to arrive from distant stars, support the idea that the universe is at least 8–14 billion years old. If radioactive age-dating is not true, then much of the basis for all of chemistry and physics is also not true.

Creationists try to disprove evolution by trying to find evidence against the succession of fossil faunas and floras. One way to do this is to discover co-occurring fossils that could not have lived at the same time, according to paleontologists. This effort has concentrated on finding fossil evidence for the presence of humans on Earth in rocks that paleontologists agree are much too old to yield such evidence. This putative evidence has commonly taken the form of fossil human footprints or sandal prints. Thus, it is claimed in Utah that a sandal print has been found in Cambrian rocks that also contain trilobites. Creationists interpret this evidence to mean that the rocks that contain these fossils were all deposited during the Noachian flood episode of the Old Testament. According to creationists all the plants and animals that we know to now be extinct were living prior to the flood and were wiped out by that flood. There was no room for dinosaurs on the ark. A Cretaceous bed in Texas has yielded what are said to be human footprints. If this is true, these occurrences would falsify our current view that man and dinosaurs could not have coexisted and that Cretaceous rocks are more than 65 million years old. Geologists who have examined these occurrences have found that the so-called footprints and sandal prints are not fossils at all but are impressions of concretions or fragments of dinosaur toe impressions in the rock.

Finally, creationists believe that one species cannot be transformed through organic evolution into another species. Thus, they attempt to discredit intermediate fossil forms that seem to bridge the gaps between major life groups. They deny that

Archaeopteryx is a transitional form between dinosaurs and birds and argue that *Ichthyostega* does not indicate the origin of amphibians from lobe-finned fish. These arguments have been countered by detailed anatomical rebuttals from paleontologists.

In summary, creationism is an unscientific approach to understanding ancient life on Earth because it cannot be falsified or modified. One either accepts creationist ideas as an article of faith or one does not. Many church leaders oppose attempts to teach scientific creationism in schools or to equate it with evolution. Gradual rather than sudden creation is accepted by some; others see life in terms of theistic or God-related evolution. An overwhelming amount of paleontological evidence and modern biological evidence contradicts creationist arguments.

In an entirely empirical way, faunal succession and age-dating of rocks are used daily by successful multibillion dollar corporations in search of fossil fuels and minerals. For instance, ancient organic reefs buried deep within Earth are prime sources of petroleum. It is difficult to imagine how they could use creationist ideas of the history of Earth to find oil and gas.

KEY TERMS

adaptive radiation	extinction	natural selection
allopatric speciation	gene	parallel evolution
analogous	genetics	phenotype
background extinction	genotype	phyletic evolution
chromosomes	homologous	punctuated equilibrium
convergent evolution	isolation	RNA
creationism	mass extinction	stasis
DNA	mutation	variation

READINGS

The processes of evolution and extinction have been written about extensively. Discussions range from the very simple to the quite complex. Each of the books listed here places emphasis on a somewhat different aspect of evolution or extinction.

Berggren, W.A., and Van Couvering, J.A. 1984. *Catastrophes and Earth History*. Princeton University Press. 461 pages. A variety of essays on different types of catastrophic events that may have shaped the history of Earth, including climate and life.

Elliott, D.K. 1986. *Dynamics of Extinction*. Wiley. 294 pages. An even-handed and thorough biological treatment of extinction.

Evolution (videotape). 1986. Hawkshill. 39 minutes. Covers the history of scientific thought on evolution, the concept of natural selection, and evidence for evolution.

Hoffman, A. 1989. *Arguments on Evolution: A Paleontologist's Perspective*. Oxford University Press. 274 pages. A series of essays on the fossil record of evolution, with emphasis on macroevolution, or adaptive radiation.

Kaufmann, E.G., and Walliser, O.H., eds. 1990. *Extinction Events in Earth History*. Lecture Notes in Earth Sciences No. 30. Springer-Verlag. 431 pages. Comprehensive discussion of all hypothesized extinction events, including cyclic phenomena.

Lewin, R. 1983. "What killed the giant mammals?" *Science* 221:1036–1037.

Nitecki, M.H., ed. 1984. *Extinction*. University of Chicago Press. 335 pages. A collection of thirteen technical papers on various aspects of plant and animal extinction.

Stebbins, G.I. 1981. *Processes of Organic Evolution*. Concepts of Modern Biology Series. Prentice-Hall. 183 pages. A short paperback with emphasis on genetic control of evolution.

Weiner, J. 1994. *The Beak of the Finch*. Vintage. 332 pages. Enjoyable book discussing scientific fieldwork documenting the existence of Darwinian natural selection.

Continental Drift and Plate Tectonics

INTRODUCTION

If you ask a man or woman on the street if the positions of oceans and continents have always been the same as they are today, the answer he or she gives will probably be "yes." The continents are the largest land masses, and because of their size, most people think they must surely be fixed in position. This has always been the traditional viewpoint, even among most paleontologists and geologists. However, in the 1960s the old idea of fixed continents and oceans was seriously challenged, and today most Earth scientists think that the continents have moved over the face of Earth.

Why should we concern ourselves with the crust of Earth in a book on paleontology? One aspect of ancient life that interests us is the past distribution of life—topics such as migration, barriers, isolation, and faunal and floral provinces. In order to discuss the geography of ancient life, also called **paleobiogeography** (see Chapter 7), we have to assume some model for the distribution of land and sea. In the 1940s we would have chosen a model of a fixed Earth crust, with oceans and continents forever in their present position. But new techniques developed to study the ocean floor have yielded new evidence and have led to a scientific revolution in our thinking about the stability of the crust. The crust is in constant motion, and the continents and oceans have not always had their present positions or outlines. Here, we will first discuss the century-old idea of continental drift and then present a synopsis of the processes responsible for drift, which is sea-floor spreading.

CONTINENTAL DRIFT

Soon after North and South America were discovered and their eastern coasts roughly mapped during the 1600s, several scientists noted that there was a similarity of outline between the eastern coasts of North and South America and the western coasts of Europe and Africa. These early scientists, including Francis Bacon, simply noted the similarity in outline on both sides of the Atlantic; they did not propose that the continents had separated, forming the Atlantic Ocean. This idea of drifting continents was first expressed in the 1850s by two European scientists, Richard Owen and Antonio Snider. However, it was not until 1912 that the theory of **continental drift** was elaborated and much supporting evidence gathered. For the most part, this work is credited to a German meteorologist, Alfred Wegener. Emergence of the theory of continental drift depended largely on geologic and paleontologic evidence that had been gathered in the southern hemisphere. The theory was strongly supported by geologists working in South America, Africa, and Australia, especially by a South African geologist, Alexander DuTroit. A number of European geologists also supported the theory, but some North American geologists were especially reluctant to accept it.

What was the evidence in this controversy for and against continental drift? The old idea about the close fit of North and South America with Europe and Africa was used, of course. The continents were not matched at their present coastlines, but at the edges of the continental crust that includes some submerged continental shelves and slopes. The best fit was obtained by matching at a depth of approximately 2,000 meters, halfway down the edge of the continental slopes. Additional support for the tearing away of the continents from each other was gained when the nature of the Mid-Atlantic Ridge was ascertained by early oceanographic expeditions. This high ridge, which only reaches the surface of the ocean at Iceland, the Azores, and two small islands in the South Atlantic, has an outline that matches the edges of the Atlantic on either side (Figure 6.1). It was suggested that the ridge records the scar left behind where the continents separated.

Rocks and fossils throughout the southern hemisphere have several unique features that suggest drift. Rocks of Permian age contain evidence for ancient continental glaciers. Ice had scoured and scratched pre-Permian rocks, then had melted away leaving behind tillite deposits. In South America the ice had moved from east to west, as determined by the directions of the grooves made by the ice, and would have had a source in the present Atlantic (Figure 6.2). In Africa, ice movement was from west to east, again from the Atlantic. Evidence for these large ancient glaciers is also present in

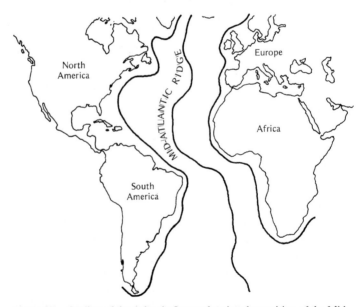

Figure 6.1   Outline of the Atlantic Ocean showing the position of the Mid-Atlantic Ridge. Heavy lines indicate the true, submerged margins of the continents; light lines indicate those parts of the continents that are dry land today.

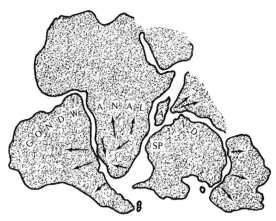

Figure 6.2 Reconstruction of Gondwana as it appeared approximately 250 million years ago. Arrows indicate directions of ice movement of ancient continental glaciers that covered most of the area. The South Pole (SP) is thought to have been situated on the coast of Antarctica.

Australia, India, and Antarctica. It was argued that large ice sheets could not have been generated in the ocean but must have accumulated on land; hence, there must have been land present where the Atlantic is now situated.

In addition to this physical evidence, there was also much paleontological evidence to support drift. Above the Permian glacial deposits in each of these areas, there are terrestrial rocks that contain many fossil plants, and in some places coal beds are formed. The most distinctive of these plants were two kinds of extinct gymnosperms, related to modern conifers. These were seed bearing and commonly called tongue ferns (Figure 6.3). Two genera, **Glossopteris** and **Gangamopteris**, are on all the southern hemisphere continents and on the peninsula of India, now in the northern hemisphere but considered by "drifters" to have migrated north after the Permian. These plants are not present anywhere in the northern hemisphere, except India, and were interpreted as a cool-climate flora because of their close association with ice deposits. This flora is referred to as the *Glossopteris* flora.

In addition to land plants, Permian rocks in these areas also contain a unique marine fauna that is characterized by distinctive kinds of brachiopods and clams. This fauna is named the **Eurydesma** fauna after one of the large, thick-shelled bivalves that

Figure 6.3 Leaves of *Glossopteris* from the Permian rocks of New South Wales, Australia. The rock slab is 15 cm across. *Gangamopteris* is similar in aspect but lacks a distinct midvein in the leaves. (Photograph by George R. Ringer, Indiana University.)

is closely associated with glacial deposits (Figures 6.4 and 6.5). Some of the beds that contain this clam also have large boulders of granite, quartzite, and other rocks that are judged to have been dropped onto the Permian sea floor by melting icebergs.

Finally, Permian rocks in South America and in Africa contain a small fossil reptile called ***Mesosaurus***. The beds that yield this fossil are freshwater in origin, and the reptile is very distinctive, quite unlike any other reptiles known from this age in the northern hemisphere (Figure 6.6). Those who opposed drift had to explain how this freshwater animal could have migrated across the Atlantic Ocean or migrated from Africa through Europe and North America to South America without leaving any trace in intermediate areas.

These lines of evidence—the Permian glaciations, Permian stratigraphy, and paleontology (the *Glossopteris* flora, the *Eurydesma* fauna, and *Mesosaurus*)—were used by the advocates of continental drift to strongly support the theory. The evidence indicated that the southern hemisphere land areas had once been joined in one giant supercontinent. This supercontinent was named **Gondwana** by Edward Seuss, a Swiss geologist who wrote a worldwide synthesis of geology in 1900. The name is taken from an area in India where some of the critical rocks are exposed. Later, in 1912, Alfred Wegener not only accepted Seuss's name for this predrift continent to the south, but he further proposed that the southern and northern hemisphere continents had all been united at one time, such that all Earth's continental areas had once been together. He called this single continental area **Pangaea**, which means single land. He thought that Pangaea had broken into two continents, Gondwana and a northern one that he called **Laurasia**. Each of these two continents, he believed, had divided to form our present continental configuration.

How were these arguments in favor of drift countered by the many geologists who did not accept the theory? The basic contention was that there was no adequate driving force known that could have caused the continents to move. The continental areas of the **crust** are composed of quite thick rocks that are lighter, or less dense, than the thinner crust that underlies the ocean floors. To move, it seemed at this time, the continents would have to have been pushed horizontally through this dense oceanic crust. The friction involved in such movement would have been enormous, and no forces were known that could have created such lateral pressures.

Figure 6.4   A large, thick-shelled marine bivalve, *Eurydesma*, found in close association with Permian marine glacial deposits of the southern hemisphere continents and India. The specimen is from Tasmania and is about 10 cm across. (Photograph by George R. Ringer, Indiana University.)

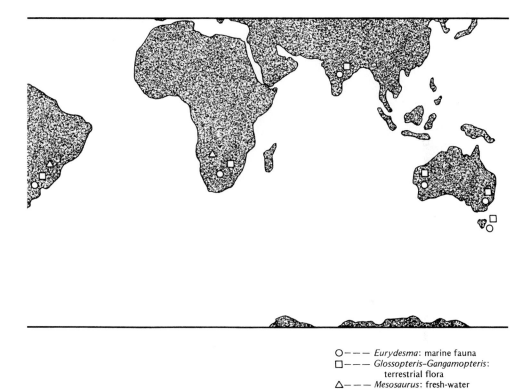

O— — — *Eurydesma*: marine fauna
□— — — *Glossopteris–Gangamopteris*:
          terrestrial flora
△— — — *Mesosaurus*: fresh-water
          terrestrial fauna

Figure 6.5   Distribution of Permian marine and terrestrial fossils associated with late Paleozoic continental glaciation in the southern hemisphere and India—areas that once constituted a single supercontinent, Gondwana, which later broke up and drifted apart.

The glacial deposits cited as evidence of drift were thought by many geologists not to have been formed by ice. They were explained as having been been formed by dense mud flows or turbidity currents carrying pebbles and boulders. Even if the deposits were accepted as glacial in origin, it was argued that there could have been several small ice caps, one on each continent, rather than a single large ice sheet. The faunal and floral evidence was dismissed by some with the argument that plants and animals have so many ways of dispersal that they may have migrated through northern areas without leaving any traces. Alternatively, some postulated that there had been a land bridge across the South Atlantic connecting South America and Africa. They contended that *Mesosaurus*, and perhaps *Glossopteris*, had migrated across this land bridge, which later sank into the ocean. Such a land bridge would have been composed of lighter continental rocks, and we now know that there is no evidence for the sinking of such a land mass in the southern oceans. It was also argued that if North America and Europe had indeed once been close together, we should find much good fossil evidence in these areas where fossils have been very extensively studied. These and other arguments convinced many geologists that drift had not occurred and that the theory

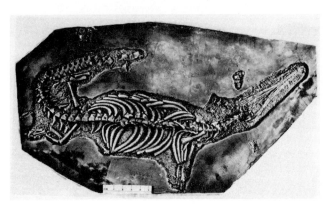

Figure 6.6 A skeleton of the Permian reptile *Mesosaurus* from Brazil. This animal was important as early scientific evidence for the former connection of South America and Africa. (Courtesy of Takeo Susuki, U.C.L.A.)

was mainly an intriguing idea that was entertained largely by a few geologists working in the southern hemisphere.

SEA-FLOOR SPREADING

This discussion brings us to the 1950s. By this time, it had been discovered that many igneous and some sedimentary rocks preserved faint traces of Earth's magnetic field. When lava is erupted and cools, tiny needlelike crystals of iron minerals become oriented in Earth's magnetic field, just as if they were many small compass needles. As the rock solidifies, the crystals become frozen into position, thereby recording where Earth's north and south magnetic poles were located at the time the lava cooled to a solid rock. Other small crystals of iron minerals that settle out of water, to form part of a sandstone bed, for instance, may also become aligned in the magnetic field. This preserved magnetism is called **remnant magnetism** because it remains in the rock. Study of remnant magnetism from many different areas and many different ages revealed either that the magnetic poles had not always been in their present positions, close to the north and south poles of Earth's rotation, or that they had always been close to their present positions, and the continents themselves had changed position (Figure 6.7). This was taken as strong evidence for continental drift by some scientists. Others argued that the magnetic field may very well have shifted position during Earth history. The major difficulty in accepting or rejecting either hypothesis was that the underlying reasons for the generation of Earth's magnetic field were not, and are still not, fully understood. Therefore, whether it could shift position or not was unclear.

Magnetic studies of rocks not only revealed that the position of the magnetic poles or the continents have shifted through time, but they also indicated that the poles have reversed their positions through time. This, called **magnetic reversals**, indicates that the north and south magnetic poles have flip-flopped back and forth through time on the average of every 400,000 years (Figure 6.8). The cause of these magnetic reversals is debated, but the history of these changes is very well documented. Rocks that have magnetic reversals can also typically be dated radiometrically, and a time scale has been developed that gives the timing of each reversal. Studies of the remnant mag-

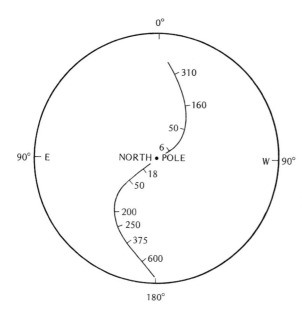

Figure 6.7   Wandering of the magnetic poles. The outer circle is at the equator, and the North Pole is in the center. The upper, curved line represents the position of the north magnetic pole for Australia as recorded in rocks from more than 310 million years ago to 6 million years ago. The lower curve is for Europe from 600 to 18 million years ago. Because the curves do not coincide, they indicate that Europe and Australia have had different relative movements over the past several hundred million years.

netism of rocks was extended from the continents to the ocean floors. It was found that igneous rocks that make up the floor of the ocean exhibit both normal polarity (North Pole in its present-day position) and reverse polarity (North Pole near the position of the present-day south magnetic pole). Reverse and normal polarized rocks occur on either side of the Mid-Atlantic Ridge and other large ridges of other oceans. The pattern of reverse and normal polarity is striped. Those rocks closest to the ridge, on both sides, have normal polarity, followed by elongate areas with reverse polarity, and so on in alternation on both sides of the ridge (Figure 6.9). These rocks were also dated and the patterns of polarity matched on either side of the ridge by age. Such study led to the conclusion that the age of the sea floor increases with distance from an oceanic ridge. The only explanation for this is that new ocean crust is formed at a ridge and is then progressively pushed farther and farther away from the ridge as younger and younger crust is formed. This process is called **sea-floor spreading**.

Much additional evidence for sea-floor spreading has accumulated. The rocks composing volcanic islands in the oceans are dated to be increasingly younger the closer they are to an oceanic ridge. Dating of the oceanic crust in the Atlantic has revealed that none of these rocks is older than about 180 million years (Jurassic Period), and therefore, the entire floor of the present Atlantic ocean has formed relatively recently. The skeletons of various kinds of floating protozoans and algae accumulate on the sea floor as they die, and their skeletons sink. These deposits, called oozes, can be dated on the basis of the contained fossils, and the deposits are oldest the farthest away from the ridge. Those deposits that overlay the ridge are quite thin and the youngest (Figure 6.10). Again, the oldest deposits in the Atlantic Ocean are Mesozoic.

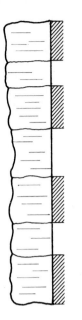

Figure 6.8   A sequence of volcanic rocks for which the polarity of the magnetic poles has been determined. Shaded intervals at the side indicate reverse polarity; white areas are normal polarity. The time scale at the right indicates a duration of 10 million years.

The demonstration of sea-floor spreading provides an adequate and sufficient mechanism for movement of the continents. It was shown that a continent did not have to plough its way through dense lower parts of the crust, like a ship through viscous tar. Instead, the continents could float on top of a continuously moving conveyor belt, or moving sidewalk, of oceanic crust that gradually moved them around. The concept of sea-floor spreading has been documented now in so many different ways that all but a very few geologists have come to accept this idea. However, we are still left with the question of what causes spreading from the ocean ridges.

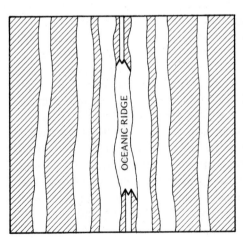

Figure 6.9   Diagrammatic map of a portion of the sea floor showing an oceanic ridge in the center and bands of rock of normal (white) and reverse (diagonal rule) polarity on either side of the ridge. This indicates that new crust is formed at the ridge and is pushed out equally to both sides.

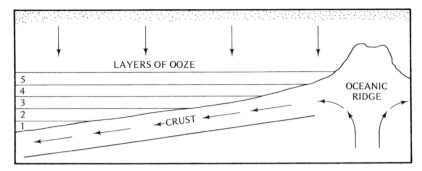

Figure 6.10   As new sea floor is formed at the ocean ridge and moves away from it to the left, successively younger layers of fossil ooze (1-5) are deposited on top of the moving crust. The skeletons of tiny marine organisms living in the surface waters sink to the bottom, forming the ooze layers. Deep-sea coring of these sediments reveals the age differences at different sites.

PLATE TECTONICS

In the preceding discussion we did not consider several questions that come to mind in light of the reality of sea-floor spreading. For instance, what happens to all the old crust if only relatively young ocean floors are known? Does it just pile up somewhere, or is it destroyed? The answer is the latter—the old crust is destroyed. Earth's crust is fairly rigid and is not easily deformed. The theory of **plate tectonics** explains sea-floor spreading. It states that the structure of Earth's crust is divided into rigid **plates** that move horizontally. These plates are rigid and are composed of the crust and the upper mantle. They interact at their boundaries. Along one edge plates have a spreading center, either active or inactive (Figures 6.11 and 6.12), where new oceanic crust is formed. On the opposite side from the ridge, there is commonly a deep ocean trench. This is where oceanic crust is destroyed. These are commonly the sites of many earthquakes, and curved chains of volcanic islands may be situated close to the trenches. Trenches are known now as the positions of the leading edges of plates. Here, the older parts of the oceanic crust, farthest from the spreading centers, are turned back down into the mantle. These trenches are called **subduction zones**; old crust is eventually destroyed and melted as it is subducted, or turned down, into Earth. The lateral edges of plates, between subduction zones and spreading centers, are areas where one rigid plate slides past an adjacent plate. These edges are marked by huge elongate systems of fractures in Earth's crust.

What happens when a plate of crust with a continent riding on it approaches a subduction zone? Because continents are composed of thick, light rocks, they are not easily turned downward. Instead, they float over the adjacent ocean floor slab; and due to the lateral forces of the subduction, the continental crust is folded and faulted into a mountain range, such as the Andes mountains of South America. Continental areas may also collide with other continental areas on an adjacent plate. In this case, neither continental area will be subducted, and the tremendous compression that results will squeeze the welded continental areas into a large mountain range. An example is the

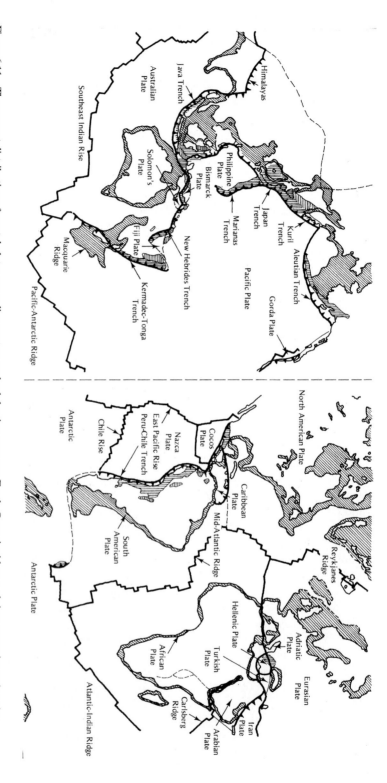

Figure 6.11   The present distribution of crustal plates, spreading centers, and subduction zones on Earth. Oceanic ridges and rises are active or inactive spreading centers. Oceanic trenches are subduction zones. Uncertain plate boundaries in Asia and southeast of South America are shown as dashed lines. (From J.F. Dewey, *Plate Tectonics*. Copyright © 1972, Scientific American, Inc. All rights reserved.)

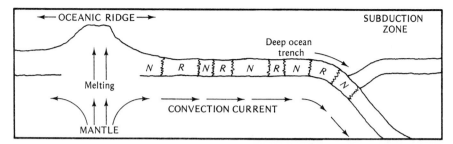

Figure 6.12   Cross section of an idealized ocean floor plate, between a spreading center (oceanic ridge) on the left and a subduction zone (oceanic trench) on the right. Convection in the mantle brings up magma under the ridge that cools and is forced to each side by newly rising magma. The N and R symbols in the crust indicate zones of rock that have normal and reversed polarity.

Himalayan mountains, where the continental mass now making up the peninsula of India moved northward and collided with the mainland of Asia.

The entire surface of Earth is divided into approximately 18 plates. Some of these are very large, one occupying virtually all the Pacific Ocean and one covering North America and the northwestern Atlantic. Others are quite small, such as the Arabian peninsula. Not all spreading centers are active, and the rate of spreading is very slow— just a few centimeters a year. Even though the spreading rate is slow, it can be measured. The fact that the Atlantic Ocean is becoming wider every year is confirmed by measurements on Iceland, where the Mid-Atlantic Ridge is exposed above sea level. The ever-increasing width of the Atlantic can be measured using satellites orbiting Earth. Subduction goes on at the same slow rate.

In the debate over the ultimate cause of sea-floor spreading, several different hypotheses have been put forward. The process obviously takes place deep within Earth, where we cannot make direct observations. The most generally accepted idea is that the material within the **mantle** is in constant motion, being heated near the **core**, moving upward under the crust, flowing sideways, and then after cooling and becoming more dense, sinking again toward the core. This process produces convection currents, much like heating soup in a pot, so that the lowest level heats first, rises to the top, cools, and sinks. The ridges mark linear areas of Earth's surface where convection cells approach the surface. Hot mantle material melts and is forced upward as volcanoes and molten lava produce new oceanic crust that is progressively pushed aside as more new crust is formed. Where this happens the ridges are uplifted by these underlying forces and are termed **spreading centers**.

The position and activity of spreading centers has changed through time. The record for the past 150 to 200 million years of Earth history is reasonably straightforward and has been worked out in considerable detail (Table 6.1). We know that the present Atlantic Ocean is quite young. The continents were all together forming a single super-continent, Pangaea, that began to break up about 220 million years ago, with spreading of the crust on either side of the Mid-Atlantic Ridge. The motions of the plates were not uniform; Africa moved north as well as east, closing a large oceanic gap between Africa and Europe. This resulted in the formation of the Alpine mountain ranges and a reduction of the ocean to the present day Mediterranean, Black, and Caspian Seas.

Table 6.1. **SEQUENCE OF EVENTS IN SEA-FLOOR SPREADING AND CONTINENTAL DRIFT. EVENTS PRIOR TO 200 MILLION YEARS ARE A SUBJECT OF DEBATE; THEY ARE NOT AS WELL SUBSTANTIATED AS ARE EVENTS AFTER 200 MILLION YEARS.**

Millions of Years Ago	Events
50	Separation of Antarctica and Australia.
100	India drifts northward.
150	Separation of Africa from India and Australia.
200	Beginning of separation of Europe and North America. Initiation of the Pangaea breakup.
300–400	Closing of Iapetus; formation of Pangaea; creation of the Appalachian Mountains.
prior to 600	Formation of proto-Altantic by breakup of an early Pangaea.

Prior to the Pangaea supercontinent, the exact position and sequence of plate configurations is less certain. There is clear evidence that earlier during the Paleozoic the continents were not together, and for the most part the continental configurations would be odd to very unfamiliar based on present-day geography. Prior to the formation of Pangaea, there was an ocean separating North America from Europe and Africa, similar to the present Atlantic Ocean. This Paleozoic sea was called the Iapetus Ocean. The continent-to-continent collision that occurred when the Iapetus Ocean closed formed the Appalachian Mountains. Deciphering continental configurations of the Paleozoic and earlier is an area of active research.

THE EFFECT OF CONTINENTAL DRIFT ON LIFE

In this chapter we have devoted considerable space to discussion of continental drift and to its underlying cause, plate tectonics. The position and arrangement of the continents has surely had a profound effect on life of different ages. Paleontologists are now investigating the relationship between these physical and biological events. We can specify several broad areas in which continental drift has had an especially important influence on life.

As the continents shifted, they changed their relative positions with respect to the poles—sometimes closer to the equator and other times closer to a pole. For instance, the United States was close to the equator during much of the late Paleozoic and has gradually drifted northward since, farther from the equator. This drift undoubtedly influenced the climate that was present over the United States, which gradually changed from tropical to subtropical to temperate. Similarly, climatic changes occurred on all the continents through time.

As the continents moved apart, came together, and were again separated, the distribution of land and ocean plants and animals was affected. When Pangaea was effectively a single large land mass, life on land was quite cosmopolitan—the same species inhabited large areas, subject, of course, to differences in climate. When the continents separated with little or no direct land communication between them, land organisms evolved separately on each continent because new ocean basins acted as a barrier to animal and plant migration. The result was distinctly different faunas in different areas with species much more restricted in their distribution. The same was true for marine organisms. With a single world ocean and Pangaea, we would expect many cosmopolitan marine species. Fragmentation into separate continents would result in many different, isolated, shallow marine areas for species evolution.

Virtually all the marine fossils we know from continental areas lived in reasonably shallow water. The total area of shallow water available to such former life would have been greatly affected by whether or not the continents were apart or together. The total shoreline of several separate continents is conspicuously greater than the shoreline of a single supercontinent. If the total number of species that can live in shallow water is proportional to the amount of area available for living sites, the overall diversity of shallow-water marine life should have been low when Pangaea was assembled and high when the continents were separate. Recent studies have shown that, indeed, this was the case.

The configuration of land areas, quite apart from their latitudinal position, also had a profound effect on world climate. A single large land mass, like Pangaea, would have had an extreme continental climate with wide fluctuations in temperature and rainfall, probably much more drastic than any climates today. Breakup of the land mass into several continents would have resulted in each smaller land mass having much more climatic influence from the surrounding oceans, resulting in less drastic extremes. Widespread deserts and salt deposits during Permian and Triassic times, when Pangaea existed, are probable indicators of climatic extremes.

It has also been suggested that movement of the continents was partly responsible for rises in eustatic sea level, forming the widespread shallow seas that have flooded the continents repeatedly during the past 600 million years. As the continents moved apart, the various mid-oceanic ridges built up. This displaced enormous volumes of ocean water, resulting in a significant sea level rise relative to land. Because large areas of the continents were low, flat lands, a slight elevation in sea level flooded immense areas of the continents. Thus, episodes of increased volcanic activity along mid-oceanic ridges may have corresponded with times when several continents experienced widespread flooding by the sea.

Continental drift helps explain several apparent anomalies in the distribution of ancient organisms. For example, we know that today coral reefs only flourish in low-latitude, tropical environments. Yet, many ancient reefs are in rocks that are at much higher present-day latitudes. Before continental drift, this apparent discrepancy was explained by appealing to widespread equitable climates. We now know that these areas were once located in low latitudes and that the continental areas on which these fossil reefs thrived have since drifted to higher latitudes. Thus, there are Silurian-age reefs in southern Canada and just north of Hudson Bay, but these reefs formed in tropical climates when North America was situated in low southern latitudes.

The distribution of ancient coal beds can be explained in the same way. Coal swamps are thought to have required warm, humid climates. The occurrence of coals today at high latitudes is interpreted to indicate that these areas were in lower latitudes during the time when the coal swamps flourished. This applies both to extensive Pennsylvanian age coals as well as the extensive Cretaceous and Cenozoic coals.

KEY TERMS

continental drift	Laurasia	plate tectonics
core	magnetic reversal	remnant magnetism
crust	mantle	sea-floor spreading
Eurydesma	*Mesosaurus*	spreading center
Gangamopteris	paleobiogeography	subduction zone
Glosopteris	Pangaea	
Gondwana	plates	

READINGS

Kearey, P. 1990. *Global Tectonics*. Blackwell Scientific. 302 pages. An advanced-level textbook dealing with all aspects of plate tectonics.

The Living Planet: The Building of the Earth (videotape). 1998. BBC, with Time-Life; Ambrose Video. 55 minutes. The formation of Earth, moving continents, and volcanoes.

Smith, A.G.; Hurley, A.M.; and Briden, J.C. 1981. *Phanerozoic Paleocontinental World Maps*. Cambridge University Press. 102 pages. A series of 88 maps showing the distribution of land and sea from Cambrian to Holocene times in two different map projections.

Wilson, T.J., ed. 1976. *Continents Adrift and Continents Aground*. Freeman. 219 pages. A selection of *Scientific American* articles by 19 different authors on continental drift, sea-floor spreading, and plate tectonics.

C H A P T E R   7

Paleobiogeography

INTRODUCTION

Fossils are distributed in rocks within both a time and a spatial framework. In this chapter the focus will be on the arrangement of fossils in space, the ancient geography of life, or **paleobiogeography**. We know that plants and animals are not evenly distributed across the face of Earth today, either in marine or terrestrial environments. We do not expect to find giraffes outside central Africa or kangaroos outside Australia, except, of course, in zoos. An example from the oceans would be that coral reefs are confined to near equatorial areas. Did life of the past have similar distinctive patterns of distribution? Yes, it did, although the patterns were very different from those of today. In this chapter we consider some of the more important examples of the geographic differentiation of life in the fossil record.

Perhaps the single most important consideration for thinking about ancient geography is that we live during a time that is quite atypical. As already stated, the marine fossil record is considerably better, more complete and more enduring, than the land fossil record. We have also noted that knowledge of fossils from rocks exposed on land is much more complete than that from rocks on the floors of the oceans. These two observations taken together must mean that many areas that are now land were once covered by the seas. Compared to the past, the present is a time when the continents stand relatively high above sea level; consequently, there are large areas of exposed land. During many past times this was not the case. Wide continental areas were flooded by shallow seas that deposited marine rocks now exposed on the continents. The continents have been alternately flooded and dry, with seas transgressing over the land and later withdrawing. During times of regression, erosion took place, destroying some of the previously deposited sediments and resulting in the interrupted and incomplete rock record that we now have available for study.

During the Paleozoic, shallow seas covered parts of the United States during each of the periods of time from the Cambrian through the Permian. At the beginning of the Paleozoic, Early and Middle Cambrian seas were confined to the borders of the continent but flooded into the interior by Late Cambrian time. Two of the greatest times of flooding were during the Ordovician and the Mississippian Periods of the Paleozoic (Figure 7.1). During the Mississippian, dry land in North America consisted of a group of large islands separated from each other by wide shallow seaways. Maximum withdrawal occurred at the close of the Permian when North America stood high relative to sea level. At this time, marine rocks were deposited only along the southern and western margins of the United States. The last great inundation of North America occurred during the Late Cretaceous, at the close of the Mesozoic Era. This

Figure 7.1 Distribution of land and sea over the United States during Mississippian time. Land areas are patterned; areas covered by seas are not. The position of the United States relative to the equator is shown.

Cretaceous seaway extended from Texas across the Great Plains region of the United States and Canada and far north into the Arctic. This seaway dried up at the end of the Cretaceous, from which time the continent has not had extensive flooding by oceans. Most of the wide seas that flooded the continents were quite shallow, probably less than 200 meters deep. This factor surely restricted currents and tidal action. If we wanted to study similar seas of the present time, we would be restricted to bodies of water such as the Baltic Sea, between Scandinavia and Europe, or Hudson Bay. However, because neither of these modern seas is in low-latitude, tropical or subtropical climates, they are not truly comparable to the epicontinental seas of the past.

FAUNAL AND FLORAL PROVINCES

We recognize the distribution of any species of plant or animal as being **cosmopolitan** (global or very, very widespread), **endemic** (confined to a small area), or someplace between. Actually, the great majority of species fall in an intermediate category. Where a relatively large area is characterized by a suite of plants or animals that are restricted to that area, the region is termed a **faunal** or **floral province**, depending on whether one is studying animals or plants. Today we can recognize such large regions as the Boreal Province, constituting land and sea that surrounds the North Pole. Australia is both a faunal and floral province with many plants and animals that are unique to that continent.

We can also recognize faunal and floral provinces throughout the fossil record, beginning at the base of the Cambrian. Prior to about 20 years ago, reconstructions of ancient provinces were mostly based on the positions of land and sea with the continents and oceans fixed in their present positions. Now much of this earlier work is being reevaluated using different arrangements of continents and oceans.

In Lower Cambrian rocks there is a distinctive assemblage of trilobites that occurs throughout western and eastern North America, except for a small region in the Maritime Provinces of Canada and in New England (Figure 7.2). In these latter areas a quite different assemblage of trilobites occurs, which is also known from various localities in western Europe and northwestern Africa. The fossils that predominate throughout North America constitute a province of ancestral North American aspect

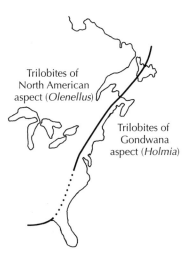

Trilobites of
North American
aspect (*Olenellus*)

Trilobites of
Gondwana
aspect (*Holmia*)

Figure 7.2   Map of eastern North America showing the boundary between the Early Cambrian trilobite provinces—the *Olenellus* province to the west, the *Holmia* province to the east. The *Olenellus* province also occurs in Europe, Norway, Scotland, and Northern Ireland, as well as in western Africa.

trilobites characterized by *Olenellus*, one of the distinctive trilobites in this assemblage. The suite that dominates the small region in the Canadian Maritime Provinces and New England, western Europe, and northwestern Africa is a province of ancestral Gondwana aspect trilobites characterized by *Holmia*.

The small area of the ancestral Gondwana province trilobites in North America is judged to represent a piece of what was the European continent during Cambrian time. During the Cambrian, ancestral North America was separated from ancestral Europe by the Iapetus Ocean, which separated these two trilobite provinces. Later during the Paleozoic, when the continents came together to form Laurasia, these two continental areas were "welded" together. When the continents drifted apart about 180 million years ago, this region in Canada and New England became part of the present North American continent. These and other distinctive trilobite provinces indicate that when marine communities with hard parts first became established during the Cambrian, life was already organized into different regions.

There are numerous other examples of marine faunal provinces. During the Devonian and Mississippian periods, marine faunas were marked by times of endemism; that is, many marine animals were confined to small areas. These intervals alternated with times of cosmopolitanism, when genera and species had a nearly worldwide distribution. What caused these differences in size, number, and distinctiveness of marine faunal provinces? One answer is based on the fact that these differences are related to the ease with which shallow marine animals could migrate or disperse from one area to another. During times with numerous barriers to migration, faunas with quite different aspects would have tended to evolve in different regions.

As discussed in Chapter 6, sharply defined faunal and floral provinces were in existence during the latter part of the Paleozoic Era, with a *Eurydesma* marine fauna and a *Glossopteris* terrestrial flora confined to **Gondwana**. These provinces were surely induced, in part, by differences in climate, with Laurasian areas near the equator and Gondwanan areas near the South Pole. The presence of continental ice sheets in the Gondwana region supports a climatic interpretation.

In Mesozoic strata, there is continuing fossil evidence for geographical differences in the distribution of plant and animal life. In this instance, there are distributions that show distinct latitudinal changes in the northern hemisphere. We can recognize various groups that have either low-latitude distribution close to the equator or high-latitude distribution closer to the pole. Mesozoic reefs, for example, are more common and larger in low-latitude areas. These reefs extend farther north, especially during Jurassic and Cretaceous times, than coral reefs today, suggesting that the temperature of those times may have been somewhat warmer than today. Some Mesozoic reefs are built by highly specialized bivalves called **rudistids** (Figure 7.3), which evolved a coral-like shape with a conical lower valve and a cap-shaped upper valve. These clams became enormous, some up to 2 meters high, and were frame-building organisms on reefs during the Cretaceous. These reefs are common in Mexico and the Caribbean, but they are rare northward in the United States where they are known mainly from Texas (Figure 7.4). Other Mesozoic fossils, such as **belemnites** (Figure 7.5), a group of extinct cephalopods, had a high-latitude distribution, being much more common in the northern United States and Canada than farther south. Belemnites had a solid, cigar-shaped, internal skeleton and were squidlike animals.

These examples of latitudinal control on the distribution of rudistids and belemnites are only part of a much more widespread differentiation with respect to latitude of various groups of marine fossils during Mesozoic time. The equatorial fauna that was latitudinally controlled goes under the general name of Tethyan. The **Tethys Seaway** was a great oceanic area of which the present day Mediterranean and Caspian seas are small, shrunken remnants. The seaway extended from the Caribbean area of the western hemisphere to the east between Africa and Europe, across the Middle East and the

Figure 7.3   Side view of a rudistid clam belonging to an extinct group of Mesozoic bivalves that built small reefs and were confined to low latitudes. One valve, shown here, is elongate and coral-like in aspect. The other valve, not shown, forms a cap on top. Length of specimen, about 20 cm. (Courtesy of Field Museum of Natural History, Chicago.)

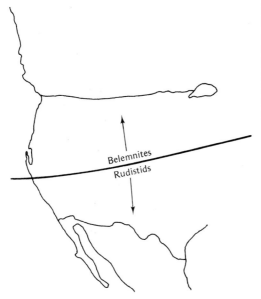

Figure 7.4   Cretaceous marine faunal provinces in western North America. Rudistid bivalves are most abundant in Mexico and Texas and become progressively rarer farther north. Jurassic and Cretaceous belemnites are most common in Canada but also occur in the northern United States.

present site of the Himalayas in India, to southeastern Asia, and into the Indonesian archipelago. The life of this seaway had a distinctively tropical aspect, with many kinds of clams, snails, ammonites, and other fossils confined to this equatorial seaway. In contrast, shallow-water marine fossils farther north in Europe, Asia, and North America lack this tropical aspect and have faunas with fewer species and fewer numbers of individuals. Thus, many lower latitude Mesozoic marine faunas are said to have a Tethyan aspect, in reference to this major low-latitude seaway. Clearly, the latitudinal control on fossil distribution during the Mesozoic reflects more fundamental controls of climate and temperature differences with latitude. Temperature was probably the ultimate controlling environmental factor that limited the distribution of these animals during the Mesozoic.

Some of the best-known ancient life provinces are recorded in terrestrial rocks of the Cenozoic Era. Fossil mammals and fossil flowering plants, both of which were dominant on land during the Cenozoic, have very distinctive patterns of distribution. Primitive groups of mammals were able to migrate to South America and Australia

Figure 7.5   The internal skeleton of an extinct belemnite cephalopod from England. The specimen is Jurassic in age and about 19 cm long. (Courtesy of Field Museum of Natural History, Chicago.)

near the beginning of the Cenozoic, when the distribution of land areas allowed for such migration. As the breakup of Pangaea continued and reduced the likelihood of migration to South America and Australia, mammals on each of these continents evolved for millions of years in isolation, thus becoming distinctive and unlike mammals elsewhere. Later, when South America was connected to North America via the Isthmus of Panama, again due to plate tectonics, South America was invaded by North American mammals that drove the vast majority of native South American mammals to extinction. A few South American mammals moved north, such as the armadillo and opossum, and these two animals are still expanding their range through the United States. A similar history occurred in Australia, where the native fauna that evolved in isolation was dominated by marsupial mammals (kangaroo, wombat, and many others). In this case, competition has largely been from human-introduced invaders, such as cats, dogs, rats, and rabbits. (This Cenozoic mammalian history is discussed in greater detail in Chapter 16.)

Floral provinces of the Cenozoic had radically different patterns of distribution earlier during the era than they do today. Subtropical plants extended far north into Alaska during the Cenozoic and were gradually pushed southward with climatic cooling during this time. As climatic cooling took place and an arid climate emerged, new assemblages of plants evolved in adjustment to these climatic changes. The widespread grasslands and prairies of the central part of the continent and the dry, desert flora of the southwestern United States evolved during this period of climatic change, resulting in a more complex and fragmented series of floral provinces than is still in existence. These floras were abruptly affected by the four major periods of continental glaciation that occurred during the Pleistocene (past 2 million years) in North America. The spreading and melting of ice sheets caused floral provinces to alternately shrink and expand, to shift south and north, as unusually sudden and extreme climate changes took place.

MIGRATION, DISPERSAL, AND BARRIERS

What are the processes that control the migration and dispersal of plants and animals and allow faunal and floral provinces to develop? Migration and dispersal occurs differently in different groups of plants and animals. For example, the dispersal stage of most plants is during the spore or seed phase of the life cycle, whereas the adult plant is stationary. Dispersal occurs by numerous mechanisms, both physical and biological. Spores and seeds are dispersed by wind and by water. In addition, spores and seeds are dispersed by innumerable animals in a variety of ways, including by accidentally attaching to the setae of an insect leg, the fur of a mammal, or being deployed to a new area in the feces of an animal. As for animals, most marine invertebrates have planktonic larvae. It is during this phase of their life cycle when the maximum dispersal occurs. Depending on the type of larva and whether it is a feeding larva or has a large attached food supply, this stage can be prolonged and allow for considerable dispersal. For large marine planktonic animals, marine nektonic animals, and terrestrial vertebrates, migration occurs in adults.

Dispersal and migration occur as far and wide as conditions allow. Limiting factors are things such as distance, climate, suitable habitat, etc. A restriction to migration or dispersal is called a **barrier**. For land animals, the oceans, large freshwater lakes, and mountains may be barriers, whereas for marine life, land and wide expanses of deep oceans may be barriers. Fresh water may or may not be a barrier to marine animals, depending on their tolerance to changes in salinity.

G.G. Simpson, a well-known vertebrate paleontologist and evolutionary biologist, described three kinds of migration routes for land animals. The concepts he formulated are also useful in thinking about migration and barriers for marine organisms. He recognized corridors, filter bridges, and sweepstakes routes (Figure 7.6). **Corridors** allow for relatively easy dispersal and migration back and forth in both directions. The steppe region of Russia, between Asia and Europe is an example of a corridor. In the case of **filter bridges**, migration and dispersal is not so easy. Generally, some animals can utilize this path, but others cannot. Thus, migration and dispersal are selective; some animals are filtered out and not able to migrate, whereas others freely move along this path. Two well-known examples of filter bridges for land animals are the Bering Strait and the Isthmus of Panama. At times when sea level was lower than today, the Bering Strait was dry land that connected Asia and Alaska. Migration was easily possible for those organisms that could tolerate the colder climates at this high latitude. Animals that lived farther south never had the opportunity to migrate along this route, so they were denied access to this bridge.

During much of the Cenozoic Era, the present day land connection between North and South America did not exist. The Isthmus of Panama developed and became dry land during the Pliocene, only 3 million years ago. Animals moved both north and south across this bridge, but many more animals from North America migrated to South America than vice versa; thus it is a filter bridge (see Chapter 16).

In a **sweepstakes route** (were G.G. Simpson to have named this today he would have called it the "lottery route"), migration and dispersal are quite difficult. This name implies that a "winning ticket" is needed, perhaps one of a million, in order to migrate successfully across such a route. Examples for terrestrial animals are islands separated from other land by wide expanses of water. The result commonly appears to be an unusual, unbalanced ecosystem because some animals, probably under quite fortuitous circumstances, made the jump, whereas most did not. Madagascar and many islands in the Pacific are examples of areas reached by animals using sweepstakes routes. Indeed, it was the unusual biological circumstances on the Galapagos Islands, a product of the sweepstakes route, that led Charles Darwin to his ideas on evolution by natural selection.

Within the last 25 million years, since the Miocene Epoch of the Cenozoic when the continents were approximately in their present positions, there have been several migration routes for land plants and animals. These include the ones already mentioned, as well as a sweepstakes route that involves island hopping across Indonesia from the mainland of Asia to Australia; the land connection between the Middle East and Africa, now severed by the Suez Canal; and possibly a land connection, which existed at certain times, across the Strait of Gibraltar at the western end of the Mediterranean Sea.

Figure 7.6   Different kinds of dispersal pathways in North America. The broad area of the Great Plains provides a two-way corridor for migration. Filter bridges occur across the Isthmus of Panama and the Bering Strait, which at times in the past was dry land. Sweepstakes routes are found between the mainland and the Caribbean islands.

The idea of corridors and filter bridges can also be applied to marine invertebrates. After the Suez Canal was completed, a former barrier became a marine filter bridge for shallow-water organisms. Some animals were able to migrate or disperse through the canal; others were not. The Suez Canal, of course, is artificial, but it is clear that changes in global sea level altered migration routes for shallow marine organisms, thus altering the ease at which dispersal can occur.

The distribution of marine animals is also influenced by physical and chemical conditions in the oceans. Temperature may provide the strongest control on shallow-water marine organisms. On both the Atlantic and Pacific coasts of North America, there are a series of shallow-water life zones from south to north, based principally on the presence and absence of marine gastropod and bivalve molluscs. Temperature, a function of latitude, seems to be the primary control, although temperature is also greatly affected by the distribution of cold- and warm-water currents. These molluscan life provinces are also recognized in Cenozoic strata along both coasts, but these provinces have shifted through time, providing strong evidence for past climatic change.

REFUGES AND LIVING FOSSILS

During the entire span of life, many groups of plants and animals enjoyed periods when they were abundant, diversified, and widespread, only to come upon hard times later. In many instances, these groups ultimately became extinct. In other cases, we find that there have been survivors, sometimes referred to as **living fossils**. Commonly, the surviving plant or animal is restricted to certain environments or to specific geographic areas where they have managed to hang on, perhaps because of reduced competition. Such restricted areas or habitats are called refuges, although they have also been termed asylums or havens.

For marine animals, the most common example is finding primitive survivors persisting in relatively deep water. These are animals that once lived in shallow water, as documented by their fossil record. One striking example is a very primitive group of molluscs, called monoplacophorans. Monoplacophorans have a single cap-shaped shell that is similar to the limpet gastropods that live attached to intertidal rocks today. However, the internal anatomy is partially segmented and very different from any gastropod or any other mollusc. Monoplacophorans are well known as fossils in Cambrian through the Early Triassic, after which they were thought to have become extinct. However, in 1957 a living monoplacophoran, *Neopilina*, was dredged from waters of the Pacific near Central America at depths greater than 3,000 meters (Figure 7.7). It is now known from deep waters in many places in the world. Here, still alive, is an animal previously thought to have been extinct for more than 230 million years!

Another example of a living fossil is the well-known fish *Latimeria* (Figure 7.8). This is a **coelacanth** fish, related to living lungfishes and to the extinct group of fish that gave rise to land vertebrates (tetrapods) during the Devonian Period. During the Paleozoic and Mesozoic, coelacanths lived in shallow-water habitats, and they were thought to have become extinct during the Cretaceous. However, in 1938 a single, live coelacanth specimen was caught by a commercial fisherman in waters several hundred meters deep off the coast of Madagascar. Since then, other specimens have been caught in the same general area. Coelacanths today seem to be largely restricted to waters deeper than 100 meters off the eastern coast of Africa.

These and other examples indicate that some animals that once lived in shallow water have moved to deeper water habitats, perhaps because they were unable to compete effectively in shallow-water communities. Shallow-water environments are an especially favorable habitat for many organisms, with abundant food supplies and diverse habitats. Presumably, competition in shallow water is more intense than in deeper water. Whether monoplacophorans, *Latimeria*, or other organisms that originally lived in shallow water and moved to deeper water became exterminated in the latter or simply migrated to deep water is unclear in most cases.

Other instances of geographic restriction of marine animals do not seem to be related to changes in the water depth at which they lived but, rather, to persistence in a small area or refuge. The bivalve *Trigonia* provides an example. During the Mesozoic, this clam was worldwide in distribution and a very common fossil (Figure 7.9). However, *Trigonia* is not known as a fossil after the Cretaceous, except in eastern Australia, where it lives today in shallow water off the eastern coast. This area of the southwesern Pacific also seems to have been a refuge area for other kinds of marine

 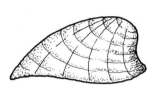

Figure 7.7   A living invertebrate that lives in a deep-water refuge. Top and side views of *Neopilina*, a living monoplacophoran mollusc. Fossil relatives are confined to Triassic and older rocks.

animals for many millions of years. Persistence of suitable climate and environments in this region could be an explanation for this phenomenon, but this is still not certain.

Similar distribution patterns are also known exclusively in the fossil record. Some marine animals developed restricted distributions, common shortly before they became extinct. During the Mississippian Period, a very distinctive kind of bryozoan evolved. It is called *Archimedes* because it had a twisted central shaft shaped like an archimedian spiral (Figure 7.10). This bryozoan had a widespread occurrence throughout North America during the Mississippian but persisted into the early part of the Pennsylvanian only in Arkansas and Oklahoma, as far as is known. During the Middle Pennsylvanian, it is known in the central United States and in Utah. It has not been found anywhere in Upper Pennsylvanian strata, but it has been reported from Alaska and Siberia in Permian rocks, after which it became extinct.

An extinct group of echinoderms called **blastoids** were cosmopolitan during the Devonian and Mississippian (Figure 7.11). They, like *Archimedes*, are found in Lower Pennsylvanian rocks only in Arkansas and Oklahoma, after which they become extremely rare until the Permian. In Lower Permian rocks of the small island of Timor in Indonesia, they are quite abundant and diverse. They also occur in Permian strata from Russia, Australia, and Indonesia, but have a much reduced distribution compared to earlier during the Paleozoic. Blastoids became extinct by the close of the Permian. Presumably both blastoids and *Archimedes* were only able to survive under a specific set of conditions that became more and more restricted, resulting in their confinement

Figure 7.8   The only living coelacanth fish, *Latimeria*. Note the thick, fleshy fins and the symmetrical tail. (Courtesy of Field Museum of Natural History, Chicago.)

Figure 7.9   External and internal views of the bivalve *Trigonia*. This specimen from Tennessee is Cretaceous in age. The shells are approximately 6 cm wide. (Courtesy of National Museum of Natural History.)

to certain small areas of Paleozoic oceans. Final elimination of these habitats at the close of the Paleozoic resulted in the extinction of both groups.

So far, our discussion has been confined to refuges of marine life, but there are also examples among terrestrial plants and animals. During the Mesozoic Era there was a group of primitive coniferous trees related to pines, called **araucarian** pines. These survive naturally today strictly in the southern hemisphere. The best known example is the Norfolk Island pine, which is now commonly planted as an ornamental tree. During the Mesozoic, araucarians were geographically widespread and abundant. For example, many of the large fossil logs in the Petrified National Forest in Arizona, Triassic in age, are this kind of conifer.

Several other trees have a pattern similar to that of the araucarians—widespread during the past but now confined to a refuge. Another example is the group of tree-sized gymnosperms called **ginkgos** (Figure 7.12), or maidenhair trees. They have fan-shaped leaves with parallel veins. Ginkgos were very common elements of the terrestrial flora during the Mesozoic but disappeared from the North American fossil record during the Miocene. Eventually, they survived only in China, where they became extinct in the wild—the last uncultivated specimens having been cut down for firewood centuries ago. However, prior to their extinction, ginkgos had been planted in Buddhist temple courtyards and cultivated for centuries. Ginkgos once again enjoy a worldwide distribution thanks to their recent dispersal by humans. Ginkgos are useful street-side trees and are common as ornamental trees.

Another tree that survives in the wild only in China is *Metasequoia* (Figure 7.13). This is a close relative of *Sequoia*, the giant redwood tree, from the coastal areas of the western United States. *Metasequoia*, initially described from fossil remains in Korea, differs from *Sequoia* in the structure and arrangement of the needlelike leaves. In 1948 the tree was discovered alive in mountainous regions of China. It too has been dispersed by humans.

An example of a terrestrial animal that might be called a living fossil is the tuatera lizard that is confined to a few small islands off the coast of New Zealand. This ani-

Figure 7.10   A small slab of limestone with two of the spiral axes of *Archimedes*. Lacy bryozoan fronds that may have been attached to the spiral axes are scattered over the surface of the rock. The slab is approximately 15 cm across and is Mississippian in age. (Courtesy of National Museum of Natural History.)

mal is not really a lizard, but rather it belongs to a group of reptiles called the rhynchocephalians, which were conspicuous during the Mesozoic. They flourished during that era but have since disappeared from the fossil record. Now they have this single small living representative.

Two other examples are mammals that live in Australia, the duck-billed platypus and the spiny echidna, or anteater. Both are unusual in that they are the only known egg-laying mammals. They have hair and mammary glands and, thus, qualify as mammals, although they are considered to be very primitive. They may be the only survivors of a quite archaic group of mammals, but unfortunately we are not sure to which group of primitive mammals they belong. Most fossil mammals are identified on the basis of their teeth, especially their molars. The teeth of these mammals are degenerate; consequently, we have great difficulty in trying to ally them with their possible ancestors.

THERMOMETERS OF THE PAST

We know that the present-day climate varies at different places on Earth's surface. At least some of these differences are caused by the angles at which the sun's rays strike Earth's surface and the length of time that these areas are exposed to sunlight. These factors must have operated throughout Earth history to produce climatic differences. Other important factors, such as the distribution of land and sea and the major oceanic currents, have changed through time due to continental drift. Thus, we can expect that ancient climates were different from what they are today. One of the most important aspects of climate is temperature. We can make inferences about ancient climate if we know the average annual temperature for an area. Fortunately, we have a chemical

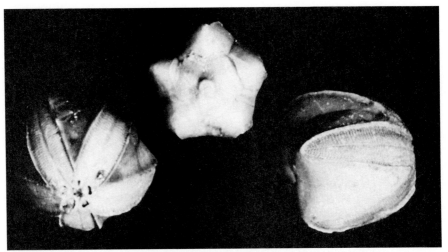

Figure 7.11   Three views of the blastoid *Pentremites*. Although this kind of blastoid is most common and diverse during the Upper Mississippian, it survived into the Lower Pennsylvanian in Oklahoma and Arkansas. The stem and food-gathering appendages are missing. The specimen is approximately 2 cm high. (Courtesy of Ward's Natural Science Establishment, Inc.)

thermometer that allows determination of temperatures for the last 130 million years. For earlier periods of Earth history, this method is less reliable, and we must also depend on indirect evidence of rocks and fossils.

The chemical method for determining ancient temperature is called the **oxygen isotope method**. As noted in discussing radioactive age dating (see Chapter 1), many elements have varieties, called isotopes, with different atomic weights. Oxygen has several isotopes including O^{16}, the common isotope that accounts for 98 percent of all oxygen atoms in nature, as well as a heavier isotope, O^{18}, that accounts for 0.2 percent of all oxygen. The ratio of these two isotopes in seawater is a function of temperature.

Figure 7.12   Leaf of the gymnosperm Gingko from Paleocene rocks in South Dakota, about 6 cm wide. (Courtesy of David Dilcher, University of Florida.)

A

B

Figure 7.13   *Metasequoia* from the John Day Formation, Oligocene, Oregon. (A) Leafy branchlets. (B) Branch. The scale on the right is in centimeters. (Courtesy of David Dilcher, University of Florida.)

Thus, warmer seawater tends to have more O^{18} than colder water. This is because O^{16} is lighter and is preferentially given up in evaporation of H_2O in warm seas. Also, many marine organisms take up oxygen when they build shells or tests with calcium carbonate, $CaCO_3$. Some organisms randomly select either O^{16} or O^{18} when constructing a skeleton of $CaCO_3$, so the two oxygen isotopes are used in the precise abundance in which they occur naturally in the water. If the shell becomes fossilized and is relatively unaltered so that the ratio of oxygen isotopes is preserved, this isotopic ratio can be measured and used to determine the water temperature when that shell was secreted. Shells that are preserved well enough to be studied using this technique are generally from rocks no older than the Cretaceous. A few belemnite shells from the Jurassic have also been studied. They are so thick that the calcite has undergone very little alteration. Because many older fossils have been chemically altered, this method cannot commonly be used on older Mesozoic or Paleozoic rocks, although some promising work on Paleozoic brachiopods has been completed. Nevertheless, the oxygen isotope method of studying paleotemperatures is a powerful tool for helping to understand the past history of Earth's climate.

The pattern of worldwide temperature fluctuations is now well understood. This history was unfolded by studying sediment cores from the major ocean basins during the deep-sea drilling project now known as the Ocean Drilling Program (commonly abbreviated ODP). ODP is a multinational scientific consortium whose primary goal is to recover information about the floor of the oceans. The project includes a large research vessel with an attached drill rig and drilling platform from which scientists drill cores into sediments at the bottom of the oceans. These data from Earth's oceans reveal that during the Jurassic and Cretaceous Periods, there was a distinct warming trend. This was followed by a drop in temperatures during the very Late Cretaceous Period and early Paleocene Epoch. A second warming trend peaked during the early Eocene Epoch, followed by a steep decline in temperature that bottomed out during the Oligocene (Figure 7.14). A second cooling trend occurred during the Miocene Epoch, which may have coincided with the formation of the Antarctic ice sheet. Certainly during this time, sea level dropped as water was withdrawn from the oceans and became tied up as ice on dry land. The third drop in temperature began during the Pliocene Epoch and culminated in the ice ages of the Pleistocene. Earth's climate during the past 3 to 4 million years is known in great detail. Although we tend to think of four major advances and retreats of continental glaciers in the northern hemisphere, in reality there were many minor advances and retreats superimposed on the larger cycle.

Many aspects of ancient life can be related to this pattern of changing climate. The spread of flowering plants and the development of the major floras during the Cenozoic reflect these changes (see Chapter 14). The distribution of mammals is related to their adjustment to colder climates. Furthermore, the pattern of coral reef development during the last 60 million years is related to change in global temperature.

OLDER PALEOCLIMATES

In older rocks we cannot use the oxygen isotope method to measure temperature. Thus, we must depend on less precise methods to determine ancient climates. Some of these involve fossils. For instance, we know that modern coral reefs do not live where temperatures are much below 20 degrees centigrade, or higher than approximately 30 degrees of latitude. If ancient reefs had similar temperature constraints, their widespread presence is an argument for tropical and subtropical conditions in a broader area than today. Therefore, the widespread development of Silurian and Devonian reefs is taken to indicate that these areas were situated in low latitudes with a generally warm cli-

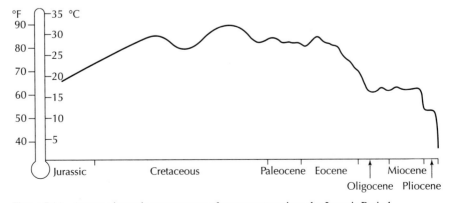

Figure 7.14   Average isotopic temperature of ocean waters since the Jurassic Period.

mate, whereas the restriction and rarity of reefs during the Permian and Triassic Periods may indicate restriction of tropical climes or extinction of reef-building organisms.

Various rock types can be used as paleoclimate indicators. For example, evaporites (e.g., rock salt and gypsum) and red beds (shales and sandstones colored by oxidized iron) are thought to reflect warm conditions and either arid or monsoonal rainfall conditions. Another rock type, coal, is considered to have formed in only warm, humid climates because the plants that produce coal beds require these conditions. Verification of this can be found in the tree trunks from Pennsylvanian coal deposits that lack growth rings, indicating that the trees lived in areas without distinct growing and dormancy seasons. Some Devonian woody tree trunks have such rings, as do Permian trunks from the southern hemisphere where glacial conditions existed.

Rocks that were clearly deposited by ice, rather than by wind or water, are also powerful climatic indicators. Such rocks are called **tillites** and are usually very poorly sorted and bedded. Tillites of various ages indicate that there have been a series of ice ages on Earth, not just the most current one that we are experiencing today, albeit from an interglacial perspective. We have discussed the Permian ice age in the southern hemisphere that played an important role in early debates about continental drift (see Chapter 6). An earlier ice age during the Ordovician Period mainly affected northern Africa, which was over the South Pole at that time. Several ice ages occurred during the late Proterozoic from approximately 1,000 to 600 million years ago, with the one occurring 800 million years ago the most extensive event. Evidence of glaciers at that time has been found on all continents, perhaps making this period the coldest time during the history of Earth.

KEY TERMS

araucarian	endemic	oxygen isotope method
barrier	faunal province	paleobiogeography
belemnites	filter bridge	rudistids
blastoids	floral province	sweepstakes route
coelacanth	ginkgos	Tethys Seaway
corridor	Gondwana	tillites
cosmopolitan	living fossil	

READINGS

Briggs, J.C. 1987. *Biogeography and Plate Tectonics*. Elsevier. 204 pages. An up-to-date review of past plant and animal distribution related to continental and ocean positions.

Wallister, O.H., ed. 1986. *Global Bio-events: A Critical Approach*. Springer-Verlag. 442 pages. This book includes technical chapters on various physical and biological events that affect all of Earth.

Whitmore, T.C., and Prance, G.T. 1986. *Biogeography and Quaternary History in Tropical America*. Oxford Monographs on Biogeography No. 3. Clarendon Press. 214 pages. Modern and ancient biogeography of Latin America, with emphasis on plants, birds, butterflies, and humans.

The Precambrian Fossil Record

INTRODUCTION

This chapter considers the Precambrian fossil record, including the oldest fossils on Earth to the base of the Cambrian. This discussion builds on Chapter 4, where the origin of Earth and the origin of life was discussed. The Precambrian comprises 88 percent of Earth history, and because the oldest fossils are 3.5 billion years old, the Precambrian constitutes the first 85 percent of time when life existed. Documentation of this early fossil record is relatively recent in paleontology and has generated much interest among all paleontologists. Prior to 1954, there were almost no fossils known at all from Precambrian rocks, and the few that were known were almost invariably questioned or regarded as non-organic. But the mystery of early life began to unravel as geologists and paleontologists began to recognize Precambrian fossils and to learn in what types of rocks these should occur. Today, more than 2,800 sites around the world are known to contain Precambrian fossils.

There are three principal types of evidence for Precambrian life, morphological fossils, stromatolites, and chemical fossils. **Morphological fossils** are the preserved remains of ancient organisms and consist exclusively of microscopic fossils for much of the Precambrian. During the later Precambrian molds and casts of larger fossils exist. Microscopic fossils are preserved in **black chert**. Chert is a microcrystalline form of silica, which in its common crystal form, quartz, is one of the most common minerals on Earth's surface. The black color is due to finely disseminated carbon, organic matter that includes the microfossils. These ancient black cherts apparently formed on the sea floor from gels of silica that were soft and sticky, allowing the cells of microscopic primitive organisms to become trapped and preserved in the gels, which were later buried and hardened into chert (Figure 8.1). Except in the case of the very youngest Precambrian rocks, in which fossils are in other kinds of sedimentary rocks, most known Precambrian fossils come from black cherts. In very young Precambrian rocks, small shelly fossils and molds and casts of larger organisms begin to dominate the fossil record.

Black cherts are commonly associated with stromatolites. **Stromatolites** are a preserved sructure that was formed by algal mats. These should be considered a type of trace fossil because typically none of the algae that made the stromatolites are preserved. Stromatolites vary greatly in shape and size. They can be tabular-, mound-, or column-shaped. Columns can be the size of a pencil or more than 1 meter in diameter. The distinguishing characteristic of stromatolites, regardless of size or shape, is that they are composed of very thinly laminated sediment, which reflects the way these algal accumulations grew. Stromatolites are still living, so the growth process is well documented. Only the upper surface of a stromatolite is alive. It contains a mixture of algal types, all of which are photosensitive. When this living mat is covered by a thin layer of sediment, algae grow through this sediment to form another living algal mat at the surface. This process is repeated many times, yielding the finely laminated structure of the stromatolites.

Chemical fossils are also a type of trace fossil. The biochemistry of life is not a random process; life leaves very definitive chemical signals. This can occur in many ways, and the two types of chemical fossils to be mentioned here are carbon isotopes and pristane and phytane. Organically derived carbon is enriched in C^{12}, the lighter isotope (relative to C^{13}), whereas inorganically derived carbon has a mixture of these isotopes that is proportional to their natural, global abundance. The **C^{12}/C^{13} ratio** is easily determined with current technology, so it is possible to ascertain whether carbon in the rock record is of organic or inorganic origin. **Pristane** and **phytane** are organic molecules that are the geologically stable products of the decomposition of chlorophyll, the pigment in green plants responsible for photosynthesis. Thus, the presence of pristane and phytane are evidence for the former existence of photosynthetic autotrophic organisms.

A few Precambrian localities are discussed below in chronological order to describe the basic pattern of evolution during this part of Earth history. This discussion will primarily involve morphological fossils, but in all cases, stromatolites and chemical fossils occur at these localities and microfossils are from black cherts.

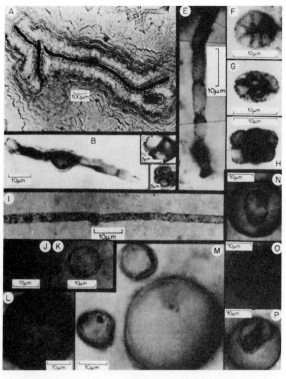

Figure 8.1 Photomicrographs of Precambrian fossils. Specimen (A) is a pseudo- or false fossil. (B-P) are preserved in black chert. Specimens (B-D) are from rocks of 2.25 bya, fossils consist of cyanobacterial filaments (B) and spherical cells (C, D). Specimens (E-H) are from 1.2 bya; (I-K), 1.4 bya; and (L-P), 0.85 bya. Magnification scale is in microns as indicated. (With permission from the *Annual Review of Earth and Planetary Sciences*, vol. 3. Copyright © 1975 by Annual Reviews. Photograph courtesy of J.W. Schopf, U.C.L.A.)

PRECAMBRIAN MICROFOSSILS

WARRAWOONA GROUP, WESTERN AUSTRALIA—3.5 BYA

The oldest uncontroversial fossils known on Earth are from the 3.5-billion-year-old **Warrawoona Group** of western Australia (slightly older stromatolites without associated microfossils are also known from both southern Africa and western Australia) (Figure 8.2). The Warrawoona assemblage of microfossils is associated with stromatolites, a light carbon isotope ratio exists, and the microfossils are from black cherts. The microfossil assemblage consists of six types of filaments that look like minute threads of beads. However, they are comparable to living cyanobacteria (prokaryotes). These are interpreted as anaerobic autotrophs. A similarly aged microfossil assemblage is also present in southern Africa.

FIG TREE FORMATION, SOUTHERN AFRICA—3.4 BYA

For more than 20 years prior to the discovery of the Warrawoona Group fossils, microfossils from the **Fig Tree Formation** were considered the oldest fossils. In basic composition, these are similar to assemblage from the Warrawoona Group in that they are composed only of prokaryotes and are associated with stromatolites, light carbon isotope ratios, and pristane and phytane.

GUNFLINT CHERT, CANADA—2.1 BYA

The **Gunflint Chert** locality is along the shores of Lake Superior, Ontario, Canada. This was the first place where convincing Precambrian microfossils were described in 1954, which started the current research activity in Precambrian paleontology. Stromatolites, black cherts, and chemical fossils are all present at this site, but what is different here is that the diversity of prokaryotes has increased dramatically. Numerous types of both cyanobacteria and bacteria exist at this locality (Figure 8.3); however, these types of life are still the only types present.

OLDEST EUKARYOTES—2.0 TO 1.8 BYA

The oldest eukaryotes, green algae, are known from single-celled fossils possibly as old as 2.0 bya and definitely present at 1.8 bya. How can the remains of one single-celled microfossil be recognized as eukaryotic? Cell size is the principal means. All living prokaryotes are composed of relatively small cells that fall within a restricted size range, typically 10 μm or smaller, but they are rarely as large as 60 μm. In contrast, the cells of eukaryotes are typically much larger, ranging to greater than 200 μm, so this basic feature of single-celled organisms can be used to infer the differences between prokaryotic and eukaryotic Precambrian organisms.

The presence of eukaryotes is especially significant because, as known from living organisms, eukaryotes require aerobic conditions for existence. This marks the apparent beginning of aerobic autotrophs. Therefore, sometime prior to 2.0 to 1.75 bya, Earth's atmosphere must have changed from an anaerobic atmosphere to one with free oxygen. Furthermore, all prokaryotic organisms reproduce only asexually, whereas sexual reproduction was introduced among the primitive eukaryotes. With the

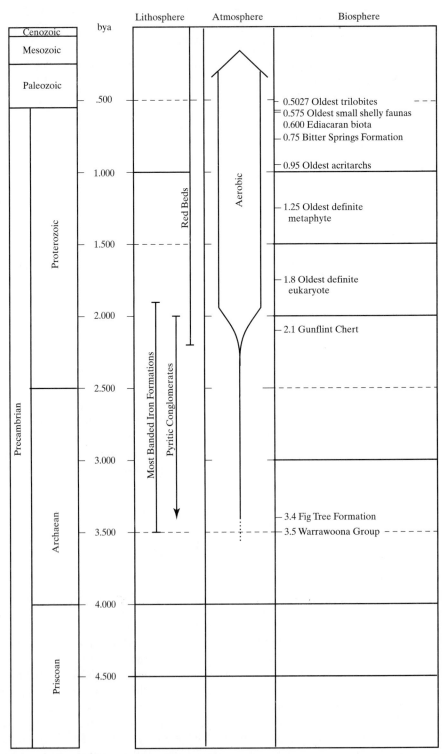

Figure 8.2 Outline of important events in the biosphere, lithosphere, and atmosphere during the Precambrian.

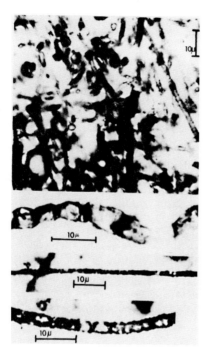

Figure 8.3   Microfossils from the Gunflint Chert, Canada, 2.1 bya. These are long filamentous strands that are the remains of cyanobacteria. Scale is indicated by 10-micron bars. (Courtesy of J.W. Schopf, U.C.L.A.)

advent of eukaryotic organisms, life began to assume a more familiar nature in some of the most fundamental features.

BITTER SPRINGS FORMATION, CENTRAL AUSTRALIA—0.8 TO 0.75 BYA

The **Bitter Springs Formation** from Australia (0.8 to 0.75 bya) is among one of the most diverse and well-described microfossil assemblages known after the first occurrence of green algae. This assemblage has cyanobacteria and bacteria, as had been present before, but green algae are also reported. The cell size of these fossils is large, consistent with eukaryotes. Furthermore, some of these cells occur in tetrads, which is consistent with meiotic cell division of eukaryotic green algae; and degraded cellular organelles, again diagnostic of eukaryotes, are reported. As with earlier Precambrian microfossil assemblages, the Bitter Springs assemblage is from black cherts, and it has associated stromatolites, light carbon ratios, and pristane and phytane.

Beginning at approximately 0.8 bya, a variety of eukaryotic microfossils occurred commonly in the rock record. Examples included chitinozoans and acritarchs (see Chapter 11) (Figures 8.4 and 8.5).

PRECAMBRIAN MACROFOSSILS

METAPHYTES—1.25 BYA

Metaphytes are macroscopic algae, such as the green or brown algae present in the oceans today. However, because macroscopic algae typically lack any hard parts, they

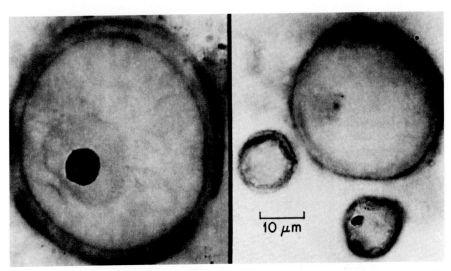

Figure 8.4   Precambrian microfossils from the Precambrian of the Grand Canyon, Arizona, 0.85 bya. These fossils are probably the remains of unicellular eukaryotes, and the dark areas inside the cells are thought to represent collapsed cytoplasm or degraded organelles of the eukaryotic cell. (Courtesy of J.W. Schopf, U.C.L.A.)

are not organisms that are likely to preserve well as fossils. Many nondescript objects from Precambrian rocks have been reported as metaphytes. Upon further analysis, paleobotanists have concluded that some of these are inorganic, some are colonial microfossils, and, indeed, some are the fossil remains of metaphytes. The oldest undisputed metaphyte fossils are from approximately 1.25 bya, after which they were relatively common fossils.

EDIACARAN FOSSILS, SOUTHERN AUSTRALIA—0.6 BYA

The remarkable **Ediacaran megabiota** was described in 1946 in the Ediacaran Hills of southern Australia, although some elements of this biota were known as early as 1908 when specimens from Namibia were discovered by German scientists. More than 30 species have been described from this assemblage. They are all preserved as molds in a fine-grained sandstone, which is quite unusual for soft-bodied organism preservation (Figure 8.6). When first described, this biota was considered to record the initial appearance of multicellular animals, metazoans, and was thought to represent the first representatives of many of the phyla, such as the coelenterates, annelids, arthropods, and echinoderms. Thus, these organisms have traditionally been considered the first record of aerobic heterotrophs. These fossils range in size from approximately 1 centimeter (0.5 inches) to the erect frondlike *Charniodiscus*, which are up to 60 centimeters (23 inches) high, so this fauna stands in striking contrast to the Precambrian microbiota that prevailed for nearly 3 billion years.

During recent years, this fauna has been discovered throughout the world during this late Precambrian interval in, among other areas, Newfoundland, China, Namibia, and Russia. Recent study of this fauna has raised many questions concerning its systematic affinities. A study intended to explain why soft-bodied organisms should be so

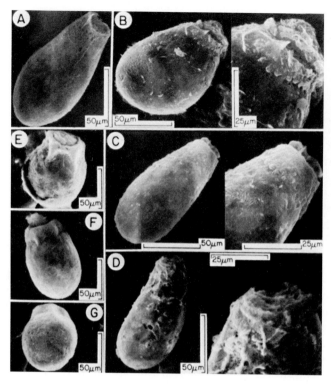

Figure 8.5   Precambrian microfossils called chitinozoans. These specimens from the Grand Canyon are from about 0.85 bya. They were released from the rock with strong acids and photographed using a scanning electron microscope. Their sizes are indicated by the bars. (Reprinted with permission from B. Bloeser, et al., *Science* vol. 195. Copyright © 1977, American Association for the Advancement of Science. Photograph courtesy of J.W. Schopf, U.C.L.A.)

well preserved as molds in the Ediacaran and not in younger faunas led the German paleontologist Adolf Seilacher to suggest that the vast majority of this fauna was not related to other metazoans. He concluded that several of the fossils were trace fossils (Figure 8.7A) and that some elements were coelenterates (Figure 8.6). However, he argued that more advanced forms did not occur in this biota. Instead of being metazoans, Seilacher concluded that the majority of this biota was part of a new, extinct kingdom (Vendobionta) of "quilted" organisms (Figure 8.7). Subsequently, it has been argued that these organisms are, indeed, part of the initial radiation of metazoans, that they are fossil lichens, and that the separate kingdom status is correct based on their different style of cell growth.

A consensus among paleontologists has not been reached on the identity of these organisms, so it is unclear whether a new kingdom is necessary for these fossils. Regardless of affinities of the entire fauna, the Ediacaran fossils record the first large organisms other than algae; and at the very least, the tissue-grade Cnidaria are present in the Ediacaran. They occur only as soft-bodied organisms just a few million years prior to the base of the Cambrian.

SMALL SHELLY FAUNAS

The first fossils with hard parts are called the **small shelly fauna** (also called the Tommotian fauna). They are first known from the latest Precambrian at appoximately

Figure 8.6 A slab from the Ediacaran beds of southern Australia, 0.6 bya, illustrating some of the difficulties of working with this material. Are these impressions a fossil coelenterate (jellyfish or medusae) or gas-pocket pseudofossils? The specimens average about 3 cm in diameter. (Courtesy of Field Museum of Natural History, Chicago.)

575 mya, 35 million years before the Cambrian and 50 million years prior to the first occurrence of trilobites. The small shelly fauna is known throughout the world, and it spans the Precambrian-Cambrian boundary.

When first described, this fauna was quite enigmatic. The fossils consist of all small, calcium-phosphate skeletal parts. These fossils are a series of cones, plates, tubes, and spicules (Figure 8.8). Some are smooth, some are ornamented, some display growth lines, but few or none seem to represent complete animals. Furthermore, they are all quite small, typically less than 2 millimeters, which undoubtedly accounts for their lack of discovery until the latter quarter of the twentieth century. Animals are being recognized from among the elements of the small shelly fauna as more and more is learned about these small shells and as more and more is learned about the complete morphology of the soft-bodied Cambrian faunas, e.g., Chenjiang fauna, Early Cambrian of China, and the Burgess Shale fauna. For example, several elements of the small shelly fauna were undoubtedly scales and spines attached to the exterior of "armored worms," such as *Wiwaxia* (see Figure 9.8D).

THE PRECAMBRIAN RECORD

THE BIOSPHERE

The prediction of the abiotic synthesis of life (Chapter 4) is that the first life forms that should have been on Earth were replicating microbes that fed on organic molecules. They formed in anaerobic conditions of the initial ocean. Thus, these earliest organisms would have been anaerobic heterotrophs. From these forms, as previously discussed, the sequence of life forms would logically be anaerobic autotrophs, aerobic autotrophs, and finally aerobic heterotrophs.

As presently understood, the Precambrian fossil record does not contain any anaerobic heterotrophs. The oldest fossils, 3.5 billion years old from the Warrawoona Group, are all anaerobic autotrophs, the second in the predicted sequence of life forms. These were all prokaryotic bacteria and cyanobacteria—the oldest fossils on Earth are

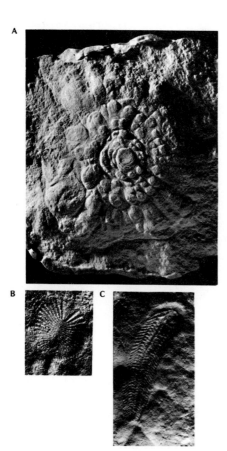

Figure 8.7 Ediacaran fossils of late Proterozoic age from southern Australia (0.6 bya). (A) *Mawsonites*, a radially symmetrical fossil, approximately 8 cm (3 in across). (B) *Dickinsonia*, a problematic fossil, 2.5 cm (1 in) long. (C) *Spriggina*, 4 cm (1.5 in) long. (Courtesy of Mary Wade, Queensland Museum, Brisbane, Australia.)

among the simplest forms of life living on Earth today! Furthermore, as predicted, the next category of life forms to evolve were aerobic autotrophs and finally aerobic heterotrophs. The sequence of fossils during the Precambrian corresponds to the prediction of the abiotic synthesis.

ATMOSPHERE AND LITHOSPHERE

A striking conclusion from interpretations of the temporal distribution of Precambrian fossils is that approximately 2.0 bya Earth's atmosphere changed. It changed from an anaerobic atmosphere, lacking oxygen and unable to support most known life, to an aerobic atmosphere that would have been poisonous to the first life on Earth. The transition from an anaerobic to an aerobic atmosphere is presumed to have occurred through accumulation of oxygen as a by-product of photosynthesis (see Chapter 4). This supposition can be tested in a number of ways because the occurrence of an oxygenated atmosphere has many consequences for Earth. Three aspects of the rock record will be discussed here in which independent preserved evidence has bearing on this question—banded iron formations, pyrite conglomerates, and red beds.

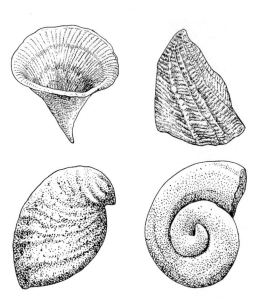

Figure 8.8   Four typical skeletal fossils of Tommotian age that are part of the small shelly fauna. All are phosphatic, and quite small (1 to 2 mm) in size. (Adapted from Matthews and Missarzhevski, 1975.)

BANDED IRON FORMATIONS

Banded iron formations are a unique rock type that consists of interbedded jasper (red quartz) and iron-rich minerals, particularly hematite (Fe_2O_3) and magnetite (Fe_3O_4). These are valuable sources of iron ore in the Lake Superior region of the United States, Montana, Wyoming, New Mexico, Labrador, Brazil, Venezuela, Australia, southern Africa, and Russia. The most unusual aspect of this distinctive rock type is that, with minor exception, it occurs globally from 1.8 to 3.5 bya worldwide, peaking in abundance approximately 2.5 bya. This restricted distribution was difficult to reconcile prior to our understanding of the Precambrian fossil record. Now it is believed that the banded iron formations basically record the "rusting" of the oceans. The initial atmosphere contained no free oxygen. In this anaerobic setting the Fe^{2+} and Fe^{3+} present in the initial oceans remained as free cations, unlike they would today in an oxygenated atmosphere. As oxygen was generated as a byproduct of early photosynthetic organisms, any free O_2 would have readily combined with iron cations to form iron oxides. This was an oxygen sink that would have persisted while iron cations existed. So, the banded iron formations represent a period of time when free iron cations and oxygen were available in the oceans. When the iron was expended, oxygen could accumulate in the atmosphere. The interval of banded iron formation existed from 3.5 to 1.8 bya, and eukaryotes, the first evidence of aerobic life, first occur at approximately 2.0 to 1.75 bya.

PYRITIC CONGLOMERATES

Pyrite, the iron sulfide (FeS_2) or fool's gold mentioned in Chapter 3, is a mineral that is unstable in an aerobic environment; it oxidizes. Conglomerate (see Chapter 3) is a sedimentary rock type in which the sediment grains are larger than 2 millimeters. **Pyritic**

conglomerates are unknown today, and they only occur in sedimentary rocks older than 2.0 bya. So based upon our understanding of the chemistry of this mineral, prior to 2.0 bya an anaerobic atmosphere is suggested and after that time an aerobic atmosphere is suggested.

RED BEDS

The timing of **red beds** is the third independent line of evidence that further corroborates the timing of this atmospheric change. Terrestrial sediments are typically called red beds because the iron oxides have oxidized. This occurs in stream, deltaic, and other terrestrial deposits that have been exposed to the atmosphere—an atmosphere with oxygen. Red beds are known from the present back to about 2.2 bya, before which red beds are not in terrestrial sediments.

DISCUSSION

These three aspects of the lithosphere, as well as others not mentioned, corroborate the atmospheric history inferred from the temporal sequence of microfossil assemblages through the Precambrian. The diverse types of information that can be gathered to determine the timing of this long past, critical change in the conditions of Earth all agree. When fossils first appeared 3.5 bya, the atmosphere was anaerobic. Photosynthetic organisms began to generate oxygen approximately 3.0 bya, but this oxygen was quickly consumed in a variety of sinks, such as the banded iron formations. Free oxygen probably began accumulating in the atmosphere approximately 2.5 bya; we can infer that the change to an oxygenated, aerobic atmosphere took place approximately 2.0 bya.

THE PROTEROZOIC DIVERSIFICATION

During at least 1 billion years of the Archean and much of the first part of the Proterozoic (Figure 8.2), life was composed of only bacteria and cyanobacteria. Approximately 1.75 bya, the first eukaryotes appeared; at 1.25 bya the first metaphytes appeared; at 0.58 bya multicellular animals first appeared; at 0.575 bya small shelly fossils were present; and a diverse shelly fauna was present at the base of the Cambrian at 0.543 bya. Assuming that this fossil record, known from more than 2,800 localities, is correct, life existed with little apparent change for 1.75 billion years, and then an incredible diversification and expansion of life occurred during the late Proterozoic. By the close of the Precambrian, the Ediacaran organisms, small shelly fauna, simple trace fossils, and metaphytes were all present in the oceans. How can this be explained? What processes could be responsible?

First, we are accustomed to noticing the evolution of morphology, but for Archean and early Proterozoic microfossils, morphological evolution was very, very slow. Rather than morphology, the primary evolutionary pressures of these organisms was biochemical. Different types of autotrophism evolved, including sulfur-reducing metabolisms, photosynthesis, and others. Still, evolution during the Archean and early Proterozoic was remarkably slow. The most likely explanation for this is reproduction.

Prokaryotes known today only reproduce asexually. This means that offspring have the exact same genetic composition as the parent; there is no mixing of genetic material from parents. Without sexual recombination of genetic material, the only source for genetic variability is mutation, which occurs at a very slow rate. As presently understood, the first organisms that could have engaged in sexual reproduction were the first eukaryotes at 1.75 bya, and the first potential evidence of sexual reproduction is cell tetrads in the Bitter Springs Formation at 0.80 bya. Clearly, the rate of evolution greatly increased at the approximate time that eukaryotes appeared on Earth, so this new style of reproduction may very well account for the accelerated pace of evolution that led to the Cambrian adaptive radiation.

Toward the end of the Proterozoic, apparently, evolutionary pressures favored larger organisms. Single-celled organisms can increase in size by growing larger, by developing multiple nuclei, by being colonial unicellular, or by becoming multicellular. Larger organisms of the first two types do exist, but there are severe size limitations for single cells, regardless of the number of nuclei. Colonial single-celled organisms also exist among algae. One of the intriguing aspects of the hypothesis that the Ediacaran biota was a separate kingdom is that this quilted construction could be yet another solution, a failed solution, to the problem of how to grow larger. Clearly, multicellularity proved the most successful way to grow larger for plants, animals, and fungi. Mulicellular organisms can grow to enormous sizes. This and the differentiation of cells into tissues and tissues into organs has allowed plants to grow into trees and animals to swim, run, and fly.

KEY TERMS

banded iron formations	Fig Tree Formation	pyritic conglomerates
Bitter Springs Formation	Gunflint Chert	red beds
black chert	metaphytes	small shelly fauna
C^{12}/C^{13} ratio	morphological fossils	stromatolite
chemical fossils	phytane	Warrawoona Group
Ediacaran megabiota	pristane	

READINGS

Barghoorn, E.S. 1971. "The Oldest Fossils." *Scientific American* 224:30–42. A popular account of the Fig Tree fossils.

Glaessner, M.F., and Wade, M. 1966. "The Late Precambrian Fossils from Ediacara, South Australia." *Palaeontology* 9:599–628. Description and illustration of this well-known locality that contains the enigmatic group of larger organisms.

Glaessner, M.F. 1985. *The Dawn of Animal Life*. Cambridge University Press. 244 pages. A biohistorical study with special emphasis on the fossils at or close to the Precambrian-Cambrian boundary and especially the Ediacaran fauna.

Schopf, J.W., ed. 1992. *Major Events in the History of Life*. Jones and Bartlett Publishers. 190 pages. A collection of papers on many critical intervals of the biosphere through time, including summaries of Precambrian events.

C H A P T E R   9

Introduction to Marine Communities and the Cambrian

INTRODUCTION TO PHANEROZOIC FAUNAS

The distributions of all known families and genera of marine organisms through time indicate that marine oceanic life is divisible into three great faunas, the **Cambrian marine fauna**, the **Paleozoic marine fauna** (from Ordovician to Permian), and the **Modern Marine Fauna** (Mesozoic and Cenozoic) (Figure 9.1). Each fauna was dominated by specific taxonomic groups that, as a whole, began with a period of low diversity, had an interval of diversification and dominance, and eventually diminished in importance. The Cambrian marine fauna was the shortest in duration and came to a close following end-Cambrian extinctions. Elements of the Paleozoic marine fauna existed during the Cambrian, but this fauna diversified during the Early Ordovician and quickly reached a diversity equilibrium. It was dominant throughout the remainder of the Paleozoic and rapidly declined during the great end-Paleozoic extinctions. Elements of the Modern Marine Fauna had been present since the Cambrian, but these organisms did not diversify to dominate marine habitats until the early Mesozoic. The Cambrian marine fauna will be treated in this chapter, but first, the ecologic structure of ocean-floor communities will be discussed. The Paleozoic and Modern Marine Faunas will be discussed in detail in Chapter 10.

PALEOECOLOGY OF MARINE COMMUNITIES

Marine organisms can be categorized ecologically in many ways—where they fit in the food chain; where they live in the water column, at the bottom, or on the ocean floor; and whether or not they move actively. Marine communities are subdivided into the familiar feeding categories present in terrestrial habitats, such as primary producer, herbivore, carnivore, omnivore, scavenger, parasite, and so on. However, two additional, very important categories exist among marine communities, suspension feeders and deposit feeders.

Primary producers are autotrophs. These are organisms that take raw materials and make their own food. Most commonly, plants and algae use chlorophyll and sunlight to manufacture sugars. Animals are consumers. **Herbivores** are organisms that feed exclusively on plant material; **predators** actively pursue and subdue their prey, which are other animals; **omnivores** eat both plant and animal tissue; **scavengers** feed

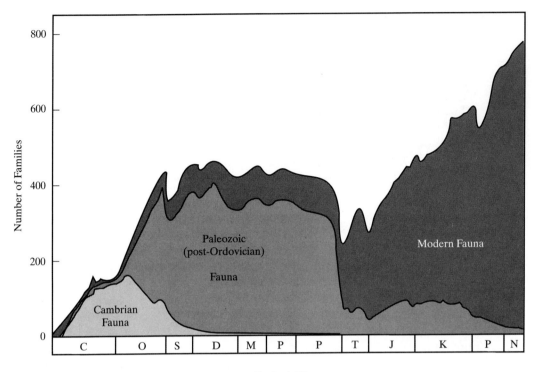

Figure 9.1   Marine evolutionary faunas of the Phanerozoic as measured by the number of families at any one time. Each fauna had a beginning with low diversity, a period of rapid diversification and stability, and a long period of decline. Note that the Paleozoic marine fauna dominated only from the Ordovician through the Permian. (Used with permission of the Paleontological Society.)

on dead animal carcasses; and **parasites** live on or within another living plant or animal (Figure 9.2). **Suspension feeders** are organisms that either pump ocean water or allow ocean currents to pass through a filter device, from which particles suspended in the water are captured and ingested. This feeding method exploits both the rich array of plankton in the ocean and other organic-rich material that is in the water. **Deposit feeders** are organisms that simply eat mud from which they extract organic materials. These are sediment feeders that either eat surface sediment or live in burrows and eat their way through the sediment.

Organisms that live in the water column, above the bottom, are called **pelagic**. If they swim, they are called **nektonic**; if they float, they are called **planktonic** (Figure 9.2). Fish, squid, and whales are examples of nekton; the Portuguese man-of-war and microscopic plants and animals are examples of plankton. The uppermost portion of the ocean waters is penetrated by sunlight and is called the photic zone. Organisms living in this zone include all the autotrophs, of which many are plankton.

Organisms living on the bottom are called **benthos**, and they can live either on the bottom, in which case they are called **epifaunal**, or they can live in the sediment below the ocean floor, and are known as **infaunal**. Many of the benthos subdivide the

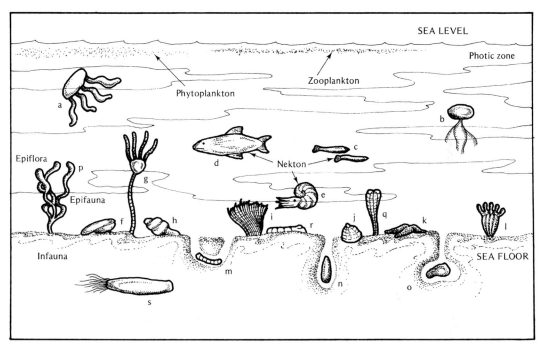

Figure 9.2   Relationship of different major groups of plants and animals to the surface of the sea and the ocean floor. Nekton: (a) jellyfish, (c) conodont animals. Carnivorous nekton: (d) fish, (e) cephalopod. Plankton: (b) graptolite. Epiflora: (p) seaweed. Epifaunal suspension feeder: (f) brachiopod, (g) crinoid, (i) bryozoan, (j) bivalve, (l) coral, (q) sponge. Epifaunal deposit feeder: (h) gastropods, (j) bivalve, (r) trilobite. Epifaunal carnivore: (k) starfish. Infaunal suspension feeder: (n) bivalve, (o) echinoid, (m) worm or crustacean trace fossil. Infaunal deposit feeder: (n) bivalve, (s) holothurian.

available resources by subdividing the available space. This partitioning of space to maximally utilize resources is called **tiering**; both infaunal and epifaunal tiering occurs.

Finally, organisms can be divided into either those that are stationary versus those that actively move about. Stationary organisms are termed **sessile**, whereas mobile organisms are called **vagile**. Thus, we can categorize organisms according to these parameters. Most common trilobites are thought to have been vagile epifaunal deposit feeders. Many worm groups are infaunal. They are either vagile infaunal deposit feeders or vagile infaunal suspension feeders. Sponges are epifaunal suspension feeders, single-celled algae in the oceans are planktonic primary producers, and sharks are vagile, nektonic predators.

THE MYSTERIOUS CAMBRIAN

Rapid adaptive radiation of shelly metazoans occurred at the beginning of the Cambrian. During the long expanse of the Precambrian, until the latest Precambrian, life was microbial and nonskeletonized. Even the larger elements of the Ediacaran biota (see Chapter 8) are nonskeletonized. Diversification of the Cambrian marine fauna filled the world's oceans with a familiar-looking, although still a bit strange, shelly fauna. Trilobites, inarticulate brachiopods, primitive molluscs and mollusc-like organisms, and primitive echinoderms were the dominant elements of the Cambrian marine fauna. However, during this time, most other phyla also evolved. Sponges, articulate brachiopods, annelids, bivalves, gastropods, cephalopods, many arthropod groups, echinoderms, and chordates, among others, first evolved during the Cambrian (Figure 9.3) and lived in and among the trilobite-dominated communities. Earlier in this century, it was thought that life evolved at the base of the Cambrian. As demonstrated in Chapter 8, we now know that this is not true.

However, the Cambrian is still a mystery. Why did oceanic life undergo such a remarkable adaptive radiation at this time? Why did so many elements of marine life nearly simultaneously develop skeletons? Another unusual aspect of the Cambrian was that, at least based on the preserved fossil record, the vast majority of marine organisms were epifaunal deposit feeders, that is, organisms that lived on top of the ocean floor and fed directly on sediment, mud, and silt. Furthermore, as a whole, Cambrian organisms lived quite close to the sediment-water interface in comparison to later life. Few of these organisms either burrowed deeply or were elevated high above the ocean floor.

The Cambrian is also unique for having an unusually high number of soft-bodied faunas. Soft-bodied faunas are fossil localities that record exceptional preservational circumstances so that organisms lacking hard-part skeletons are preserved, typically

Ordovician — Middle		
Ordovician — Lower	BRYOZOA	
Cambrian — Upper	Cephalopoda FORAMINIFERA Polyplacophora	
Cambrian — Middle	CHORDATA GRAPTOLITHINA	
Cambrian — Lower	ARCHAEOCYATHIDA ARTHROPODA Bivalvia BRACHIOPODA ECHINODERMATA	Gastropoda Monoplacophora PORIFERA
Proterozoic	?ANNELIDA CNIDARIA Ediacara Biota	

Figure 9.3 Times of first appearances of major groups of marine animals having preservable skeletons. Phyla and the Ediacaran biota are indicated in uppercase letters; those classes listed are in upper- and lowercase letters. Questionable occurrence of Annelida based on trace fossils; depending on affinities of Ediacaran biota, some additional phyla may have their origins during the Late Proterozoic.

through carbonization. Soft-bodied faunas are known throughout the Phanerozoic, but many, many examples are known from the Cambrian. The best known example is the Middle Cambrian Burgess Shale from British Columbia, Canada. Other examples of this type of fauna occur in many places around the world, including a fairly recent discovery, the Early Cambrian soft-bodied Chengjian fauna from China.

THE CAMBRIAN MARINE FAUNA

The base of the Cambrian is now globally recognized as the first occurrence of the trace fossil *Phycodes pedum*, from approximately 543 mya. From the base of the Cambrian, the fossil record is continuous until the present day. For approximately the first 10 million years of the Cambrian, the only skeletonized fossils were the small shelly organisms.

The first **trilobites** occurred approximately 527 mya. Trilobites dominated the skeletonized component of Cambrian communities, typically comprising more than 90 percent of the fauna (Figure 9.4). Inarticulate brachiopods are another 5 to 7 percent of fossils with skeletons, and all other fossils comprise the remainder.

Trilobites are an extinct class of **Arthropoda**, so they are related to crabs, lobsters, and shrimps. Arthropods are among the more advanced and complex of any phyla. The first trilobites had large eyes and long antennae (Figure 9.5). They clearly had a well-developed nervous system. They possessed many appendages that were used for walking, feeding, and even, perhaps, swimming. Although trilobites are primitive in many respects as far as arthropods are concerned, they are quite advanced and

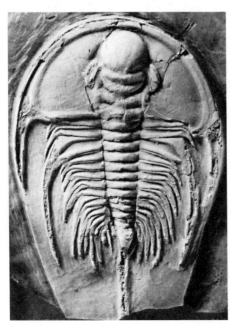

Figure 9.4   An Early Cambrian trilobite, *Olenellus*, from southern California. The large head and eyes and small tail are typical of many Early Cambrian trilobites. The specimen is 12 cm long. (Courtesy of Takeo Susuki, U.C.L.A.)

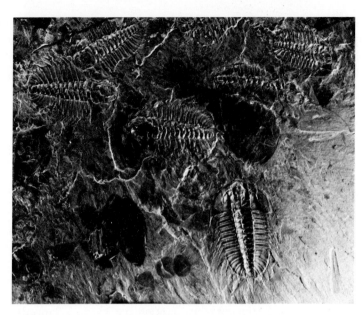

Figure 9.5   A slab of Middle Cambrian shale containing numerous trilobite specimens of *Olenoides*. Carbon-film impressions of several other fossils are also on the slab. The trilobites are approximately 7 cm long. (Courtesy of National Museum of Natural History.)

complex in comparison with other phyla and in comparison to the small shelly fauna of the late Precambrian.

Inarticulate brachiopods are a class of the Brachiopoda that have reasonably simple shells, most of which are phosphatic rather than calcareous, as in other brachiopods. Otherwise, inarticulate brachiopods were shelled and fed using a lophophore, like the articulate brachiopods. Primitive molluscs that occur in Cambrian strata include **monoplacophorans**. These are the partially segmented molluscs that have a single cap-shaped shell, and the story of the discovery of the living *Neopilina* was discussed in Chapter 7. A few rare specimens of echinoderms are also known from the Lower Cambrian. Echinoderms are the spiny-skinned invertebrates whose living relatives include starfish and sand dollars. Their skeletons are composed of many calcite plates. Cambrian echinoderms are quite unusual in morphology, like the football-shaped helicoplacoids (Figure 9.6). Presumably, they could expand and contract in a spiral fashion, resembling spiral concertinas. Another group of early Cambrian animals are the **archaeocyaths**. These are an extinct phylum that is thought to be closely related to the sponges (Figure 9.7). The skeleton looks much like a small, porous, double ice-cream cone and is composed of calcite. These animals lived solitarily or in dense aggregations and built small moundlike reefs on the sea floor.

Apparently, Cambrian communities were dominated by trilobites and other arthropods. These other organisms were locally important. Despite this dominance by trilobites, nearly all other phyla of animals, including chordates, evolved during the

Figure 9.6   Specimen of helicoplacoid from Lower Cambrian rocks of California, enlarged several times. The mouth was at the top, and the spiral rows of small plates could be expanded and contracted when the animal was alive, thus changing the shape. (Courtesy of J.W. Durham, University of California, Berkeley.)

Cambrian (Figure 9.3). The Cambrian was a reasonably short period of geologic time, but it witnessed remarkable change in the inhabitants of Earth. Trace fossils parallel this history. Undisputed Precambrian trace fossils from metazoans (excluding stromatolites) do extend back to approximately 700 mya but consist of only quite simple types that typically burrowed very shallowly (1 centimeter or less). At the beginning of the Cambrian, the number and diversity of trace fossils increased dramatically. Trace fossils were larger and lived substantially deeper. There were many, many more kinds, both globally and

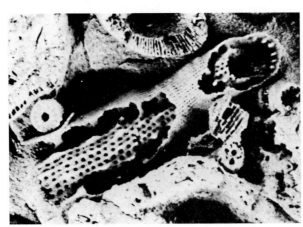

Figure 9.7   Several partial specimens (enlarged) of an extinct Cambrian fossil, archaeocyaths, a phylum related to sponges. The specimens are from Australia. They have been replaced by silica and partially etched from the surrounding rocks by acid. (Courtesy of Ward's Natural Science Establishment, Inc.)

within single communities. During the Early Cambrian, trace fossils become identifiable as being made by trilobites and other arthropods.

THE BURGESS SHALE FAUNA

The Burgess Shale fauna is the best known representative of the numerous and diverse soft-bodied faunas of the Cambrian. Soft-bodied faunas are an extra-special glimpse at the typically unpreserved organisms of the past. However, this special preservation usually occurs due to unusual environmental circumstances, so it tells us more about the organisms living in or deposited in unusual environmental circumstances than about organisms living within the communities of the skeletonized organisms that we commonly collect as fossils. The Cambrian is an exception because these soft-bodied faunas are part of the typical Cambrian trilobite-dominated communities, so, indeed, we are able to obtain a fairly accurate understanding of the entire fauna in these communities.

The **Burgess Shale** fauna is one of the most famous fossil localities known. It is high on the side of Mount Field in the Canadian Rockies of British Columbia. The shale is black and platy, and on split surfaces there are flattened films of many kinds of soft-bodied organisms (Figure 9.8). All of these lacked hard skeletons or had only very lightly calcified skeletons, including algae, jellyfish, sponges, several kinds of worms, numerous arthropods, chordates, and several groups of uncertain affinities. Burgess Shale soft-bodied organisms are preserved as carbon films coated with clay minerals. Organisms with hard parts in this fauna include inarticulate and articulate brachiopods, molluscs, hyoliths, trilobites, and echinoderms. Examples illustrated here are *Ottoia*, a priapulid worm; *Pikaia*, a chordate; *Waptia*, an early arthropod; and *Wiwaxia*, an animal of unknown affinities, but presumably a complete example of an animal that produced the skeletal elements of the small shelly fauna (Figure 9.8).

More than 170 species, including prokaryotes, protists (algae), and animals, are recognized from the Burgess Shale. The highest diversity is among arthropods. Eighty-six percent of the organisms in the biota are without hard parts. Trilobites, the dominant element of normal Cambrian faunas, have a species diversity of 22 in the Burgess Shale. This biota includes plankton, nekton, vagile epifaunal benthos, sessile epifaunal benthos, and infauna. Some of these animals are rare, being known from a single specimen, but others are extremely abundant, like *Marrella*, the "lace crab," known from more than 15,000 specimens. Furthermore, similar soft-bodied faunas are now known from more than 30 Early and Middle Cambrian localities throughout the world—no other time period has so many.

The Burgess Shale fauna is significant in many ways. Most important, it dramatically illustrates the incredible rapidity of morphological diversification that occurred by the Middle Cambrian. Nearly all living phyla evolved during the Cambrian (Figure 9.2). At least twelve phyla of living animals occur in the Burgess Shale, from sponges to chordates. More than 20 unusual Burgess Shale creatures cannot be classified into existing phyla and may require their own. Even within phyla, especially among the arthropods, a great mophological diversity is evident. These soft-bodied faunas present us with a glimpse of Cambrian ocean life that is unparalleled at any other time during Earth history other than the present day.

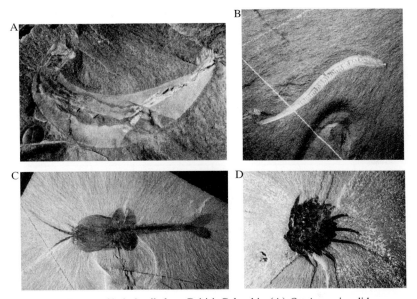

Figure 9.8   Burgess Shale fossils from British Columbia. (A) *Ottoia*, a priapulid worm with small skeletons in the gut. (B) *Pikaia*, an early chordate. (C) *Waptia*, an early arthropod. (D) *Wiwaxia*, an animal of unknown affinities, composed of multiple small spines and plates. (Courtesy of Derek E.G. Briggs, University of Bristol.)

PALEOECOLOGY OF CAMBRIAN COMMUNITIES

As previously mentioned, the Cambrian fauna is unique from later times because it was apparently dominated by deposit feeders. It is estimated that 90 percent of skeletonized Cambrian fossils are trilobites, and most benthic trilobites are interpreted as epifaunal deposit feeders. Even with consideration of Cambrian soft-bodied faunas, deposit feeders dominated. Many of the non-trilobite arthropods, including the abundant *Marrella* are thought to have been deposit feeders.

Other feeding types do exist, but only suspension feeders would be recognized if the soft-bodied fauna were not known. Examples of Cambrian suspension feeders are sponges, brachiopods, and echinoderms. Some swimming arthropods and worms were probably also suspension feeders. Arthropods and worms were probably also scavengers, and predators include *Anomalocaris* and the priapulid worms (see Chapter 12). The priapulid worm *Ottoia* illustrated in Figure 9.8A has preserved prey in its intestines.

Despite the revisions that include accomodation of the soft-bodied fauna, the balance of feeding types is skewed toward epifauna deposit feeders and probably shallowly tiered infaunal deposit feeders (Figure 9.9). The deposit-feeding aspect of the Cambrian fauna and its low levels of tiering stand in sharp contrast to ocean-floor communities later in Earth history (see Chapter 10). Suspension feeders or suspension and deposit feeders dominated during the remainder of the Paleozoic, and tiering complexity increased dramatically after the Cambrian.

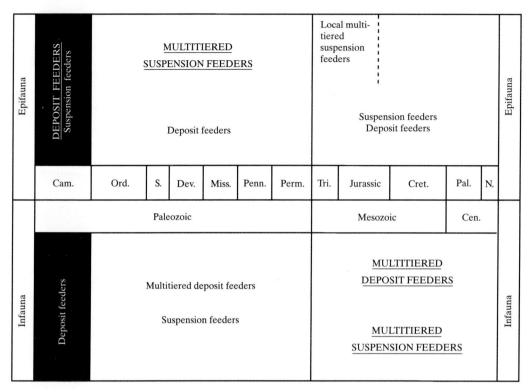

Figure 9.9   Phanerozic history of the important types of feeding among organisms on the ocean floor. Distinction is made between infauna and epifauna. Feeding types underlined and in uppercase letters were the dominant types. Feeding types of the Cambrian marine fauna are highlighted.

THE MYSTERY OF THE CAMBRIAN

We now understand that the base of the Cambrian does not represent the origination of life; indeed, it does not even represent the origination of skeletons. However, the Cambrian does represent 1) a remarkable adaptive radiation of multicellular animals, 2) a remarkable diversification of the morphology of animals, 3) the global expansion of these animals in the oceans, and 4) the nearly synchronous development of hard-part skeletons in many phyla of animals. In order to understand the processes responsible for this unique time, it is essential that the data are well constrained. Is the fossil record accurate—do the fossils that we have correctly record the diversification of major phyla at or near the base of the Cambrian or is there a long history of soft-bodied multicellular organisms not recorded in the fossil record? Is the skeletonization of animals simultaneous with their diversification or did their diversification take place much earlier?

In some ways, it is more satisfying to invoke a poor fossil record and claim that during a long, unpreserved history the variety of metazoan body forms developed. This gives us a comfort zone of time for the formation of this fantastic diversity of morphology. Even recent analyses of molecular data support this notion. However, various other types of data do not; they suggest that the fossil record is basically an accurate history of the timing of the evolution of multicellular animals. These data include the following: 1) Precambrian microfossils do not become larger and, presumably, eukaryotes until 1.75 bya, 2) macroscopic algae first appeared 1.25 bya, 3) the Ediacaran biota contains, at best, either only cnidarians or cnidarians and primitive members of other metazoa (600 mya), and 4) trace fossil diversification began approximately at the Precambrian-Cambrian boundary (543 mya). The entire morphological fossil record, as presently known, occurs in the correct sequence. If the diversification of metazoans (multicellular animals) occurred considerably earlier, what are the odds that the proper sequence would occur at the wrong time? Furthermore, if a long history of nonskeletonized organisms occurred earlier during the Precambrian, there should be a record of this left as abundant trace fossils; this is absent. Instead, the trace fossil diversification parallels the diversification of the skeletonized fauna.

One alternative hypothesis is that a long history of microscopic metazoans precedes the Precambrian-Cambrian diversification. Such microscopic forms could have been planktonic or could have lived among sand grains.

Pending additional data to the contrary, the fossil record can be presumed basically correct. So, the base of the Cambrian probably represents the global adaptive radiation and expansion of metazoans. They were largely evolving into unoccupied shallow-water habitats, which promoted the evolution of many, many body plans, thus creating many phyla and many classes within phyla. This adaptive radiation was remarkable, but was it substantially more remarkable than the adaptive radiation of flowering plants during the Early Cretaceous (see Chapter 14) or the adaptive radiation of mammals during the early Cenozoic (see Chapter 16)? In these evolving Cambrian ecosystems the development of efficient predation must have had a substantial impact, and perhaps the development of skeletons across many phyla records the protection solution to increased predation.

The origination and early evolution of multicellular organisms is an active area of research by numerous individuals, from molecular biologists to paleontologists. This is an evolving field of study with new data and new interpretations presented regularly. Continuing development in this area makes it one of the most important and fascinating areas of paleobiologic research.

KEY TERMS

archaeocyaths	infaunal	predator
Arthropoda	Modern Marine Fauna	primary producer
benthos	monoplacophorans	scavenger
Burgess Shale	nektonic	sessile
Cambrian marine fauna	omnivore	suspension feeder
deposit feeder	Paleozoic marine fauna	tiering
epifaunal	parasite	trilobites
herbivore	pelagic	vagile
inarticulate brachiopods	planktonic	

READINGS

Briggs, D.E.G.; Erwin, D.H.; and Collier, F.J. 1985. *The Fossils of the Burgess Shale*. Smithsonian Institution Press. 238 pages. An account of the geology and an atlas of the Middle Cambrian Burgess Shale fossils.

Conway Morris, S., and Whittington, H.B. 1985. *Fossils of the Burgess Shale*. Geological Survey of Canada Miscellaneous Report No. 43. 31 pages. This report features large black-and-white photographs of these important Middle Cambrian fossils.

Conway Morris, S. 1982. *Atlas of the Burgess Shale*. Palaeontological Association. 24 plates. Photographs of the important fossils from this famous locality.

Gould, S.J. 1989. *Wonderful Life: The Burgess Shale and the Nature of History*. W.W. Norton. 347 pages. A readable account of the important Burgess Shale fauna with Gould's special view of the history of study and importance of these fossils.

McMenamin, M.A.S., and Schulte McMenamin, D.L. 1990. *The Emergence of Animals, The Cambrian Breakthrough*. Columbia University Press. 217 pages. Account of the evolution of animals during the late Precambrian and into the Paleozoic.

Raff, R.A. 1985. *The Shape of Life, Genes, Development, and the Evolution of Animal Form*. University of Chicago Press. 520 pages. A detailed but readable analysis of the evolution of animal morphology with consideration of the fossil record and our understanding of genetics and development.

Whittington, H.B. 1985. *The Burgess Shale*. Yale University Press. 151 pages. A complete discussion of the origin and significance of the Burgess Shale fauna.

Paleozoic, Mesozoic, and Cenozoic Ocean-Bottom Faunas

INTRODUCTION

Life on Earth changed dramatically during the Ordovician. In the oceans, the subject of this chapter, the Cambrian fauna, composed dominantly of trilobites, was replaced at the start of the Ordovician by the Paleozoic fauna that persisted to the end of the Permian (Figure 10.1, also see Figure 9.1). Also, the first evidence of life on land is from the Ordovician, although true land plants do not appear until the Silurian (Chapter 13).

In the oceans, the Paleozoic fauna was typically dominated by **epifaunal** suspension-feeding organisms, especially corals, bryozoans, brachiopods, and stalked echinoderms. Communities of this feeding style stand in marked contrast to those of the Cambrian described in Chapter 9. Epifaunal suspension feeders dominated in most shallow-water ocean settings from the Ordovician to the end of the Paleozoic (Figure 10.1), for more than 250 million years.

THE PALEOZOIC MARINE FAUNA

RADIATION OF THE ORDOVICIAN FAUNA

As with the Cambrian fauna, the increase in diversity was remarkably rapid. Following Late Cambrian extinctions, marine diversity quickly became greater than it had been at any time during the Cambrian, and it reached an apparent equilibrium high by at least the Late Ordovician (Figure 10.1). The rise of epifaunal suspension feeders is more than just the diversification of new organisms. It was a fundamental shift in the dominant method of food procurement by ocean-floor organisms. During the Ordovician, the predominant style of feeding was suspension feeding above the substratum, i.e., epifaunal suspension feeding. Various filtration devices evolved and developed—tentacles of corals, lophophores of bryozoans and brachiopods, gills of bivalves, brachioles and arms of stalked echinoderms, and others.

The cause of such a major shift in the biosphere is commonly difficult to identify, but one particularly plausible idea has been proposed for the radiation of Ordovician suspension feeders. During the Early Ordovician, the carbonate composition of the oceans shifted so that the carbonate mineral calcite was precipitated from seawater

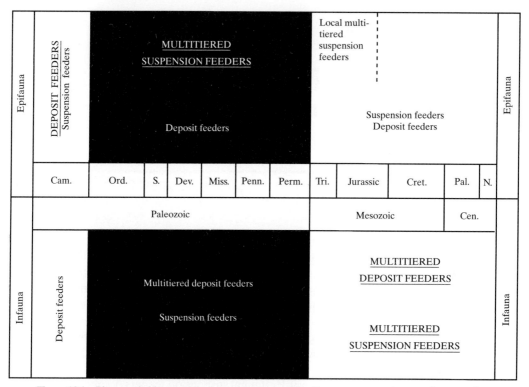

Figure 10.1   Phanerozoic history of the important types of feeding among organisms on the ocean floor. Distinction is made between infauna and epifauna. Feeding types underlined and in uppercase letters were the dominant types. Feeding types of the Paleozoic marine fauna are highlighted.

rather than the carbonate mineral aragonite (these two minerals are of the same composition, $CaCO_3$, but have different molecular structures). This resulted in extensive cementation of carbonate ocean floor sediments, making what geologists call **hardgrounds**. Hardgrounds were not suitable habitats for many Cambrian organisms, because the cemented carbonate mud could not be a source of nutrients for deposit feeders. Alternatively, many suspension feeders have attachment structures that can easily adhere to a hardground surface. Thus, a change from unconsolidated (loose) carbonate sediments to hardgrounds during the Early Ordovician resulted in a substantially different ocean-floor habitat that was rapidly exploited by epifaunal suspension feeders. In turn, epifaunal suspension feeders adapted to other habitats and, then, dominated in most settings throughout the remainder of the Paleozoic.

Presumably, the ocean bottom could be filled with suspension feeders, and, perhaps at least locally, suspended particulates could be depleted. Two strategies seem to have been in place that diminished competition among suspension feeders. First, organisms subdivided the space above the sea floor, which is termed **tiering**. From the viewpoint of an organism in benthic settings with currents, food moves past "horizontally." Therefore, organisms at different heights above the ocean floor will not sample the same parcels of water and not directly compete. Additionally, the food particle size varies among different groups of suspension feeders. Some cnidarians ingest whole fish, whereas the mouth of a living bryozoan is generally less than 100 µm in diameter. This **size selection for food** must also have played a role in differentiating food resources among suspension feeders.

During the Ordovician, epifaunal tiering among suspension feeders changed dramatically (Figure 10.2) from relatively low heights and minimum tier subdivisions inherited from Cambrian communities to the much higher multitiered Ordovician communities. The increase in tiering paralleled the rapid radiation of the Paleozoic marine fauna. Stalked echinoderms routinely reached heights perhaps as high as 50 to 70 centimeters above the ocean floor, and the characteristic maximum level of tiering was 100 centimeters by at least the Middle Silurian. This tiering occurred on level ocean bottoms, and any structure on the bottom, such as reefs, further enhanced tiering. It is important to note that epifaunal suspension feeders were not the only feeding group. Most communities were composed of many other feeding types, such as deposit feeders (not on hardgrounds), scavengers, grazers, parasites, and predators. Infaunal organisms included both suspension feeders, which initially, during the Ordovician, lived only in shallow burrows, and deposit feeders.

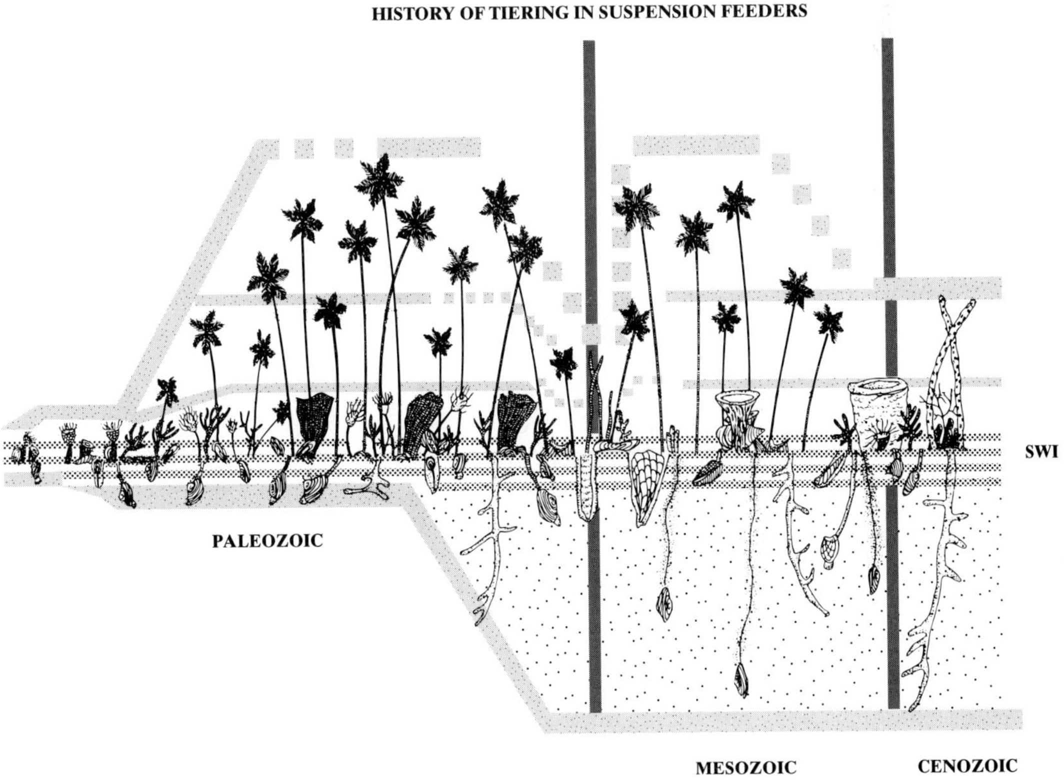

HISTORY OF TIERING IN SUSPENSION FEEDERS

SWI

PALEOZOIC

MESOZOIC   CENOZOIC

Figure 10.2   Diagramatic Phanerozic history of the tiering of suspension feeders both above and below the sediment surface (epifaunal and infaunal). (Used with permission of the *Journal of Geological Education* vol. 39. Copyright © 1991 National Association of Geology Teachers.)

A typical Ordovician or Silurian ocean floor community was composed of bryozoans, brachiopods, corals, stalked echinoderms, sponges, bivalves, gastropods, ostracods, graptolites, trilobites, starfish, cephalopods, and conodonts. The basic community structure was dominated by epifaunal tiering of the suspension feeders. Tiering among organisms was undoubtedly not compartmentalized, but for discussion purposes, four tier subdivisions are recognized. During the Silurian, the highest two tiers were composed nearly entirely of stalked echinoderms, with the tallest crinoids comprising the highest tier (~50 to 100 centimeters); and crinoids, rhombiferans, diploporans, and other stalked echinoderms making up the second highest tier (~15 to 50 centimeters) (Figure 10.3). The next to lowest tier was composed of a large diversity of solitary and colonial organisms and reached from between ~5 to 15 centimeters above the bottom. This tier included erect bryozoans, corals, sponges, graptolites, and perhaps some larger brachiopods and bivalves. The lowest tier fed immediately above the sediment and was composed of brachiopods, encrusting bryozoans, bivalves, some corals, and numerous other organisms (Figure 10.3). In addition to suspension feeders, other members of Ordovician communities included deposit feeders (trilobites, bivalves, worms), algal grazers (gastropods), scavengers (perhaps ostracods and various worms), parasites (worms), and predators (asteroids and cephalopods). Although trilobites declined rapidly after the Cambrian, they were still locally important in some younger habitats (Figure 10.4).

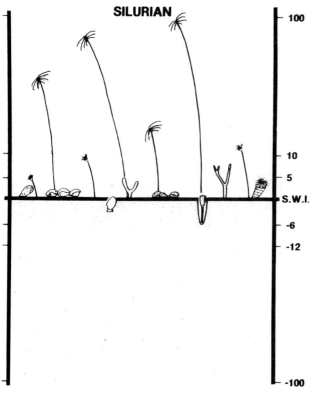

Figure 10.3   Diagram illustrating the composition and tiering of a typical Silurian ocean-floor community. (Used with permission of the *Journal of Geological Education* vol. 39. Copyright © 1991 National Association of Geology Teachers.)

Figure 10.4   A slab of Ordovician limestone containing many specimens of a large trilobite, *Homotelus*, characteristic of that period. The block is about 40 cm long. (Courtesy of the Los Angeles County Museum of Natural History. Photograph by Lawrence S. Reynold.)

Although considerable evolution and devastating mass extinctions dramatically changed the composition of the Paleozoic Fauna between the Ordovician and the Permian, the basic fauna always remained one composed dominantly of epifaunal suspension feeders in most settings. As mentioned above, brachiopods, bryozoans, corals, and stalked echinoderms were dominant on the ocean floor. Each of these are treated below.

BRACHIOPODS

The earliest **brachiopods** had a shell composed of chitin and apatite, and the two valves enclosing the soft parts were not hinged together (**inarticulate brachiopods**). This kind of brachiopod dominated during the Cambrian but was replaced abruptly during the Ordovician by calcareous brachiopods that had the two shells hinged together with a tooth-and-socket arrangement (**articulate brachiopods**). Although the inarticulate brachiopods persist to the Holocene (Figure 10.5), they were never conspicuous fossils after the Cambrian.

The calcareous brachiopods displayed several trends during the Paleozoic, when they were a dominant suspension feeder. These included a tendency in some groups for the hinge line to change from wide to short, with an accompanying change from a quadrangular to an oval shape. The mode of attachment to the sea floor was also subject to several modifications. Ordovician brachiopods typically had a wide hinge with coarse ribs (Figure 10.6). The shell was attached by a horny stalk called a pedicle that projected from the back of the shell through an open triangular hole (Figure 10.7). As the hinge became shorter in some groups, the hole for the pedicle became partly filled by calcareous plates, or it changed in shape from triangular to circular. In still other lineages, a functional pedicle was lost altogether, the hole being completely closed in adults. These more advanced types either rested loose on the bottom or developed an alternative mode of fastening themselves to the ocean floor. Some had part of the shell buried in the soft sediment, which served as an anchorage. Others were cemented to other shells by the beak of the shell, much as oysters are today (Figure 10.8). Still other

Figure 10.5   A specimen of a Holocene *Lingula*. The chitinophosphatic shell consists of two valves above, with a long fleshy stalk called the pedicle extending below from between the two valves; the pedicle anchors the shell in a soft bottom. (Courtesy of the National Museum of Natural History.)

groups developed spines, either along the hinge or over both valves, that penetrated into soft sediments and served as anchorage.

In addition to changes in shape and attachment, brachiopods also experienced progressive modification of the calcareous internal supports of their food-gathering structure, the **lophophore**. The lophophore consists of two coiled, fleshy arms covered with cilia that beat to direct food-laden water currents into and out of the shell. In early calcareous brachiopods, the lophophore was supported by only two short, straight rods at the base. In more advanced groups, the lophophore became more fully supported by internal structures. In two groups, the spiriferids and atrypids, the lophophore was supported by two helically coiled ribbons of calcite. In still others, the terebratulids, the calcareous loop was bent back onto itself.

Water was drawn into the shell on both sides anteriolaterally and expelled at the center of the shell opposite the hinge and, at least in an ovoid shell with a short hinge, also posteriorly. Many features of the shell are related to this aspect of brachiopod biology. The center part of the shell is set off from the two sides, resulting in a raised fold on one valve and a depressed sulcus on the other. This serves to separate incoming and outgoing water currents, helping to ensure that the brachiopod does not recycle already filtered water. The coarse ribbing on many shells results in a zigzag pattern where the valves meet when closed. This pattern produces an increased area through which water can pass so that the shells need not open too wide, which might allow unwanted larger particles into the shell interior. Some brachiopods have small, closely spaced spines that project between the opened valves, acting like a fence, to further keep out intruders.

Figure 10.6   A slab of Ordovician limestone containing many specimens of wide-hinged brachiopods that have been replaced by silica. Both interiors and exteriors of the valves are shown. A bryozoan colony is near the center of the picture. The specimen is reduced somewhat. (Courtesy of National Museum of Natural History.)

The brachiopod shell is built by a thin layer of tissue called the mantle that lines the interior of both valves. Growth proceeds along the edges of the shell and is called accretionary growth. Thus, the outer edges of the shell, where it opens, are the most recently formed part of the valves. Most of the space within the shell is called the mantle cavity, within which the lophophore is housed. The gut is reduced in size. In inarticulate brachiopods the gut is complete, having a mouth and anus, but in the more advanced articulates, the gut has only a single opening that serves both for ingestion of food and for ejection of waste products. The fecal material is pseudofeces because there is no true anal opening. To expel accumulated pseudofeces from the mantle cavity, brachiopods rapidly clap the two valves together, causing water currents to wash the pseudofeces away.

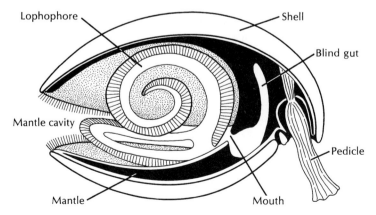

Figure 10.7   Cross section of a living brachiopod showing major morphological features.

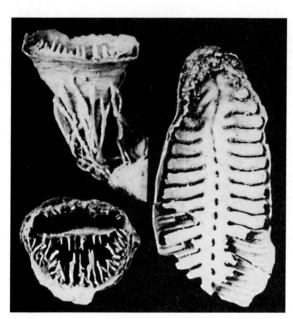

Figure 10.8 Two advanced brachiopods from Permian-age reefs in western Texas. The one on the right has a much reduced upper valve and lived an oysterlike existence attached to the reef. The one on the left is shown in both side and top views. This brachiopod adopted a coral shape and was attached by spines. The specimens are reduced in size and are silicified, having been etched from limestone. (Courtesy of National Museum of Natural History.)

During each period of the Paleozoic, one or another brachiopod group tended to predominate in most oceanic settings. The orthids and strophomenids (Figure 10.9) held sway during the Ordovician (Table 10.1). These brachiopods had wide hinges, an open triangular pedicle opening or lacked an opening, and coarse ribbing. The lophophore supports were stubby and primitive. The pentamerids were especially common in many Silurian limestone settings on or near reefs. These large egg-shaped brachiopods had a very short hinge and a smooth or faintly ribbed shell. The Devonian was predominated by strophomenids, spiriferids, and atrypids. Spiriferids had exceedingly long hinges, so that the shell appears to be winged, whereas the atrypids had a very short hinge and a round pedicle opening. Most were ornamented with ribs and had a conspicuous fold and sulcus along the midline of the shell (Figure 10.9). A conspicuous aspect of both of these forms is that they have a spirally coiled lophophore support structure, but these evolved independently. In the spiriferids it coils parallel to the direction of shell growth, whereas in the atrypids it coils perpendicular to shell growth. During the Mississippian spiriferids and atrypids remained common, but two groups of strophomenid brachiopods also increased in relative dominance, the chonetids and productids. Both groups had lost a functional pedicle, had spines, and had concavo-convex shells. The chonetids were small brachiopods with spines only along the hinge line. The productids were much larger and generally had spines on one or both valves. Spines served to anchor the brachiopod onto the bottom. One valve was generally highly convex, whereas the other (uppermost in life position) was flat or gently concave, forming a lid over the body cavity. Chonetids and productids dominated Pennsylvanian and Permian communities (Figure 10.9). By the close of the Paleozoic, most of these various groups of brachiopods, so conspicuous for millions of years, had become extinct.

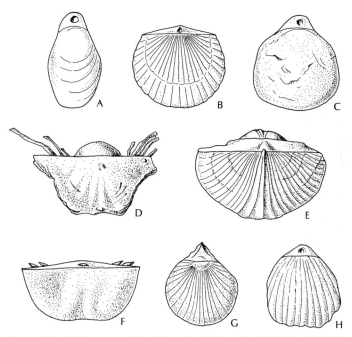

Figure 10.9   Major groups of brachiopods, all oriented with the brachial valve in front, pedicle valve behind, and posterior up. (A) Terebratulid. (B) Orthid. (C) Pentamerid. (D) Productid. (E) Spiriferid. (F) Chonetid. (G) Strophomenid. (H) Rhynchonellid.

Table 10.1.   DOMINANT GROUPS OF BRACHIOPODS AND BRYOZOANS DURING THE PHANEROZOIC. *SPECIALIZED GROUPS OF STROPHOMENID BRACHIOPODS.

	Dominant Brachiopods	Dominant Bryozoans
Cenozoic	Rhynchonellids Terebratulids	Cheilostomes
Mesozoic	Rhynchonellids Terebratulids	Tubuliporans
Permian	Chonetids* Productids*	Fenestrates
Pennsylvanian	Chonetids* Productids*	Fenestrates
Mississippian	Atrypids Chonetids* Productids* Spiriferids	Fenestrates Cystoporates
Devonian	Atrypids Spiriferids Strophomenids	Cystoporates Fenestrates Trepostomes
Silurian	Pentamerids Strophomenids	Trepostomes
Ordovician	Orthids Strophomenids	Trepostomes

BRYOZOANS

Virtually any outcrop of Paleozoic marine rocks yields fossil **bryozoans**. In many such rocks, bryozoans are the single most common group present. Clearly, they were dominant animals in many Paleozoic marine communities (Figure 10.10). The individuals that make up each bryozoan colony are tiny; they usually cannot be easily seen without a hand lens or microscope. In order to identify a bryozoan to its major group, genus, or species, it is commonly necessary to examine the details of the colony under a microscope, preferably in **thin section**. Thin sections are thin slices of rock or fossil that are mounted on a glass slide and ground so thin that light can pass through the rock or fossil. Thus, the internal details of the bryozoan colony and skeletal structure are examined with a microscope.

Like brachiopods, bryozoans exhibit a succession of dominance in different groups through the Paleozoic (Table 10.1). Bryozoans evolved during the Ordovician and underwent an adaptive radiation, becoming conspicuous, dominant members of ocean-bottom communities. Trepostomes are the dominant Ordovician bryozoans and are commonly referred to as "stony" bryozoans because these colonies are typically massive, ranging in shape from an irregular lump to a large, flat sheet to a bush form with thick, cylindrical branches (Figure 10.11). They are important rock formers in many Ordovician limestones and shales. Trepostomes were also dominant during the Silurian and Devonian, but by the Devonian, a variety of less conspicuous, more delicate bryozoans were also abundant in many settings. The more delicate forms include the fenestrates, which form colonies that have an open meshwork, much like that of a window screen. The skeleton of a fenestrate is called the frond, and the entire colony is typically conical in shape.

During the late Paleozoic, fenestrates continued to be varied and abundant. Other bryozoans include both the "stick" bryozoans, delicate cylindrical branches that formed bushy colonies, and bifoliate bryozoans that were again bushlike in colony

Figure 10.10 A bryozoan-rich slab of limestone from the Rochester Shale of Silurian age from New York. Several different kinds of bryozoans are shown, mostly twiggy, ramose forms, as well as brachiopods and other fossils. About natural size. (Courtesy of National Museum of Natural History.)

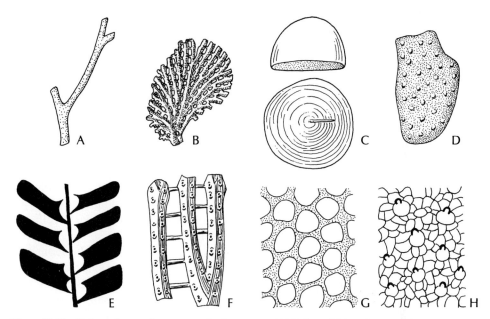

Figure 10.11   Colony shapes of representative bryozoans. (A) Twiggy. (B) Fan-shaped. (C) Massive or stony. (D, E) Bifoliate or laminar (E in cross section). (F) Fan-shape magnified. (G) Enlarged trepostome section. (H) Enlarged cystoporate section.

form, but the branches had a ribbon morphology composed of two layers of bryozoan animals back-to-back. A specialized fenestrate form was *Archimedes* that had the fronds attached to a solid spiral core of skeletal tissue (see Figure 7.10).

All bryozoans are suspension feeders and were an important aspect of multi-tiered epifaunal suspension-feeding communities. Like brachiopods, bryozoans possess a lophophore, and cilia on the lophophore are used to capture microscopic suspended food particles from the water (Figure 10.12). In many instances, bryozoans were clearly

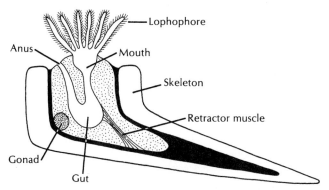

Figure 10.12   Magnified cross section of an individual bryozoan showing the relationship between the skeletal and soft-part anatomy. Highly magnified.

the dominant animals in Paleozoic marine communities, although their abundance is easily overlooked because many are commonly preserved as small, broken fragments. A casual fossil collector must be careful not to ignore bryozoans completely in favor of the more obvious brachiopods, trilobites, and corals.

The most conspicuous Paleozoic groups of bryozoans became extinct by the end of the Permian or were greatly reduced at the end of the Paleozoic and became extinct by the close of the Triassic.

SPONGES

Sponges are another group of suspension feeders. They filter water through their bodies and feed on tiny food particles, such as detritus, zooplankton, and phytoplankton, that are carried into the body cavity with water. With a celullar grade of organization, different cell types perform different functions. Cells with whiplike flagella beat in unison to force water currents through holes in the wall of the sponge (Figure 10.13A). The water then exits through a large hole in the top of the body. The sponge skeleton is composed of many small spicules that are composed of either silica, calcium carbonate, or organic material (spongin) as in the bath sponge (Figure 10.13B, 10.13C). The spicular skeleton commonly falls apart at the death of an animal; therefore, individual spicules are common as fossils, but whole sponges are rare. The siliceous or glass sponges (Figure 10.13D) were common Paleozoic fossils in shallow-water communities, but today, glass sponges live only in very deep water.

CORALS

In many books **corals**, phylum **Cnidaria**, are treated as predators. However, because they are always sessile, fixed to the bottom, and capture suspended food particles, they are considered here to be selective suspension feeders. It is true that corals ingest only

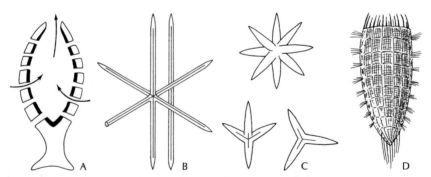

Figure 10.13   Fossil sponges. (A) Cross section of a simple sponge showing water currents entering the central cavity through pores and exiting through the large upper opening. The dark inner lining indicates the position of food-gathering cells. (B) Enlarged siliceous sponge spicules. (C) Enlarged calcareous sponge spicules. (D) Complete Devonian siliceous sponge with anchoring root tufts of spicules at the bottom (Modified from R.C. Moore, C.G. Lalicker, and A.G. Fischer, 1952 *Invertebrate Fossils*, McGraw-Hill).

small nektonic and planktonic animals, but this alone does not qualify them as predators. In this book predators are restricted to carnivorous animals that seek out and capture their prey.

The oldest well-known corals are Ordovician. Like calcareous brachiopods and bryozoans, corals very quickly became a dominant aspect of marine communities during the Ordovician. From the Ordovician to the present, corals have been a dominant group. We presume that ancient corals played many of the same roles in ancient marine communities as they do in living communities.

Modern corals catch small crustaceans, fishes, and other nektic animals by using tentacles equipped with stinging cells that poison and subdue the prey. Sea anemones, cnidarians that lack a calcareous exoskeleton, have the same feeding habits. Obviously, the size of the coral partly determines the size of the prey that a coral can capture.

When we think about corals living today, images of coral reefs and tropical isles immediately come to mind. In shallow-water tropical areas, corals, along with calcareous algae and many other organisms, build **reefs** that are enormous structures above the sea floor composed of the skeletons of these organisms. The diversity of life on reefs is amazing; they are the marine equivalent of tropical rain forests in terms of numbers of species present. The corals that build reefs are called **hermatypic corals**. Within their soft tissue, hermatypic corals contain small, symbiotic, photosynthetic, unicellular protists called **zooxanthellae algae**. The relationship between the coral and the protists is mutually beneficial; neither thrives without the other. Organic reefs are raised above the sea floor, and active reef growth is confined to shallow, sunlit waters because the zooxanthellae require sunlight for photosynthesis. Living corals that lack zooxanthellae are called **ahermatypic**. Although they may live on reefs, they do not contribute to the rigid skeletal framework that causes the reef to grow upward. Ahermatypic corals are more common in deep water. Organic reefs are common throughout the fossil record and have a long history. However, not all ancient reefs were built by corals and calcareous algae, like modern reefs. Moreover, we have little direct evidence that fossil corals contained zooxanthellae, a relationship that seems vital to the formation of living reefs.

In the sense of a rigid organic framework raised above the sea floor, reefs extend into the Precambrian as stromatolites; archaeocyathids formed reefs during the Cambrian, as discussed above. Small algal reefs and reefs composed of corals, bryozoans, and algae first occurred during the Ordovician. During the Silurian, reefs become very large and common. Some Silurian reefs in Illinois, Indiana, and Ohio are several hundred feet thick and cover more than one square mile in area. The primary builders of these Silurian reefs are algae; **stromatoporoids**, a group of sponges (Figure 10.14); and **tabulate corals**, an extinct group of corals. Tabulate corals are exclusively colonial. The skeletal tube that housed each coral animal in the colony is small, rarely more that a centimeter in diameter (Figure 10.15). Each tube has a series of horizontal partitions in it, called tabulae (hence the name *tabulate*), that served as a floor for the soft coral polyp at successive growth stages (Figure 10.16). Tabulate corals were common participants in reef growth during the Silurian and Devonian. Large oil-producing Devonian reefs are known in western Canada. Like Silurian reefs, Devonian reefs are build principally of algae, stromatoporoids, and tabulate corals.

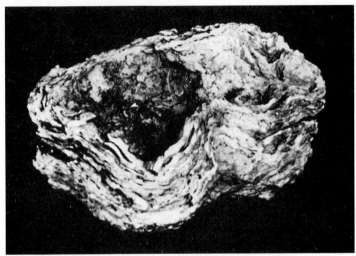

Figure 10.14 Skeleton of an extinct group of sponges called stromatoporoids. These were important reef builders during the Silurian and Devonian. The specimen is several centimeters across and consists of concentric layers of silica, which has replaced the original calcite skeleton. The specimen is shown upside down to better illustrate the concentric growth layering. (Courtesy of Ward's Natural Science Establishment, Inc.)

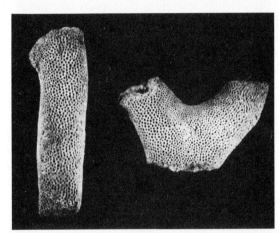

Figure 10.15 Two specimens of the tabulate coral *Favosites*, of Silurian age. Each specimen is a portion of a much larger colony; each hole in the colony housed an individual coral polyp when the animals were alive. The specimens are about 4 cm in maximum dimension. (Courtesy of Field Museum of Natural History, Chicago.)

Figure 10.16 Magnified detail of the individual tubes in a Devonian colony of tabulate coral, *Favosites*. Portions of the horizontal cross partitions that served to support the soft polyp in life can be seen. (Courtesy of Ward's Natural Science Establishment, Inc.)

In addition to tabulates, the other major group of Paleozoic corals is the **rugose corals**. These include both solitary individuals and colonial forms (Figure 10.17). The individuals are conspicuously larger than those of tabulates. Some solitary rugosans may be up to 1 meter long, and colonies may be up to 1 meter in diameter. In addition to lacking tabulae, rugose corals differ further from tabulates in having vertical partitions called **septa** (singular, septum) that divide the soft polyp into a series of compartments (Figure 10.18). The internal structure may be quite complex and is used as a basis for classifying families, genera, and species.

Interestingly, rugose corals virtually never took part in forming organic reefs. They may have lived on such reefs, as do ahermatypic corals today, but they did not contribute to the rigid organic framework. Some beds in the Paleozoic are crowded with rugose corals, such as the famous Middle Devonian beds of the Falls of Ohio, near Louisville, Kentucky. However, these corals form a horizonal bed; they do not bind together into an organic framework raised above the sea floor. It is at least possible that tabulate corals may have had zooxanthellae and that rugosans did not. There is little direct evidence to support such a contention, as there is no obvious morphological means to distinguish a skeleton of a modern hermatypic coral from one that lacks zooxanthellae.

During the Late Paleozoic, reefs were composed of a variety of organisms, but neither tabulate nor rugose corals were conspicuous. Sponges, bryozoans, calcareous algae, and even some kinds of brachiopods contributed to making rigid reef frameworks. Rugose and tabulate corals became extinct at the close of the Permian.

A

B

Figure 10.17   Solitary and colonial rugose corals. (A) Two different solitary rugose corals shown from the side and somewhat enlarged. Notice the prominent radial septa showing at the top of each specimen. (B) A colonial coral of Devonian age shown in both top and bottom views. The skeleton for each coral in the colony is tightly compressed against its neighbors. (Courtesy of National Museum of Natural History.)

Figure 10.18   Polished section through a colonial rugose coral showing the internal structure of the skeleton. The specimen is of Devonian age. Thin, dark walls separate each individual from adjacent corals. Many radiating septa are present within each individual. Notice two small individuals along the left edge that have budded from adjacent ones. (Courtesy of National Museum of Natural History.)

STALKED ECHINODERMS

Important components of many Paleozoic ocean-floor communities were the **stalked echinoderms**. Like brachiopods and bryozoans, this phylum is unusual because it is represented in the fossil record by many more kinds of extinct forms than living forms, especially during the lower and middle Paleozoic. A total of twenty-one classes are commonly recognized within the Echinodermata. Echinoderms underwent an extensive adaptive radiation during the Cambrian and Ordovician, developing many different body plans. Many of these new types were only successful for a relatively short period of time, but others persisted. Of these, only five are still living today, leaving sixteen extinct major groups. Many of these extinct classes are confined to either Cambrian or Ordovician times. Some groups are known from only a handful of specimens that are assigned to a few genera and species. Yet, these fossils are so distinctive that they have been separated at the class level.

Some of these primitive echinoderms could move about. Others were attached by their bodies to the bottom. Still others had bodies elevated above the sea floor on stalks or stems.

The most common and diverse stalked echinoderms were the crinoids, blastoids, rhombiferans, and diploporans. The **blastoids** were most common in Silurian through Mississippian rocks (Figure 10.19) and became extinct during the Permian. **Rhombiferans** and **diploporans** were most common during the Ordovician and Silurian, and both became extinct during the Devonian. In blastoids, rhombiferans, and diploporans the stalk was typically short, generally only a few centimeters. The vital organs are encased within a theca composed of a few to many calcareous plates. From this theca, fine food-gathering appendages, brachioles, extended upward to filter food from the water. Each of these groups has characteristic pore structures that were presumably used for respiration.

Crinoids were a more successful group than the blastoids, rhombiferans, and diploporans combined. They range from Ordovician to Holocene and have been very diverse; nearly 6,000 fossil species are known (Figure 10.20). Their basic structure is the same as other stalked echinoderms, except for differences in the arrangement of calyx

Figure 10.19 A nearly complete blastoid, *Orophocrinus*, with a short stem, external growth lines on the thecal plates, and delicate, slender food-gathering appendages called brachioles above. This is unusually good preservation. The specimen is Mississippian in age, from Iowa, and about 4 cm high. (Courtesy of D.B. Macurda, Jr., University of Michigan.)

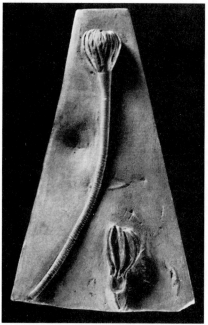

Figure 10.20 Two stalked crinoids (large, *Taxocrinus*; small, *Dichocrinus*) from Mississippian rocks of Iowa. The stem is not complete on either specimen. Notice the array of food-gathering arms around the top of the body on each specimen. The slab of rock is about 22 cm high. (Courtesy of Field Museum of Natural History, Chicago.)

plates and differences in the structure of the food-gathering appendages, or arms. Crinoids generally have a longer and more robust stem than other stalked echinoderms. They were able to arrange their arms into a feeding fan through which water currents could pass. Crinoids evolved numerous types of arm branching that yielded several styles of filtration fans, each adapted to capture different-sized food particles. Crinoids were the dominant high-tier organisms throughout the Paleozoic (Figure 10.2). They typically occurred in reasonably high-diversity communities and contributed greatly to the tiering complexity in these communities. Crinoids reached their peak of abundance and diversity during the Mississippian Period (Figure 10.21).

PALEOZOIC RADIATIONS AND EXTINCTIONS

The Paleozoic (Ordovician to Permian) fauna radiated rapidly after the end-Cambrian extinctions, and it came to an end at the great end-Permian extinction when as many as 84 percent of the genera of ocean life became extinct. This was the most devastating mass extinction ever. During the 250-million-year reign of the Paleozoic marine fauna, two additional major mass extinctions occurred, the end-Ordovician extinction and the Frasnian-Famennian extinction (near the end of the Devonian)(see Figure 5.7). The end-Ordovician extinction was the second most severe extinction in Earth history (57 percent generic extinction), and 50 percent generic extinction occurred during the Devonian mass extinction. These significant extinction events made dramatic changes in ocean-floor life, but neither changed the basic epifaunally tiered structure of ocean-bottom communities.

The end-Ordovician extinctions were caused by a global ice age. Extensive continental glaciers developed at the Ordovician South Pole, which is in present day northern Africa. Evidence for this glaciation is preserved in African rocks, including glacial grooves and glacial till (rock deposits left by a retreating or melting glacier). During this or any continental glaciation, a huge volume of water is transferred from the oceans to glacial ice spreading over the land. Thus, sea level drops, shallow continental

Figure 10.21 A slab of Mississippian limestone containing examples of complete crinoids. Numerous long stems that supported the flowerlike head and arms are preserved. The slab is from LeGrand, Iowa. The paper clip at right indicates size. (Courtesy of National Museum of Natural History.)

seas dry up, and even the shallow continental shelves may be drained. It is estimated that during the Late Ordovician glaciation, sea level may have dropped by as much as 100 meters. Thus, shallow-water ocean habitats became very scarce, and climatic cooling occurred associated with a global ice age. For example, during the Middle Ordovician, shallow continental seas were extensive across North America, but the only very late Ordovician shallow-water deposits in all of North America are on Anticosti Island in the mouth of the St. Lawrence Seaway. This near elimination of shallow-water habitats, as well as an undoubted global cooling, caused the end-Ordovician extinctions. Trilobites, stalked echinoderms, and brachiopods all experienced considerable extinctions during this event. Remarkably, as mentioned above, these extinctions did not bring about a fundamental change in the organization of marine communities. Crinoids and brachiopods experienced Early Silurian adaptive radiations, and multitiered epifaunal suspension-feeding communities still dominated in most shallow-water settings.

The end-Frasnian extinctions (the Frasnian is a series of geologic time during the Late Devonian) were also quite significant, and are generally ranked among the five most significant mass extinctions (see Figure 5.7). Like the end-Ordovician extinctions, the basic ecological structure of most marine communities was unaffected by the end-Frasnian extinctions. However, one habitat in particular, shallow-water tropical reefs, was eliminated during these extinctions. As discussed above, Silurian and Devonian reefs were extensive worldwide in shallow-water, tropical seas. The framework of these reefs was composed largely of tabulate corals, stromatoporoids, and algae. Many, many other organisms populated these reefs, similar to reefs today. However, this Paleozoic reef habitat perished abruptly during the end-Frasnian extinctions, and multitaxonomic coral reefs did not return in Earth's oceans for another 140 million years, during the Late Triassic. Corals, stromatoporoids, trilobites, crinoids, ammonoids, bryozoans, brachiopods, and fishes experienced substantial end-Frasnian extinctions. The cause of the end-Frasnian extinction is not known, although numerous proposals have been made. Potential causes include global cooling or a sudden overturning of ocean waters to bring anoxic waters to the surface layer of the ocean, but problems exist with both of these scenarios. A meteorite impact is commonly proposed as the mechanism for global cooling at this time.

END-PERMIAN MASS EXTINCTION

Life in the oceans was nearly eliminated by the end-Permian mass extinctions. This extinction was so pervasive that it changed the composition and organization of ocean-floor communities. The multitiered, epifaunal suspension-feeding communities of the Paleozoic fauna never really returned. Complete extinction of many groups occurred, including, among others, trilobites, blastoid echinoderms, fusulinid foraminifera, rugosan corals, tabulate corals, some bryozoan groups, and some brachiopod groups. Other significant extinctions included presumably all but one genus or species of crinoid echinoderms, most brachiopod orders, most bryozoan orders, many ammonoids, and reptiles.

Again, numerous explanations have been offered for this, the largest, extinction; but no consensus has been reached. Indeed, it is not even clear whether the extinction was gradual, stepwise, or one sudden episode. This was an unusual time in Earth history because Pangaea was assembled into a single supercontinent, sea level was quite low, and climates were relatively warm. Some of the more probable causes that have been suggested include a global drop in sea level, climate change, constriction of tropical habitat space around a single supercontinent, and increased levels of oceanic and atmospheric carbon dioxide. It is also possible that the unprecedented end-Permian extinctions were brought about by the coincidence of two or more of these potential extinction factors.

THE MODERN FAUNA

TRIASSIC AND JURASSIC OCEAN-BOTTOM FAUNAS

When seas reflooded the continents during the Early Triassic, a depauperate fauna was all that remained following the end-Permian extinctions. Low diversity, but commonly high abundance, faunas were composed of stromatolites, bivalves, brachiopods, and crinoids. From this beginning, ocean-floor communities that were either dominantly epifaunal suspension feeding, infaunal suspension feeding, or deposit feeding developed.

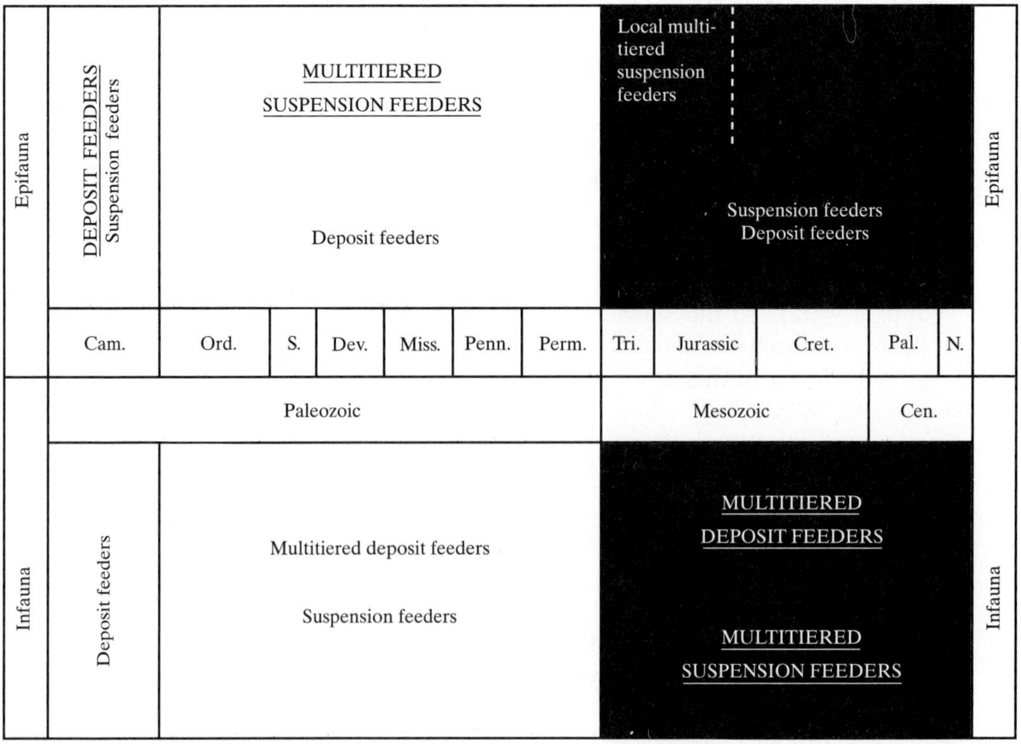

Figure 10.22 Phanerozoic history of the important types of feeding among organisms on the ocean floor. Distinction is made between infauna and epifauna. Feeding types underlined and in uppercase letters were the dominant types. Feeding types of the Modern marine fauna are highlighted.

Most significant at this time was the expansion and radiation of **infaunal** suspension feeders and infaunal deposit feeders (Figure 10.22). Infaunal suspension feeders, such as bivalves, worms, and crustaceans, had occupied infaunal tiers during the middle and late Paleozoic, but Paleozoic communities were only locally dominated by infauna. During the Triassic and Jurassic, siphonate heterodont bivalves evolved. The **siphons** are extensions of the mantle that act as a "snorkel" for infaunal bivalves, supplying them with oxygen and/or food from water just at the ocean floor. Siphons can easily be extended into position or retracted into the shell. This morphological innovation was an adaptive breakthrough for infaunal bivalves, and as a consequence, they began radiating. Adaptive radiations occurred in both suspension-feeding and deposit-feeding bivalves.

Along with deposit-feeding and infaunal suspension-feeding communities, multi-tiered epifaunal suspension-feeding communities were also present. The highest tiers were still typically maintained by crinoids during the Triassic and Jurassic. A typical epifaunal suspension-feeding community from the Jurassic would have a ~50 to 100 centimeter tier composed of crinoids; a ~20 to 50 centimeter tier composed of sponges, soft corals, and crinoids; a ~5 to 20 centimeter tier composed of sponges, corals, bryozoans, soft corals, bivalves, and crinoids; and a 0 to ~5 centimeter tier composed of numerous organisms (Figure 10.23). Infauna were also important with suspension-feeding tiers of ~0 to –6 centimeters, ~–6 to –12 centimeters, and ~–12 to –100 centimeters. Bivalves, gas-

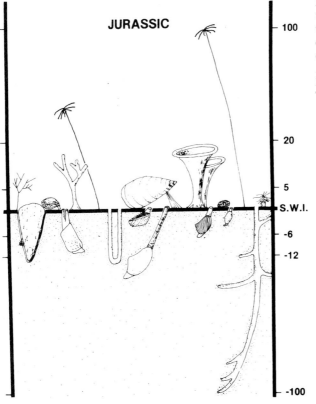

Figure 10.23   Diagram illustrating the composition and tiering of a typical Jurassic ocean-floor community. (Used with permission of the *Journal of Geological Education* vol. 39. Copyright © 1991 National Association of Geology Teachers.)

tropods, various worms, and decapod crustaceans were important infaunal suspension feeders, with decapods the only common organisms below –12 centimeters. Deposit feeding was rapidly becoming the dominant trophic mode in many Mesozoic settings. Deposit feeders also displayed tiering, with specialized feeders at the sediment surface and others at progressively deeper levels. Important deposit feeders included annelid worms, bivalves, crustaceans, holothurians, and echinoids.

CRETACEOUS AND CENOZOIC OCEAN-BOTTOM FAUNAS

During the Cretaceous, stalked crinoids left shallow-water environments and became restricted to the deep sea. Today stalked crinoids live only in water deeper than 100 meters. This coupled with the continued expansion of the infauna resulted in another substantial change in ocean-floor communities—only infaunal deposit feeders and infaunal suspension feeders dominated from the Cretaceous to today in most open shallow-water settings away from reefs. These communities are infaunally multitiered (Figure 10.2). Deposit-feeding communities occur in settings with more mud, whereas suspension-feeding communities dominate in sandier habitats.

A typical Cenozoic infaunal community would be dominated by either suspension feeders or deposit feeders. A shallow-water suspension-feeding community away from reefs would be dominated by the infauna. A diverse example would typically have a ~20 to 50 centimeter tier with sponges and soft corals; a ~5 to 20 centimeter tier with sponges, soft corals, bryozoans and bivalves; and a 0 to ~5 centimeter tier with sponges, corals, soft corals, bryozoans, bivalves, and others. Infaunal suspension-feeding community tiers are crowded, with subdivisions at ~0 to –6 centimeters, ~–6 to –12 centimeters, and ~–12 to –100 centimeters (Figure 10.24). Various bivalves, worms, decapod crustaceans, and gastropods form the infauna with increased depth penetration by all groups. Bivalves are no longer confined to the shallower tiers but live down to 100 centimeters.

Similarly, the depth and intensity of infaunal deposit feeding increased substantially by the Cenozoic and was organized into tiers. Worms, bivalves, crustaceans, holothurians, and echinoids remained the dominant deposit feeders.

Although certain aspects of the ocean biosphere were greatly affected by the end-Cretaceous extinction event, the basic organization of ocean-floor communities remained the same. Along with dinosaurs on land (Chapter 15), ammonoid cephalopods, reef-forming rudist bivalves, and large marine reptiles became extinct. Other groups greatly affected by the end-Cretaceous extinctions were corals, echinoids, bryozoans, sponges, and planktonic foraminifera.

BIVALVES

If you go shell collecting along any sandy beach, you will find mostly gastropod and bivalve shells. In the shallow water of our time, **bivalves** are the dominant suspension feeders and gastropods and bivalves are the principal deposit feeders with preservable skeletons. The great majority of Paleozoic bivalves were epifaunal, attaching themselves to the bottom by tough horny threads, similar to some living bivalves. Others moved along using their muscular foot and fed from detritus on the ocean floor (Figure 10.25). Most Paleozoic bivalves were poor burrowers, and generally had all or part of the shell exposed at the sea floor. Such bivalves are refered to as nestlers,

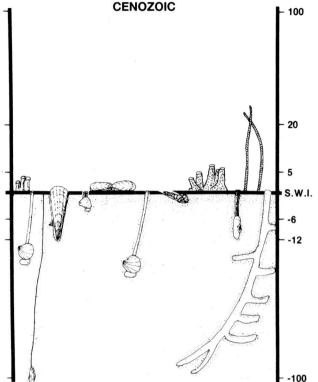

CENOZOIC

100
20
5
S.W.I.
-6
-12
-100

Figure 10.24   Diagram illustrating the composition and tiering of a typical Cenozoic ocean-floor community. (Used with permission of the *Journal of Geological Education* vol. 39. Copyright © 1991 National Association of Geology Teachers.)

because they simply settle down into soft sediment until they are nearly covered. They cannot burrow deeper, or they will suffocate.

As mentioned above, this all changed during the Mesozoic with the evolution of siphons. The presence of siphons allowed bivalves to be better and more efficient burrowers, and the abundance and diversity of bivalves increased during the Mesozoic and Cenozoic. Although some Mesozoic bivalves retained an epifaunal habit (Figure 10.26), siphonate, infaunal bivalves have been a Mesozoic and Cenozoic success story. Suddenly, they became one of the most common organisms in most marine settings (Figure 10.27).

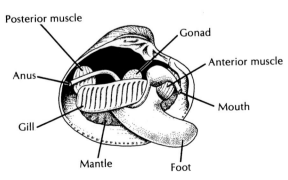

Posterior muscle
Gonad
Anterior muscle
Anus
Mouth
Gill
Mantle
Foot

Figure 10.25   Gross anatomy of bivalves. Gills circulate water in the mantle cavity for food and oxygen. The foot can be withdrawn when the shell is closed or extended for locomotion.

Figure 10.26 A giant epifaunal, oysterlike bivalve, *Inoceramus*, from the Cretaceous chalks of western Kansas. The upper valve, with many small oysters attached to it, is almost 1 m high. This particular kind of clam became extinct during the end-Cretaceous extinctions. (Courtesy of Sternberg Museum of Natural History, Fort Hays State University, Hays, KS.)

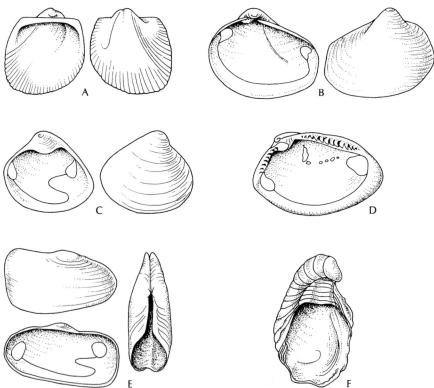

Figure 10.27 Major groups of fossil bivalves. (A) Hinge with many small teeth and sockets. (B) Hinge with single large tooth and socket. (C) Advanced tooth and socket arrangement (heterodont). (D) Nuculid, deposit-feeding bivalve. (E) Burrowing bivalve with large gape and internal notch along mantle line where long siphons can be retracted. (F) Oyster with large single muscle scar. (Modified from R.C. Moore, C.G. Lalicker, and A.G. Fischer, 1952. *Invertebrate Fossils*, McGraw-Hill.)

Burrowing allowed bivalves to escape predation from fishes and crustaceans that were also undergoing extensive radiation at this time. This new adaptation involved conspicuous changes in the bivalve morphology. The long, tubular siphons are rolled up portions of the mantle, the fleshy tissue that surrounds the body and secretes the shell (Figure 10.25). The anterior edges of the mantle became fused so that unwanted mud and sand could be kept from the body cavity. Fossil bivalves with long siphons can be identified because they have a deep pallial sinus, an indentation in the anterior connection of the mantle to the shell (Figures 10.27D, 10.27E). Also, some bivalves developed permanent gapes between valves in the anterior in order to accomodate large siphons.

GASTROPODS

Gastropods, or snails, have a long history and have been common fossils in some beds since the Ordovician (Figure 10.28). Many of the Paleozoic forms were probably herbivores, living in shallow water and grazing on various types of algae. Others moved over soft bottoms and were epifaunal deposit feeders, and some may have been predators. During the radiation of the Modern marine fauna, gastropods became very important, and in many settings they were either co-dominant with bivalves or the dominant faunal element (Figure 10.29).

Two gastropod lineages also became very efficient Mesozoic and Cenozoic predators, the naticids and muricids. These are shell-boring gastropods, which principally attack bivalves by boring through their shells.

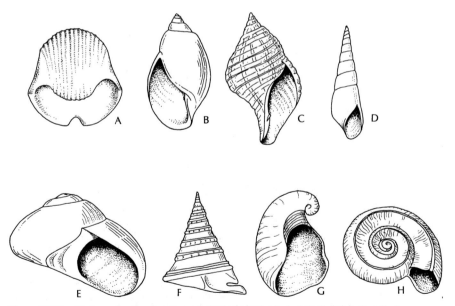

Figure 10.28   Various shapes of gastropod shells. (A) Planispiral shell. (B) Left-coiled (sinistral) shell with the aperture on the left; all others have an aperture on the right (dextral coiling). (C) Advanced shell with elongated siphonal notch. (D–F) Different styles of coiling. (G) Gastropod shell normally attached over the anus of Paleozoic crinoids. (H) Flat coiled shell.

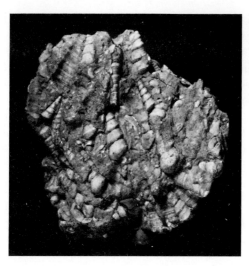

Figure 10.29   An aggregation of a high-spired gastropod, *Turitella*, from Eocene rocks in Virginia. Molds where the shells have been removed are on the upper-left part of the block. (Courtesy of National Museum of Natural History.)

ECHINOIDS

The mobile echinoderms called **echinoids** date back to the Ordovician. During the Paleozoic, they were exclusively epifaunal, moving on the bottom probably as algal grazers, deposit feeders, or perhaps scavengers. These are all called **regular echinoids**, which are echinoids that are more or less spherical or flattened with the mouth and anus at opposite poles of the animal (up and down) (Figure 10.30). Regular echinoids have five ambulacral grooves running from the mouth region to the anal region, and they typically have long spines. During the Mesozoic, irregular echinoids evolved. Parallel to the development of siphons in bivalves, the **irregular echinoids** achieved a

Figure 10.30   Regular echinoid viewed from the top. The prominent bosses on the test supported large spines that were used for movement and protection. In life, the mouth, not visible in the photograph, is next to the sea floor. About natural size. (Courtesy of National Museum of Natural History.)

morphological breakthrough that allowed them to invade the infaunal realm. However, echinoids underwent conspicuous changes in their hard-part morphology in this transition. Irregular echinoids lost their spherical five-part symmetry and became bilaterally symmetrical. They became either ellipsoidal sea biscuits or flat sand dollars (Figure 10.31). Spines became more numerous and smaller, giving a hairlike aspect to the animal. The anal opening shifted to the side of the test, which is the posterior, and in extreme examples even moved to the oral surface. The ambulacra are confined to the upper oral surface and are called petals (Figure 10.31).

These changes reflect ecological changes from an organism living on the surface to one living as an infaunal deposit feeder or suspension feeder. In infaunal echinoids spines are no longer needed for protection, so the short spines served as sediment-moving devices. For sanitation purposes, the anus moved to a posterior position so that waste material is left in the burrow behind the animal rather than on top of itself. Irregular echinoids are very effective infaunal deposit feeders that plow their way through soft mud or sand.

MESOZOIC AND CENOZOIC REEFS

Beginning during the Triassic, a new group of corals, the **scleractinian corals**, arose, perhaps from the rugose corals or, more probably, from soft corals. They differ from rugosans in their pattern of skeletal growth, which produced a different symmetry of septal arrangement within one individual. They are also composed of the carbonate mineral aragonite rather than calcite. The earliest scleractinians were small and solitary. Reefs did not exist during the Early Triassic. However, they lived on Middle Triassic reefs, but were not frame builders until the Late Triassic. Middle Triassic reefs generally had a sponge-

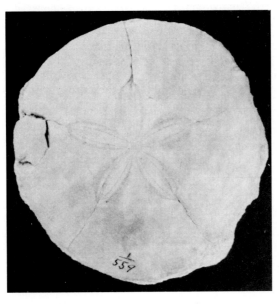

Figure 10.31 Top view, about natural size, of an irregular echinoid of Neogene age from the island of Malta. The animal was partly buried in the sediment when alive. Both the mouth and the anus are on the underside. The five areas shaped like flower petals were for respiration. (Courtesy of National Museum of Natural History.)

algal framework. By the Jurassic, larger colonial scleractinians had evolved, and they began to become important reef builders. They continued in that role through the Cenozoic to the present, interrupted only during the Cretaceous when rudistid bivalves were the dominant reef frame builders. Presumably, scleractinians did not inherit zooxanthellae from their rugosan ancestors; neither rugosans nor the earliest scleractinians are frame builders. Therefore, the intimate relationship between the protists and these corals must have developed sometime during the Late Triassic or Jurassic.

MESOZOIC AND CENOZOIC BRACHIOPDS AND BRYOZOANS

Articulate brachiopods were locally abundant again during the Mesozoic but never as diverse as before. Two groups of Paleozoic articulate brachiopods are still alive, the terebratulids and the rhynchonellids (Table 10.1). Both have an oval shell, circular pedicle opening, and a functional pedicle.

Among bryozoans, the tubuliporans were present throughout the Paleozoic but never dominated. This group survived the end-Permian extinctions and end-Triassic extinctions and diversified to become the dominant Jurassic bryozoans. The cheilostomes arose during the Jurassic and diversified to become the dominant Cenozoic bryozoans. These are delicate, and the colonies are commonly small. Many of them encrust on shells, seaweed, rocks, or other hard objects. Although they are quite diverse, with many families and genera, they are rarely as abundant and important a part of marine communities as bryozoans were during the Paleozoic.

KEY WORDS

ahermatypic	gastropods	septa
articulate brachiopods	hardgrounds	siphons
bivalves	hermatypic corals	size selection for food
blastoids	inarticulate brachiopods	sponge
brachiopods	infaunal	stalked echinoderms
bryozoans	irregular echinoids	stromatoporoids
Cnidaria	lophophore	tabulate corals
corals	reefs	thin section
crinoids	regular echinoids	tiering
diploporans	rhombiferans	zooxanthellae algae
echinoids	rugose corals	
epifaunal	scleractinian corals	

READINGS

Ausich, W.I., and Bottjer, D.J. 1991. "History of Tiering among Suspension Feeders in the Benthic Ecosystem." *Journal of Geological Education* 39:313–319. An outline of the composition and patterns of suspension-feeding communities through the Phanerozoic.

Boardman, R.C., and others. 1987. *Fossil Invertebrates*. Blackwell Scientific. 687 pages. A detailed account of the morphology and classification of all invertebrate fossils. Individual chapters vary somewhat in content and detail.

Brusca, R.C., and Brusca, G.J. 1990. *Invertebrates*. Sinauer. 922 pages. A modern textbook on living invertebrate animals.

Erwin, D.H. 1993. *The Great Paleozoic Crisis, Life and Death in the Permian*. Columbia University Press. 327 pages. A thorough consideration of the causes and consequences of the end-Permian mass extinctions.

Hess, H.; Ausich, W.I.; Brett, C.E.; Simms, M.S.; and Kindlimann, R. 1998 (in press). *Fossil Crinoids*. Treatment of the most important occurrences of fossil crinoids throughout geologic time, and discussion of morphology, paleoecology, and taphonomy.

McGhee, G.R., Jr. 1996. *The Late Devonian Mass Extinction, The Frasnian/Famennian Crisis*. Columbia University Press. 303 pages. A thorough consideration of the causes and consequences of the end-Frasnian mass extinctions.

McKinney, F.K., and Jackson, J.B.C. 1989. *Bryozoan Evolution*. Unwin Hyman. 238 pages. A comprehensive treatment of fossil and living bryozoans, their coloniality, biology, and evolution.

The Fossil Record of Plankton and Nekton

INTRODUCTION

The foundation of the marine food web is in the **phytoplankton**, the microscopic plants and photosynthetic protists that float near the surface of oceans. Their conversion of simple inorganic substances into complex organic molecules takes place only in surface waters where sunlight can penetrate, that is, in the photic zone of the ocean. Phytoplankton account for a significant percentage of primary production on Earth.

Marine phytoplankton have undoubtedly been in existence for a very long time, well back into the Precambrian. As discussed in the previous chapter, early Paleozoic faunas were dominated by animals that are suspension feeders. They captured live or dead organic particles from the water. In order to support such an extensive suspension-feeding fauna, there surely had to have been a rich phytoplankton base. However, there is a limited preserved record of Paleozoic phytoplankton.

Here we will consider some of the preserved primary producers, the phytoplankton, and some of the nonphotosynthetic plankton. Also, we will consider some of the evidence for fossil **nekton**, the swimming animals that lived immediately above the bottom or higher up in the water column (Figure 11.1). Other swimmers are higher-level consumers—predators—that will be considered in Chapter 12.

MARINE PHYTOPLANKTON

By dragging a net through the upper water column of oceans, one finds that the majority of marine phytoplankton consist of microscopic single-celled organisms that do not have hard parts. This has probably always been the case and, undoubtedly, accounts for the generally poor fossil record of phytoplankton. However, we do have a long fossil record for elements of the phytoplankton that were preservable.

ACRITARCHS

During the late Precambrian and continuing into the Cambrian and younger Paleozoic rocks, the primary record of phytoplankton consists of a group of microscopic fossils grouped into the **acritarchs** (Figure 11.2). These are composed of highly resistant organic material, cellulosic in composition, that enables them to be preserved in ancient rocks. Acritarchs range in age from the late Precambrian to the present. They are quite abundant in lower and middle Paleozoic rocks throughout much of the central United States, where the rocks have not been buried very deeply and have not been subjected to the heat and pressure of mountain building.

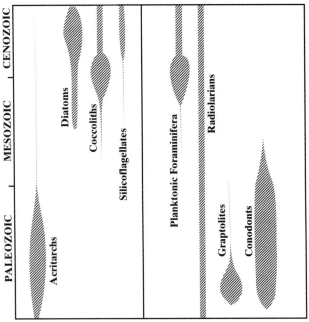

Figure 11.1   Relative prominence of major marine phytoplankton, zooplank ton, and nekton in three geologic eras. Not to scale.

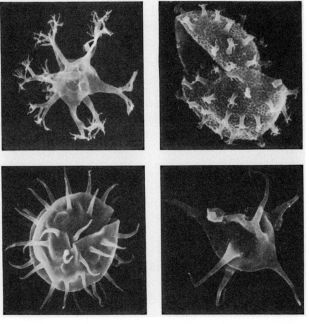

Figure 11.2   Photomicrographs of acritarchs, made with a scanning electron microscope, display variation in shape and ornament. These are from Devonian rocks of Ohio and are highly magnified. (Courtesy of E. Reed Wicander, Central Michigan University.)

Although acritarchs are sensitive to heat and pressure, they are very resistant to chemical change. They can be released from rocks by treating the rocks to a series of strong chemicals (mainly acids). This progressively dissolves away the mineral composition of the rocks, leaving behind a residue of highly resistant material, including acritarchs.

The affinities of acritarchs is problematic, and, indeed, the name, which means "old things that are confusing," was specifically coined to reflect this uncertainty. The acritarch consists of a spherical or polygonal central sac or bag. This may have a variety of different kinds of spines projecting from it—some very long and fragile, others very short and hairy. The spines may branch or have small secondary spines. A tear is present along one side of the sac or bag. The range of variation both in size and in character of the central sac and the spines is considerable; thus, a great many genera and species of acritarchs have been named.

These fossils are regarded as being the remains of reproductive cysts of one or several kinds of marine algae. Many marine algae have a complicated life cycle. After fertilization (union of two gametes), they may enter a dormant stage during which the zygote (first cell of the new generation) divides into a number of spores. In the dormant stage, the zygote is enclosed in a highly resistant organic-walled **cyst** that is comparable, but not identical, to many of the fossil acritarchs. The cyst ruptures, whereupon the spores are released and proceed to develop into new algal adult bodies. Presence of a cyst stage has considerable survival value in cases where adverse conditions may prevail for a period of time. One of the best-known living groups of algae that have such organic-walled cysts are the dinoflagellates (officially named the Pyrrophyta). These have a very distinctive and complex arrangement of organic material that makes up the cysts; they are divided into numerous discrete areas. The patterns on the cysts of dinoflagellates do not match those on acritarchs. Although the two groups may be related, at least in part, we cannot firmly place acritarch fossils within this or other categories of living organisms.

In early and middle Paleozoic rocks, acritarchs are very common fossils. They undoubtedly record an abundant and diverse presence of oceanic phytoplankton. In some samples, tens of thousands of specimens can be obtained from just a few grams of rock. During the Cambrian, acritarchs are relatively simple spheres and are not especially large or ornamented with complicated spines. During the Ordovician and Silurian, they reach their peak of diversity, abundance, and complexity. Many genera and species are known, which has proven very useful in biostratigraphy; fossil zones have been established based on acritarchs.

Although still common during the Early Devonian, acritarchs are not as diverse, and in younger and younger Devonian rocks, they become increasingly less common and have fewer types. By the beginning of the Mississippian, they have virtually vanished from the scene, leaving us with a puzzle for the remainder of the Paleozoic: What were the principal kinds of phytoplankton during the late Paleozoic from the Mississippian through the Permian?

Two answers have been put forward. First, it is possible that there was a real crisis in the oceans, i.e., that the decline in acritarchs records a serious decline in oceanic productivity that should have had a profound effect on all marine life. Not only would a decline in the phytoplankton erode the base of the oceanic food chain, but it could affect the balance of oxygen and carbon dioxide in the atmosphere. Much of the photosynthesis on Earth is accomplished by marine phytoplankton. A substantial decrease in these organisms could have produced a dramatic decrease in oxygen and a corresponding increase in carbon dioxide. Carbon dioxide is an important greenhouse gas, and a substantial increase would be expected to increase global temperatures. A second effect would have been on the suspension-feeding animals that depend on plankton for food. At the close of the Paleozoic, many important groups of marine invertebrates became extinct. Virtually all of those—bryozoans, brachiopods, corals, stalked echinoderms, and others—were suspension feeders.

The alternate answer is that no crisis occurred in the oceans. There was an ample supply of phytoplankton through the late Paleozoic; however, the dominant groups simply did not have an encysted stage and, thus, are not recorded as fossils. There is no evidence to support this answer, making it very difficult to either prove or disprove. One criticism of the first hypothesis is that the acritarchs essentially disappeared at the beginning of the Mississippian, yet major extinctions of bottom-dwelling suspension-feeding invertebrates did not occur until the close of the Permian—a time lag of nearly 120 million years. Why should it have taken so long for a crucial decline in productivity to have affected the remainder of the food chain? Based on this reasoning, the second hypothesis is favored.

DIATOMS AND COCCOLITHS

Not only is the record for phytoplankton poor during the latter part of the Paleozoic, but the paucity of these fossils continues into the Triassic at the beginning of the Mesozoic. By Jurassic time, fossil phytoplankton appear again in the record. At first they are rare and poorly preserved, but they quickly become more abundant and diverse. The Jurassic and Cretaceous phytoplankton are the same groups that dominate the oceans today, including the diatoms and coccoliths. **Diatoms** belong to the algal phylum Bacillariophyta. They have a silica skeleton composed of two valves that fit together like two parts of a pillbox or petri dish. There are many different kinds important today in both marine and fresh waters (Figure 11.3). The oldest definite diatoms are Cretaceous, and they reached their peak in abundance during the middle part of the Cenozoic, the Miocene, when there were thick deposits composed largely of their skeletons. These beds are called **diatomites**, or diatomaceous earth, and are of commercial value. This sediment is used in insulating material, abrasives, ceramics, and in filtering and filling materials.

Coccoliths belong to the algal phylum Haptophyta. They are very tiny spherical organisms that first appeared in the record during the Upper Triassic. In fact they are so small that they cannot be clearly examined with even the most high-powered light microscopes. It was not until the invention of electron microscopes that careful study of this group of organisms could begin. The spherical body of these organisms is composed of a series of small calcareous platelets (Figure 11.4). These platelets are produced throughout the life of an individual. As new platelets are formed, the old ones are forced off and drop to the sea floor. An individual may have all the platelets alike, or it may have two or three different kinds. Therefore, the study of loose plates from the ocean floor is not a reliable index of how many species produced the plates. Coccoliths were especially abundant and diverse during the Cretaceous Period, in

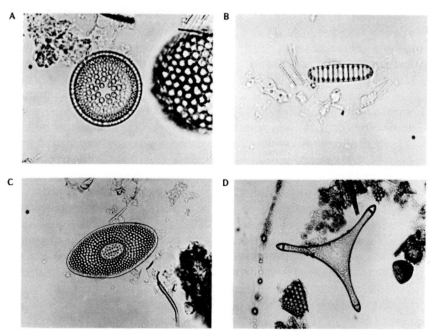

Figure 11.3 The siliceous skeletons of marine diatoms, highly magnified. (A) *Actinocyclus* and (B) *Denticula* are of Miocene age. (C) *Coscinodiscus* and (D). *Trinacria* are Eocene in age. (Courtesy of John A. Barron, U.S. Geological Survey.)

Figure 11.4 The surface of a small fragment of Cretaceous chalk from western Kansas, viewed at high magnification with a scanning electron microscope. The horizontal bar indicates scale in microns. Numerous individual coccolith platelets are scattered across the surface, forming a significant portion of the chalk. (Courtesy of Donald E. Hattin, Indiana University. Photograph by George Ringer.)

which their remains are the major component of **chalk**. One cubic centimeter of chalk may contain several billion of these tiny platelets. It is hard to believe that these ultra-small fossils comprise the chalks of the extensive white cliffs of Dover, England, which are Cretaceous in age. Extensive Cretaceous chalk deposits also occur in France and other parts of Europe, western Kansas, Nebraska, Texas, and Alabama. After the Cretaceous, coccoliths dwindled somewhat in variety, although they are still very important primary producers in the oceans today.

In addition to these two groups, there are other, less important groups of phyto-plankton that have a fossil record. These include **silicoflagellates**, with an open lattice-work kind of skeleton composed of silica (Figure 11.5). These belong to the algal phylum Chrysophyta and occur from the Cretaceous to the present day.

In summary, the fossil record of phytoplankton is incomplete and not fully under-stood. Acritarchs were apparently the dominant algal group(s) preserved during the late Precambrian and early and middle Paleozoic. We know very little about primary production from the Mississippian through the Triassic. After the Triassic, there is an essentially modern aspect to the phytoplankton, with peaks of diversity reached by the coccoliths during the Cretaceous and by diatoms during the Miocene.

MARINE ZOOPLANKTON

In addition to photosynthetic organisms that float in the water, there are many kinds of nonphotosynthetic protists and animals that have the same habit. Many of these organ-isms, called **zooplankton**, do not have preservable skeletons, similar to phytoplankton, so they do not appear in the fossil record. For paleontologists there are two major groups of zooplankton. One consists of microscopic protists, such as radiolarians and foraminifera. The second group consists of larger animals, such as the extinct groups of organisms called graptolites and conodonts.

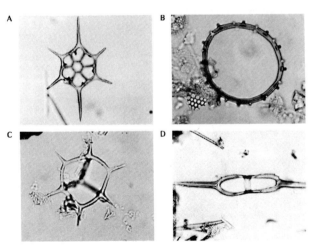

Figure 11.5   Various fossil silicoflagellate skeletons, all of Cenozoic age from California. The open meshwork of silica is typical of these microfossils that are highly magnified here. Specimen (A) is Pliocene, (B) is Miocene, (C) and (D) are Eocene. (Courtesy of John A. Barron, U.S. Geological Survey.)

RADIOLARIANS AND FORAMINIFERA

Radiolarians have a highly symmetrical, intricate skeleton composed of silica (Figure 11.6). They are assigned to the phylum Actinopoda. The silica, like that of diatoms, sponges, and other silica-producing organisms, is opaline silica, which is not very stable under heat and pressure much greater than that at the surface of Earth. Nevertheless, radiolarians have a remarkably good fossil record, having been found without question in Cambrian rocks. The record is spotty, but radiolarians have been found in abundance in some formations throughout the Phanerozoic. This may be explained by reference to the present-day distribution, which tends to have concentrations of radiolarians in offshore waters that are from 3,000 to 4,000 meters deep. Thus, they occur far from shore and are rare in coastal waters. Because our fossil record is primarily from rocks deposited in quite shallow water over or at the edge of the continents, one might expect radiolarians to have a spotty fossil record.

Foraminifera are a phylum whose members are much more common as fossils than radiolarians, but we are concerned here only with certain members of the "forams"—those that were planktonic. Although all radiolarians have been planktonic, foraminifera have been benthonic (bottom-dwelling) for the greater part of their history (Figure 11.7). Foraminifera are treated here as zooplankton, but it should be noted that some forams today have symbiotic photosynthetic algae within their tissue, so they behave both as zooplankton and phytoplankton. Benthonic forams first occur in Cambrian rocks, but planktonic representatives did not evolve until the Triassic. The oldest forams built their skeletons from foreign material; they gathered up bits of silt, sand, or other debris with their pseudopods and glued this all together with an organic cement. Such a covering is said to be agglutinated (glued together) or arenaceous (composed of sand particles). Some forams began to secrete

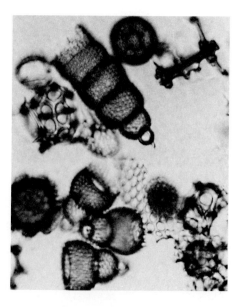

Figure 11.6 Several radiolarians of Miocene age, from the southwestern Pacific, showing the intricate siliceous skeleton. Highly magnified. (Courtesy of Joyce R. Blueford, U.S. Geological Survey.)

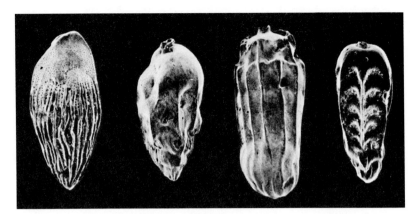

Figure 11.7   Four different kinds of calcareous benthonic (bottom-dwelling) foraminifera, highly magnified, from the Miocene of the Atlantic coast. (Courtesy of Thomas M. Gibson, U.S. Geological Survey.)

a calcareous skeleton during the Ordovician, but calcareous forams do not become common until the middle Paleozoic. All forams continued to be benthonic until the Triassic, when some of the highly globular, calcareous forms evolved into floating or planktonic organisms. These planktonic foraminifera evolved rapidly, producing many new genera and species (Figure 11.8). They reached a peak of diversity during the Cretaceous, but at the close of that period many forms became extinct. Floating forams continued through the Cenozoic to the Holocene, although reduced in number and variety. They are very useful index fossils for correlation and biostratigraphy because species evolved rapidly and they are widely distributed in ocean waters.

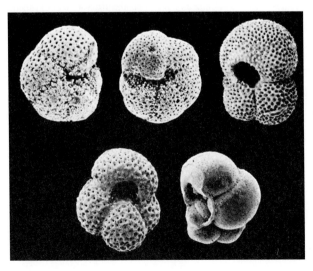

Figure 11.8   Planktonic foraminifera of Miocene age from the Atlantic coast, highly magnified with a scanning electron microscope. Highly inflated, globular chambers formed the skeleton. (Courtesy of Thomas M. Gibson, U.S. Geological Survey.)

GRAPTOLITES

In addition to the microscopic zooplankton just discussed, another group of much larger floating animals, the **graptolites**, flourished during the Paleozoic. These were colonial animals assigned to the phylum Hemichordata, with proteinaceous skeletons, and most are preserved through carbonization. Both planktonic and benthonic graptolites were important during the Paleozoic. Graptolite colonies consisted of one to many branches, each built of a chain of thecae, or small tubes that housed the individual animals (Figures 11.9, 11.10).

The oldest graptolites are from the Middle Cambrian. They became extinct during the Pennsylvanian but had their heyday during the Ordovician and Silurian, when they were very common and underwent conspicuous evolutionary changes. Three aspects of graptolites are of special interest to us: 1) their change in habitat and evidence for such change; 2) their relationship to living animals; and 3) the nature of evolutionary changes in the group.

The oldest graptolites are thought to have been benthonic. They evolved as a bottom-dwelling organism, but certain lineages later evolved to a planktonic life style. Most benthonic graptolites colonies have many branches that in some forms are connected at intervals by thin crossbars (Figure 11.11). Shrubby or mosslike in appearance, they have a basal attachment to the substratum. The colony is composed of two

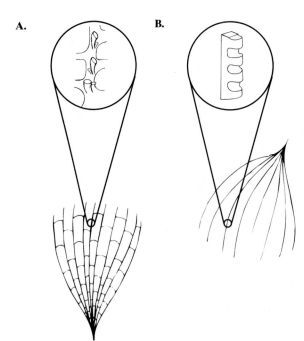

A.   **B.**

Figure 11.9   The two major kinds of graptolites: (A) The primitive benthonic type, with many branches and crossbars, that lived on the sea floor. Enlargement shows polymorphic individuals. (B) An advanced graptolite that was planktonic, with eight branches and no crossbars. Enlargement shows that all individuals are alike.

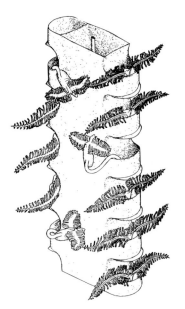

Figure 11.10 Reconstruction of a portion of a graptolite colony with two branches back-to-back.(Modified from P. Crowther and B. Rickards, 1977 *Geologica et Palaentologica*).

types of thecae that differ in size and shape; hence they are said to be dimorphic (having two forms). By Ordovician time a new group of graptolites that is thought to have been exclusively planktonic had evolved. These have fewer branches in the colony, all the thecae along a branch are the same, and the crossbars linking the branches have disappeared (Figure 11.9B). These more advanced graptolites evolved rapidly. Many individual species and genera have wide geographic distribution, and some are present on several continents. Dark shales and mudstones are the most common rocks that preserve graptolites. Because of their wide distribution, a characteristic of planktonic organisms, and their occurrence in rocks that commonly represent bottom conditions unsuitable for most life, these graptolites are considered to have been planktonic. In

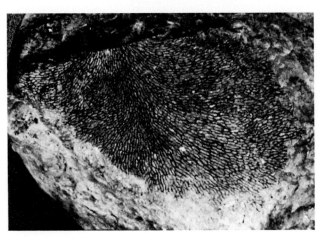

Figure 11.11 The skeleton of a benthonic graptolite colony, *Dictyonema*, from the Devonian of New York. The specimen shows the presence of many branches and crossbars connecting branches, magnified. (Courtesy of Ward's Natural Science Establishment, Inc.)

addition, small vane- and bulb-like structures are found on some planktonic grapto-lites that may have aided in floating.

During the Ordovician and Silurian, graptolite evolutionary changes included a reduction in the number of branches in the colonies (Figure 11.12). Also, some special-ized spiral-shaped colonies evolved. Whereas branch reduction and other general colony characteristics were once considered important for understanding graptolite phylogeny, it is now recognized that the structure in the earliest formed part of the colony has primary importance for determining evolutionary relationships. **Branch reduction** occurred throughout the Ordovician and Silurian. Some Early Ordovician forms are multiramous; the basic stocks of planktonic graptolites commonly have eight or fewer branches. From this starting point, many lineages had a progressively reduced number of branches. A typical Silurian single-branched graptolite is *Monograptus*, meaning "one branch" (Figure 11.13). Because these changes occurred more or less simultaneously in more than one lineage of graptolites, this is an excellent example of **parallel evolution**. Some environmental selection pressure, most likely related to the planktonic mode of life, had a constant impact on the graptolites, bringing about these repeated, parallel evolutionary changes of branch reduction. Another morphological trend in graptolite colonies was the development of specialized, spiral-shaped forms.

Planktonic graptolites continued into the Early Devonian, when they became extinct. Their bottom-dwelling counterparts persisted longer, into the Pennsylvanian, when they became extinct.

Because the planktonic graptolites have widespread geographic distribution and because they evolved very rapidly, they are ideal biostratigraphic or guide fossils. Furthermore, graptolite-bearing facies are common in the United Kingdom, where biostratigraphic subdivision of the lower Paleozoic rocks was first established. These factors have placed graptolites in a position of primary significance for defining bios-tratigraphic zones of the lower Paleozoic, especially for the Ordovician and Silurian.

Figure 11.12   A four-branched planktonic graptolite colony from the Lower Ordovician, somewhat enlarged. The specimen is preserved as a black carbon film on the rock slab. (Courtesy of National Museum of Natural History.)

Figure 11.13   A single-branched graptolite colony, *Monograptus*, from the Silurian of New York. Each individual theca has a prominent projection. (Courtesy of Ward's Natural Science Establishment, Inc.)

Finally, there has been a great debate concerning the affinities of graptolites. In the past, it was suggested that they were coelenterates, related to a group of this phylum called hydrozoans, and specifically to a group of hydrozoans called siphonophores. These include the Portuguese man-of-war, *Physalia*, a floating colony. In *Physalia*, one individual is highly modified into a gas-filled float, and other polymorphic individuals occur in tentacles below the float. Some individuals are specialized for feeding, whereas others are specialized for reproduction or as stinging cells. Although this may appear to be a good very general description of some aspects of graptolite morphology, evidence now suggests other affinities. The majority of graptolites, preserved in dark shales, are very thin, flattened carbon films in which details of the microstructure of the thecae have been obliterated. However, preserved graptolites occur in three dimensions in limestones and chert nodules. These can be released from the rock by dissolving the rock in acid, thus revealing exquisite microscopic detail of the thecae. Some of these well-preserved structures matched those of the living animal *Rhabdopleura*, which belongs to the phylum Hemichordata. Once considered closely related to the vertebrates, hemichordates are now thought to be more distant relatives to chordates, but the gill slits and dorsal nerve cord in some hemichordates suggest an evolutionary connection to chordates. The relationship between hemichordates and graptolites is now generally accepted by most paleontologists.

MARINE INVERTEBRATE NEKTON

Although many different marine animals live on the bottom, either sluggishly moving about or fixed in one position, and many protists, plants, and animals float up in the water, relatively few animals actively swim. The dominant swimmers are fish, and these have been an important part of marine communities for hundreds of millions of years. There have also been marine reptiles, and there are still mammals today that actively move through oceanic waters. Some reptiles—sea turtles and snakes—are also still oceanic swimmers. All these groups are primarily predaceous in habit. Because they are higher level consumers (predators), these and other carnivores, along with the cephalopod molluscs, will be considered in Chapter 12. Other swimmers in today's oceans include the nudibranch gastropods, which lack a shell, and various worms, but these have virtually no fossil record. For our purposes here, we will consider two additional groups of larger nekton that have reasonably good to excellent fossil records: the jellyfish, which, surprisingly, have a long fossil record despite their gelatinous body, and a group of extinct animals called conodonts that were important nekton during the Paleozoic and early Mesozoic.

JELLYFISH

Of marine animals, one would expect that **jellyfish** (phylum Cnidaria) would be among the least likely to ever be preserved as a fossil. The body consists of more than 90 percent water, and there are no hard parts (Figure 11.14). However, jellyfish have a remarkable fossil record; the oldest ones are probably from the Ediacaran fauna (late Precambrian) and from the Cambrian soft-bodied faunas, such as the Burgess Shale (Middle Cambrian). Apparently, when a jellyfish dies, it undergoes dehydration, so that as water is lost the tissue becomes more dense and compact. Under the right conditions and in the absence of scavengers, ancient jellyfish were preserved mostly as either molds or impressions showing the outline of the body and tentacles. Despite this long and impressive history of fossil jellyfish, the record is predictably rather spotty. Among the most diverse and well-known localities for fossil jellyfish are two famous soft-bodied faunas: the Mazon Creek from Pennsylvanian strata of Illinois and the Solenhofen Limestone from the Jurassic of Germany. Mazon Creek jellyfish are preserved as carbon films in ironstone concretions. Presumably, these animals were blown into a near shore setting, buried, and incorporated into concretions. Similarly, the Solenhofen jellyfish, along with other spectacular fossils discussed later in this book, were probably blown into a non-normal marine setting where special preservation conditions were present. Some specimens from the famous Solenhofen Limestone have traces of the original organic material and imprints of muscles preserved.

Figure 11.14 A fossil jellyfish preserved as an impression in a concretion from the Pennsylvanian of Illinois. The body, or float part, is at the left and several clusters of tentacles are to the right. The specimen is about 10 cm in diameter. (Courtesy of Field Museum of Natural History, Chicago.)

CONODONTS

Conodonts have been among the most enigmatic of fossils. Their fossil remains consist of tiny toothlike elements composed of calcareous apatite, similar to the mineral of which bone is made (Figure 11.15) Despite this mineralogic affinity, conodonts have historically not been thought to be directly related to the vertebrates.

These small fossils commonly occur in rocks as single, isolated elements, but rare specimens show that, in fact, each conodont had a complex skeleton composed of numerous elements arranged into a bilaterally symmetrical apparatus (Figure 11.16). The elements of this apparatus grew from the center outward and, thus, are thought to have been enclosed in soft tissue, at least when they were not in use. Because the skeletal elements of conodonts are found in all sedimentary rock types from black shales to conglomerates, they were probably parts of animals that swam.

The biological affinity of conodonts has been a matter of considerable research and debate. In recent years, a few soft-bodied fossils preserved as carbon films have been recovered with conodont elements inside. In each case, these were heralded as complete conodonts. The first such claim was based on a specimen from the Mississippian of Montana (Figure 11.17). However, rather than being a conodont, this

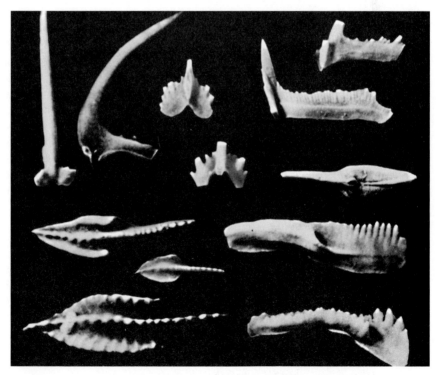

Figure 11.15   A variety of different kinds of conodont elements of Early Mississippian age from Indiana. These individual fossils are about 0.5 mm long. (Courtesy of Carl B. Rexroad, Indiana Geological Survey.)

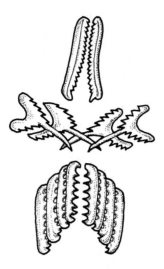

Figure 11.16 An assemblage of conodont elements, schematic and highly magnified, showing the close association of several kinds of discrete elements, with bilateral symmetrical arrangement. (Modified from Rhodes, 1954.)

fossil is now considered to be an unknown organism that ate a conodont, with the individual skeletal elements situated in the gut of this soft-bodied creature.

More or less complete conodonts have apparently been discovered in Lower Carboniferous rocks in Scotland (Figure 11.18). Several examples of this fossil are known. The best specimen is approximately 40 millimeters (1.5 inches) long, very slender, and seemingly divided into segments with the assemblage of skeletal elements located in the head. The elements are thought to have served as teeth or as grasping structures. Two possibilities seem most plausible for the affinities of these animals. First, it is possible that the Scottish specimens represent a group of animals that is different at the phylum level from all other known animals. Alternatively, these fossils might be assignable to an existing animal group. The latter hypothesis is accepted by many today, and this small elongate organism containing small phosphatic elements is thought by many to be a chordate because of its general shape, segmentation, fins, and several features of its internal anatomy. However, some of the diagnostic chordate characters are lacking, perhaps due to preservation. If conodonts were chordates, they represent a parallel development to the jawless agnathan fishes.

Figure 11.17 A carbon-film impression of a conodont-eating animal approximately 6 cm long. (Courtesy of W.G. Melton, Jr., University of Montana.)

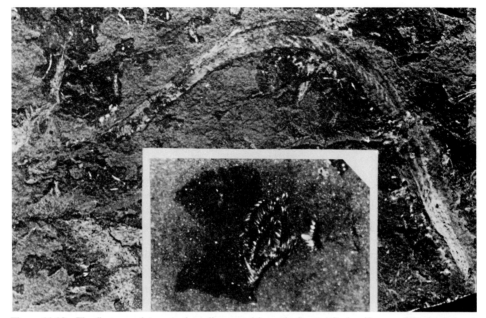

Figure 11.18   The first complete conodont discovered from the Lower Carboniferous of Scotland. Inset shows the opposite side of the head with position and orientation of the skeletal apparatus. The animal is approximately 40 mm (1.5 in) long. (Courtesy of R.J. Aldredge, University of Leicester.)

Conodont elements are very abundant in many marine rocks, which suggests that these animals were relatively common during the Paleozoic. The oldest conodonts known, broadly defined, lived during the Cambrian Period, and conodonts became extinct at the end of the Triassic.

Conodonts are exceedingly important fossils. They are among the best biostratigraphic index fossils for the Paleozoic and Triassic, and their elements are reliable "geothermometers." Conodonts have all the necessary characteristics of ideal biostratigraphic fossils. Because they were nektonic, they are geographically widespread, and they evolved rapidly through their history. Furthermore, the acid-resistant calcareous apatite composition of their skeletal elements means that they can be easily isolated from limestone and dolomite samples. These rocks can be dissolved in acid, and conodont elements remain as an insoluble residue. Their small size means that they can be easily collected from relatively small samples, and conodonts can even be expected from rock chips recovered from drilling a well. As a consequence of these characteristics, the Upper Cambrian through Triassic is subdivided into biostratigraphic zones based on conodonts.

Conodont elements also act as geothermometers because their calcareous apatite elements include films of carbon that change color (in the sequence amber-gray-black-glass-clear) with increased temperature. This color change is understood in well-calibrated temperature intervals, so the historical heating of a rock unit can be determined easily. This is useful for interpreting the geologic history of an area and for determining whether the oil and gas has been baked out of a particular rock unit.

KEY TERMS

acritarchs	diatoms	parallel evolution
branch reduction	diatomite	phytoplankton
chalk	foraminifera	radiolarians
coccoliths	graptolites	silicoflagellates
conodonts	jellyfish	zooplankton
cyst	nekton	

READINGS

Briggs, D.E.G.; Clarkson, E.N.K.; and Aldridge, R.J. 1983. "The Conodont Animal." *Lethaia* 16:1–14. This is the initial report on the fossil that is now believed to be the conodont animal.

Haq, B.U., ed. 1983. *Calcareous Nannoplankton.* Benchmark Papers in Geology. Hutchinson Ross. 338 pages. A series of articles on very small marine plankton with calcareous skeletons, primarily coccoliths.

Moore, R.C., ed. 1953–1997. *Treatise on Invertebrate Paleontology.* University of Kansas Press and Geological Society of America. This important series of monographs includes descriptions and illustrations of all known invertebrate and protist fossils. Numerous volumes that have appeared so far include revised volumes. Each book includes general chapters on evolution, ecology, and classification. Those volumes of interest in connection with this chapter include Part C on foraminifera; Part 2 on radiolarians; Part F on coelenterata; Part V on graptolites; and Part W on Miscellanea, which has four chapters on conodonts.

Ocean Drilling Program Proceedings. National Science Foundation. This series is a continuation of the reports of the Deep Sea Drilling Program (now the Ocean Drilling Program). More than 120 volumes of results have been published over the past two decades. This includes many papers on all skeletal marine plankton, plant and animal. This is by far the most comprehensive and up-to-date series on fossil plankton and nekton.

Palmer, D., and Rickards, B. 1991. *Graptolites, Writing in the Rocks.* The Boydell Press. 182 pages. Well-illustrated book to introduce graptolites to the amateur and professional paleontologist.

Sweet, W.C. 1988. *The Conodonta, Morphology, Taxonomy, Paleoecology, and Evolutionary History of a Long-Extinct Animal Phylum.* Oxford Monographs on Geology and Geophysics No. 10. 212 pages. A thorough account of the evolutionary history of the conodonts.

CHAPTER 12

Marine Predators

INTRODUCTION

In this chapter we will primarily discuss three major groups of animals that were dominant predators, occupying second and third consumption levels within marine communities. Predator is used here to refer to animals that actively pursue their prey. These groups are arthropods, cephalopods, and fish. We will also discuss, very briefly, two other groups of vertebrates that evolved from the terrestrial environment back to an aquatic existence: marine reptiles and marine mammals. These were important predators during the Mesozoic and the Cenozoic, respectively. Every species of each group was probably not actively engaged in predation of larger animals. Earliest fish were undoubtedly suspension or deposit feeders, and other fish were surely scavengers, feeding on dead organisms. Nevertheless, the majority of species of these groups were predaceous and were at the highest trophic level of their marine ecosystem.

PREDATOR HISTORY

The Phanerozoic history of marine predation can be subdivided in two ways: the predators and the inferred predation pressure. Five major predator phases are identified (Figure 12.1). Cambrian predators are relatively poorly known, with the exception of *Anomalocaris*, because they were soft-bodied arthropods, worms, and arthropodlike organisms known only from exceptionally preserved faunas. Nautiloid cephalopods and eurypterids were the major Ordovician to Silurian predators. The radiation of jawed fish during the Devonian ushered in the third phase (Devonian to Permian), which was dominated by fish and ammonoid cephalopods. Through the entire Mesozoic, bony fish, ammonite cephalopods, sharks and rays, ichthyosaurs, plesiosaurs, mosasaurs, and crustaceans (crabs and lobsters) formed the fourth phase of predators. By the close of the Cretaceous, ammonite cephalopods, ichthyosaurs, plesiosaurs, and mosasaurs had all became extinct. The final marine predator phase was composed of bony fish, sharks and rays, crustaceans (crabs and lobsters), whales, and dolphins (Figure 12.1).

In addition to these changes in dominent predators, paleontologists have identified three phases of increased predation from the Cambrian until today, based on a variety of indirect measures (successful predation leaves no evidence in the fossil record). The first phase, from the Cambrian through the Silurian, was the interval with the lowest predation intensity. Ammonoid cephalopods and fish were apparently much improved predators, so that the intensity increased substantially during the Devonian to begin the second phase. This lasted through the Permian. Finally, the third phase of increased predation pressure began during the Triassic and includes the radiations of ammonite cephalopods and the continuing radiations of bony fish and crabs and lobsters.

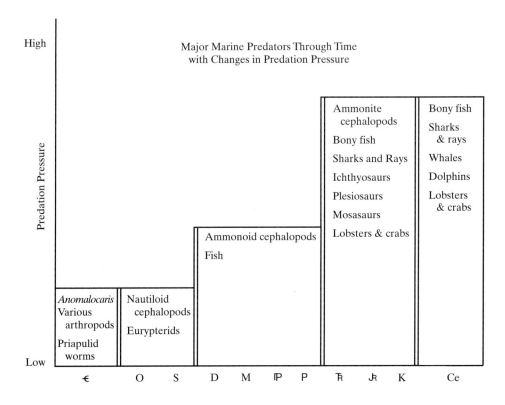

Figure 12.1 Temporal history of phases of predation through the Phanerozoic, which indicates relative changes in predation intensity through time.

ARTHROPODS

ANOMALOCARIS, PRIAPULIDS, AND OTHER ARTHROPODS

The predator ***Anomalocaris*** is probably the most celebrated of the Burgess Shale organisms (see Chapter 9). This animal is sufficiently large (up to 50 centimeters) that individual body parts were originally described as different organisms. The front grasping appendages (Figure 12.2) were named *Anomalocaris* and considered the body of an unusual shrimplike arthropod for which a head had never been recovered. Laggania was the name given to the nondistinct body, which was believed to be a holothurian echinoderm; and the unique mouth structure was named *Peytoia* and regarded as an unusual jellyfish. It was not until 1985 that the true relations of these separate pieces were sorted out. The complete *Anomalocaris* must have been a formidable predator. Arcuate scars on Cambrian trilobites that nearly became *Anomalocaris* meals undoubtedly record the predator's circular mouth structure.

Interpretation of other Burgess Shale organisms can be conjectural, but it certainly appears that **priapulid worms** (at least five species recognized from the Burgess Shale) were predators during the Cambrian, as they are today. Specimens of *Ottoia* are preserved with hyoliths in the gut (see Figure 9.8A), convincing evidence of predation. Arthropods like *Sidneyia* and *Sanctacaris* were probably predators, as were organisms

Figure 12.2   Reconstruction of *Anomalocaris*, the Cambrian swimming predator. Note the grasping arms and mouth structure that were once thought to be separate organisms. (D. Collins. 1996. *Journal of Paleontology*. Used with permission of the Paleontological Society.)

whose phylum affinities are not known, such as *Opabina* and, of course, *Anomalocaris* discussed above.

EURYPTERIDS

Eurypterids were Paleozoic (Ordovician to Permian) predaceous arthropods, although they were only thought to be dominant predators during the Ordovician and Silurian (Figure 12.1). The cuticle of eurypterids was not calcified, so their preservation was dependent upon suitable conditions. Eurypterids may look like scorpions, and some of the smaller specimens are approximately the same size—10 centimeters or less. Typical eurypterids are 10 to 20 centimeters (4 to 8 inches); however, the giant eurypterids reached lengths as long as 1.8 meters (5.8 feet). Eurypterids are thought to have been benthonic animals that could either walk on the sea floor or swim in the water column. Appendages vary from paddlelike devices for swimming to claws for grasping prey (Figure 12.3, also see Figure 3.16).

CRUSTACEANS

Crustaceans evolved during the Paleozoic, but it was not until the Jurassic that lobsters and crabs developed claws and underwent an adaptive radiation. From this time onward, crustaceans were formidable, shell-crushing predators that played an important role in marine communities. Many morphological features of Mesozoic and

Figure 12.3　Large, partially disarticulated eurypterid, *Truncatiramus*, from the Middle Silurian of northwestern Europe. Magnified view of grasping appendage, the chelicera. (Courtesy of R.E. Plotnick, University of Illinois, Chicago.)

Cenozoic gastropods have been attributed to an antipredatory function. Presumably, shelled invertebrates evolved ever more sophisticated defenses against crabs and lobsters, which became more effective predators.

Unlike many Paleozoic arthropods, crabs and lobsters have heavily calcified exoskeletons, so they are well preserved.

CEPHALOPODS

Among the best known and most common shelled predators in the fossil record are the **cephalopods**, represented today by squids, octopuses, and one living representative with an external shell, the chambered nautilus. Although cephalopods are known from the Late Cambrian, they are very rare. Beginning during the Ordovician, **nautiloid cephalopods** underwent an extensive adaptive radiation, evolving many different groups, some of which included forms that reached very impressive sizes, up to 10 meters long. They mostly had a long, conical, tapering shell that was divided into a series of chambers. Others had a slightly curved shell or one with small, tapering early chambers and inflated adult chambers (Figure 12.4). Although very diverse and common, most of these early nautiloids were probably rather sluggish swimmers that swam near the sea floor. They may have preyed on a variety of bottom-dwelling invertebrates, such as trilobites, brachiopods, and echinoderms.

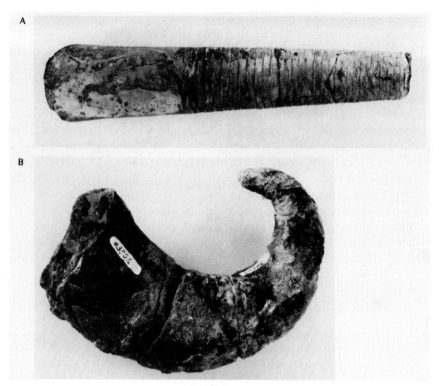

Figure 12.4   Two fossil nautiloid cephalopods of Paleozoic age. (A) Specimen from the Ordovician of Ohio that has a straight shell. The final living chamber is at the left, and watch-glass-shaped partitions (septa) divide the shell into chambers that can be seen at the right. The edges of these simple septa are visible because the outer shell is not preserved. (B) Silurian-age specimen is openly coiled rather than straight; the chambers cannot be readily seen. Both specimens are considerably reduced in size. (Courtesy of National Museum of Natural History.)

By Devonian time, the nautiloids had already passed their peak of evolution. Many of the Ordovician and Silurian varieties became extinct. The Devonian witnessed the origin of another group of cephalopods that evolved from the nautiloids, the **ammonoid cephalopods**. These rapidly became the most common fossil cephalopods; they were so diverse during the Mesozoic that it is sometimes called the Era of Ammonoids. Ammonoids differ from nautiloids mainly in the nature of the partitions that divide the shell into chambers. In nautiloids these partitions, or septa, are simple with straight edges. In ammonoids the edges of the septa become fluted and are arranged in a series of waves, so that the edge of the septa, as seen from the outside with the shell removed, traces a series of folds. The functional significance of this evolutionary change has been much debated. The most generally accepted theory is that the fluted septa edge helps strengthen the shell, avoiding crushing or implosion of the shell if the animal changes its living depth rapidly. The chambers provided buoyancy, being partly gas filled, and external water pressure on them can be severe. Most ammonoids and some nautiloids, especially advanced ones, have the chambers coiled so that they are above the living chamber. This helps keep the animal upright in water, with the lighter gas-filled living chamber located above the heavier animal.

Although ammonoids were small and not very common during the Devonian, they rapidly increased in size and abundance through the remainder of the Paleozoic. Their most conspicuous change was the increasing complication of septal shapes. The wavy septa developed small, secondary crinkles, first on every other fold and then on each fold. The primitive simple type is called a **goniatite septum** (Figure 12.5), the one with secondary crinkles on alternating folds is called a **ceratite septum** (Figure 12.6), and the most complicated forms are called an **ammonite septum** (Figure 12.7). By Permian time, most ammonoids had evolved the ammonite septum. If the prime impetus for this evolutionary change was to strengthen the shell, perhaps some of the more advanced forms must have dived to increasingly greater depths. The ammonoids underwent a crisis at the close of the Permian. Most of the forms that had been common during that period of time became extinct, and only a few genera survived to provide the ancestral stock for Triassic ammonoids. All the ones with complicated ammonitic septa became extinct; the survivors had ceratite septa. More than 300 genera of ceratite-type cephalopods are known from the Triassic, and ammonite-sutured forms evolved again from these forms. Ammonoids underwent another extinction crisis at the close of the Triassic; all cephalopods with ceratite septa became extinct at that time. A few forms with the advanced ammonite septa persisted into the Jurassic, resulting in another burst of ammonoid evolution. Several hundred genera are known from the Jurassic.

During the Jurassic and Cretaceous, ammonoids reached their peak of abundance, diversity, and rapidity of evolution. They are used for biostratigraphic correlation of marine rocks of these ages on a worldwide basis. The Jurassic Period is divided into about 74 zones based on fossils, and intervals of about 800,000-years duration can be discerned with their use. During the Cretaceous, some ammonoids became exceptionally large, their coiled shells reaching 2 meters in diameter. Another group of Cretaceous ammonoids came "unstuck," so to speak. In some lineages, the regular coiled shell changed to one of several other varieties. Some of these ammonoids were coiled, then straight, then had a half turn at the end, producing a canoe-shaped shell with the animal directed back toward the early part of the shell (Figure 12.8). Others had a helically coiled shell like that of the gastropods, and still others had the shell in the form of an irregular knot. These types are especially common in Cretaceous rocks, although a few unrelated forms do occur during the Jurassic. These unusually coiled ammonoids are called **heteromorphs** because they depart from the normal coiled shell. For some reason, the normally closely controlled growth and coiling program that built ammonoids went completely awry in heteromorphs.

The ammonoids became extinct at the very end of the Cretaceous. Why they became extinct has been a long- continuing puzzle in paleontology. Until they disappeared, they were abundant and diverse, yet they vanished abruptly from marine rocks. Because some of the youngest ammonoids are highly ornamented or heteromorphic, some paleontologists have suggested that these ornaments signal the racial senescence, or old age, of the ammonoids. Yet, these heteromorphs were surely successful animals during their time. They were common, and they survived for millions of

Figure 12.5   A primitive ammonoid, *Tornoceras*, from the Devonian, has a few very simple wavy septa dividing the shell into chambers. These are goniatite septum. As in other photographs, the outer shell is not preserved; the specimen is 4.4 cm high. (Courtesy of Takeo Susuki, U.C.L.A.)

Figure 12.6   A typical Triassic ammonoid, *Submeekoceras*, with ceratite septa. Alternate waves of the septa have secondary crenellations on them. The specimen is from the Lower Triassic of Idaho. (Courtesy of Takeo Susuki, U.C.L.A.)

Figure 12.7   An ammonoid with complex ammonite septa. This Late Cretaceous specimen of *Placenticeras* is from South Dakota and is 22 cm high. (Courtesy of Takeo Susuki, U.C.L.A.)

Figure 12.8   One of the heteromorph ammonoids, *Hamites*, from the Cretaceous of England. The early growth stages are to the right (partly broken away on this specimen), and the mature part is to the left. Length along the outside curve is approximately 90 cm. (Courtesy of Field Museum of Natural History, Chicago.)

years. Another more plausible explanation for ammonoid extinction is that first nautiloids, then ammonoids were unable to compete successfully with fish. It is certain that true fish underwent an extensive adaptive radiation during the Mesozoic, and they may have competed directly with ammonoids by preying on them and indirectly by competing for the same food supply. However, if such competition did adversely affect the ammonoids, it would be more likely that they would have slowly dwindled away instead of suddenly disappearing after having been so common. Another possibility is that the cause of dinosaur extinctions at the close of the Mesozoic (discussed in Chapter 15) also caused the final, complete extinction of the ammonoids.

FISH

An event that was to have enormous consequences on the nature of marine communities was the appearance of the fish during the Cambrian. The first fish were small, only a few centimeters long, and harmless animals at best. They were not predators but, rather, suspension or deposit feeders, and so they did not compete directly with the large cephalopods. Our knowledge of Cambrian and Ordovician fish is scanty. Cambrian fish remains consist entirely of isolated plates and spines. Ordovician fish are also known primarily from isolated plates, although a few partial specimens have been found. Ordovician fish are best known from western North America and Australia. The most famous of these localities is the Harding Sandstone in the Rocky Mountains, from which thousands of plates have been recovered. More complete specimens are known from younger rocks.

The earliest fish are jawless. The front end of the gut, the mouth, was simply a round, immovable hole. These animals apparently sucked water, mud, or both into the pharynx and expelled the water through numerous gills that were separate external openings. The gills presumably served as a filter or screen to strain small food particles out of the water or mud; they also absorbed oxygen from the water. Thus, gills were initially dual-purpose structures used for both feeding and respiration. The early fish were small (2 to 6 inches) and generally had flattened bodies with a heavy external armor of bony plates. They lacked an internal bony skeleton.

The early fish are the oldest known vertebrates (phylum Chordata, the phylum of animals characterized by a dorsal nerve cord). This jawless group belongs to the class **Agnatha** (*A* for without, *gnatha* for jaws) (Figure 12.9). The armored agnathans are known from the Ordovician through the Devonian, toward the end of which they became extinct. Before their extinction, however, another group of jawless vertebrates evolved from them. This group lacks any bony skeleton, having an internal skeleton of soft cartilage that is very rarely preserved as a fossil. Two kinds of these soft, jawless fish are still living: the lamprey eel and the hagfish, both of which are parasitic on other fish. In addition, lampreys are known from carbon-film impressions in Pennsylvanian strata.

During the Late Silurian and Devonian, the major groups of jawed fish all appeared in the fossil record. In fact, the Devonian Period was a time of rapid evolution of fish and is sometimes called the Age of Fish. The **placoderms**, generally regarded as the most primitive of the jawed fish, are first known from the Lower Devonian. These armored fish were active predators, although still small during the Early Devonian and no match in size for the cephalopods. The presence of movable, biting jaws is a clear indication of their predaceous life style. How did these jaws, which characterize all the rest of the vertebrates both in water and on land, come about? We mentioned that in the agnathans there are numerous gills, each with a separate opening (Figure 12.10). Each of the fleshy walls between the gills contains a series of small bones that supports and strengthens the partitions between the gill chambers. These bones are called **gill arches**. Modern fish have only five pairs of gills, but agnathans have seven or nine gill pairs. The bony gill arch of one of the front gill pairs gradually moved forward until it was positioned in the side of the face just behind the mouth and ultimately became incorporated into the rim of the mouth. The bones of each arch are hinged near the center so that the lower half of each arch became one half of the upper jaw. The mouth could then open and close on this hinge. Gradually, bony teeth developed on these bones that had become jaws. Thus, the typical bony jaws, teeth, and mouth evolved from agnathans. Detailed comparative anatomy suggests that differently jawed fish may have evolved separately.

In placoderms the evolution of jaws signaled other changes in the skeleton that are related to changes in feeding methods. The heavily armored agnathans were undoubtedly slow, sluggish bottom swimmers; they did not need to be active to obtain their diet of microscopic food particles or sediment. Alternatively, the placoderms took

Figure 12.9   Bottom view of an articulated specimen of a Devonian agnathan, *Cardipeltis.* The heavy, bony scales provide an armor over the body and tail. (Courtesy of Field Museum of Natural History, Chicago.)

Figure 12.10   Bottom view of a model of a primitive agnathan, *Hemicyclapsis*. The body is covered with heavy, bony scales; the mouth is a simple hole, lacking jaws; and there are numerous separate gill openings around the margin of the head. The peculiar lateral fins are unlike those of more advanced fish. Original specimens are about 20 cm long. (Courtesy of Field Museum of Natural History, Chicago.)

up an active life of predation. Placoderms changed from a flattened body shape suitable for slow movement over the sea floor to a more streamlined, fusiform, or torpedo-like body shape. The heavy armor became progressively reduced, allowing the fish to swim faster. There were also changes in the structure of the tail, the main propulsion organ of fish, which permitted more efficient swimming motions. The fins on the sides, top, and bottom underwent a series of changes that allowed increased efficiency in balancing, braking, and making sharp turns in water. All these changes were advantageous to active predators such as the placoderms, and some of these fish were quite formidable (Figure 12.11).

In addition to placoderms, two other major groups of fish evolved during the Devonian, both with internal skeletons. These are the **Chondrichthyes** (cartilaginous fish, or sharks and rays) and the **Osteichthyes** (the advanced bony fish), which together comprise virtually all modern fish. **Sharks** and **rays** are cartilaginous because their

Figure 12.11   Reconstruction of the skull and neck region of the giant placoderm *Dunkleosteus*, best known from the Middle Devonian of Ohio. Complete specimens reached 10 m in length. The peculiar jaws, with tusks and shearing edges, are quite unlike the jaws of other fish. (Courtesy of Field Museum of Natural History, Chicago.)

internal skeletons are composed exclusively of cartilage, not bone. Their only bones are their teeth, in spines that support the leading edges of some fins, and in tiny plates embedded in the skin, called **dermal denticles**. (These give a sandpaperlike feel to the skin of living sharks.) When a shark died, only these small parts of the animal were likely to be preserved. The oldest sharks are Upper Silurian in age. The Silurian and Devonian record of sharks is based almost exclusively on isolated teeth and spines. Sharks and rays diversified during the upper Paleozoic and later. Still, the bony teeth, spines, and dermal denticles are the most common chondrichthyes fossils, but rare complete specimens are known from sites of exceptional preservation.

Osteichthyes (**bony fish**) have been the most diverse fish throughout the Mesozoic and Cenozoic, and primitive bony fish are actually the oldest jawed fish. Bony fish differ from placoderms in a variety of ways. Bony fish lack the heavy external armor, having a series of small, lighter weight scales instead. In addition, they have an internal skeleton of bone that is lacking in most placoderms and chondrichthyes. The structure of the tail is consistent, and the number and arrangement of lateral, top (dorsal), and bottom (ventral) fins is fixed.

Among the advanced bony fish, there are two main groups. The most common and diverse are the **ray-finned fish** that have many thin, parallel bones supporting the fins. The second group, the **lobe-finned fish**, includes the living **coelacanth** (see Figure 7.8), the lungfish, and the Paleozoic rhipidistians. The lobe-finned fish have thick, fleshy fins with a central axis of larger bones that are the primary support of the fins.

The lobe-finned fish were never very diverse or common. However, they are especially significant because the rhipidistians evolved directly into the first tetrapod (four-legged) vertebrate animals that lived on dry land, i.e., amphibians.

The ray-finned fish constitute all the common living fish with which you are familiar, except for sharks and rays. They have been and are today enormously successful animals (Figure 12.12). They underwent a series of evolutionary changes during the late Paleozoic and especially during the Mesozoic. The tail became shorter and reduced in size. The bones that supported the lateral fins were reduced in number, and the scales became thinner with fewer bony layers. The most conspicuous changes occurred in the nature of the mouth and in the position of the lateral fins. Older bony fishes had a mouth that was long and slitlike, parallel to the main axis of the body. The mouth progressively shortened, with the jaw hinge shifting downward, so advanced types always look a little "down in the mouth." The front pair of lateral fins shifted upward in position, closer to the backbone, and the hind pair of fins moved forward until they were directly under the front fins or even slightly ahead of them. By the close of the Cretaceous, most modern groups of fish had appeared in the fossil record. Holdovers from the more primitive early Mesozoic grades of fish include living sturgeons (the source of caviar) and paddlefish. Intermediate grades include freshwater garpikes and the bowfish, both of which are rapidly approaching extinction in major rivers of central North America.

The fish added a new dimension to the complexity of marine communities. They rapidly took over as the dominant carnivores of the sea—a role they still play. Apparently, the nautiloid cephalopods were not able to stand up to this direct competition, so that by the end of the Devonian, they had started to dwindle in both variety

Figure 12.12   A giant bony fish, *Xiphactinus*, approximately 3 m long, from the Cretaceous chalk of western Kansas. A small bony fish is preserved in the rib cage. (Courtesy of Sternberg Museum of Natural History, Fort Hays State University, Hays, KS. Photograph by E.C. Almquist.)

and abundance. The success of Devonian fish is epitomized by the giant placoderm of the Middle Devonian, *Dunkleosteus*, that reached a length of 10 meters (Figure 12.11). Eventually, the ammonoid cephalopods also suffered from this competition.

LATER OCEANIC PREDATORS

Before we conclude this discussion of marine predators, we must mention two groups that came on the scene quite late in the fossil record. These are the marine reptiles of the Mesozoic and the marine mammals of the Cenozoic. Not all these animals were active predators—for example, marine turtles are herbivores—but the great majority were.

MARINE REPTILES

Marine reptiles evolved from terrestrial reptiles during the Mesozoic, acquiring an aquatic existence as a secondary adaptation. The giants of the Mesozoic seas were reptiles, as were the land giants of that time, the dinosaurs. Several groups were represented. Large marine turtles were present, especially in the Cretaceous chalk beds of western Kansas and Nebraska. Some of these turtles reach 4 meters (13 feet) in length. Another conspicuous group is the **ichthyosaurs** (Figure 12.13). These were dolphinlike reptiles (or perhaps we should say that the dolphins are ichthyosaurlike mammals). They were streamlined, probably highly active predaceous animals. Mostly small ichthyosaurs are found in Triassic strata, although some in Nevada are 15 meters long. The peak of diversity and abundance was during the Jurassic, when they reached lengths of 7 meters. Ichthyosaurs are probably best known from Jurassic beds of Germany. Other marine reptiles include large marine lizards, **mosasaurs**, with large heads and short necks that reached 10 meters in length. The giants of the Cretaceous seas were the **plesiosaurs**, most of which had very long necks and tails but small heads (Figure 12.14). Some reached up to 16 meters in length. The limbs of these latter reptiles were modified into large, paddlelike fins in which the finger bones, the phalanges, were multiplied over and over again, so that there might be as many as 50 bones in a paddle. Plesiosaurs used these large fins to effectively "fly" through the water.

Figure 12.13 An ichthyosaur specimen of Jurassic age from Germany, much reduced. Notice the large number of bones that support the lateral fins and the way the vertebral column extends into the lower lobe of the tail. (Courtesy of National Museum of Natural History.)

During the Mesozoic, these and other reptiles underwent an extensive radiation in the marine environment; these groups independently took up a fully aquatic or a semiaquatic existence. Many of these reptiles must have been at or very near the top of the marine food chains of the Mesozoic. They probably fed mainly on fish and ammonites. However, along with their reptilian counterparts on land, all these large marine reptiles, with the exception of the turtles, became extinct prior to the close of the Mesozoic.

MARINE MAMMALS

Several groups of marine mammals that are active predators today have a fossil record from the Cenozoic Era. Seals actively feed on fish and squid. Some sea otters eat sea urchins as a major part of their diet. Walruses are also marine predators. However, the most diverse group of predatory marine mammals are the whales and their relatives. Dolphins and porpoises are active swimmers that feed mostly on fish and squid. Among whales, sperm and bottlenose whales feed mostly on squid, and killer whales feed on large squid. These are all toothed whales. In contrast, whalebone whales mainly strain seawater through the long strands of baleen in their mouths and feed on

Figure 12.14 A short-necked plesiosaur about 3 m long from the Cretaceous chalk of western Kansas. A large number of bones form the oarlike paddles. (Courtesy of Sternberg Museum of Natural History, Fort Hays State University, Hays, KS.)

zooplankton as well as small fish and crustaceans up to 4 centimeters long. The right whale simply opens its mouth and lets water flow through the baleen as it swims forward. The other type of whalebone whale, the rorqual, feeds in just the opposite way; it takes in a mouthful of seawater and then expels it, thus trapping phytoplankton in its baleen.

KEY TERMS

Agnatha
ammonite septum
ammonoid cephalopods
Anomalocaris
bony fish
cephalopods
ceratite septum
coelacanth
Chondrichthyes

crustaceans
dermal denticles
eurypterids
gill arches
goniatite septum
heteromorph
ichthyosaurs
lobe-finned fish
mosasaurs

nautiloid cephalopods
Osteichthyes
placoderms
plesiosaurs
priapulid worms
ray-finned fish
rays
sharks

READINGS

Carroll, R.L. 1988. *Vertebrate Paleontology and Evolution.* W.H. Freeman and Company. 698 pages. A comprehensive survey of the phylum Chordata, with many illustrations and considerable detail.

Schultze, H.P., ed. 1986. *Handbook of Paleoichthyology.* Gustaf Fischer. A series of 10 planned volumes, each on a major group of fossil fishes. Very well illustrated.

Wiedmann, J., and Kullman, J. 1988. *Cephalopods Present and Past.* E. Schweitserbart. 765 pages. A series of papers by various authors on living cephalopods, especially *Nautilus*, and reconstruction of soft-part anatomy of fossil cephalopods.

Origin and Early Evolution of Terrestrial Communities

INTRODUCTION

Tremendous differences exist between Cambrian deposit-feeding communities dominated by trilobites, epifaunal suspension-feeding communities during the remainder of the Paleozoic, and mollusc-dominated communities of today. In tracing this record through more than 600 million years, one aspect is outstanding: the marine record, especially for benthonic life, is virtually continuous. For nearly every time interval, sedimentary rocks somewhere on Earth reveal what sea-floor life was like during that time. The greatest gap in our knowledge of marine life occurs close to the Paleozoic-Mesozoic boundary.

In sharp contrast to the marine record, the fossil record for terrestrial life is discontinuous; it contains gaps, even in younger rocks. There are a number of reasons for this. First, the area where terrestrial sediments are deposited is much smaller than the area of marine deposition. Also, once sediments are deposited on land—for example on a floodplain or in a lake—they are much more likely to be eroded than are marine sediments. Even after terrestrial sediments and their fossil content become rocks, they are gradually elevated, on average, to a somewhat higher position than marine rocks of the same age. Later during Earth history, if there is a period of uplift and erosion, the terrestrial rocks are eroded first or may be eroded selectively, leaving behind marine rocks at a lower elevation. For these reasons, our fossil record of terrestrial rocks is not as complete as the marine record.

Despite these problems of preservation for the terrestrial record, paleontologists have assembled a remarkable understanding of the appearance and evolution of plants and animals on land from scattered river, flood-plain, and lake deposits. We will begin our discussion of ancient life by considering separately the origin of land plants and the origin of animals. These events are separated in time; furthermore, each deserves separate treatment because of the complex evolutionary changes necessary for life to make the transition from submersion in an aqueous environment to existence in air.

In this chapter we will discuss the origins of these communities. In subsequent chapters, primary producers (plants), primary consumers (herbivores), and, finally, secondary consumers (carnivores) will be considered.

ORIGIN AND EARLY EVOLUTION OF VASCULAR PLANTS

THE TRANSITION FROM WATER TO LAND

The problems that are involved in making the transition from water to land are formidable for any organism. There are two possible pathways by which such a change can take place. One is a direct transition from marine waters to dry land—an approach that might be called "over the beaches." The other way is to first adjust from salt water to fresh water and then to make the transition from fresh water to dry land. Both pathways have been followed, but the latter approach is generally considered easier for most plants and animals because it lessens the major difficulty of having to accommodate the differences in the osmotic pressure of cells and tissues from salt to fresh water. If that difficulty can be handled while an organism is immersed in water—for example, by adapting to fresh water along an estuary, then adjusting to rain water as a source of moisture on land—the transition to land is more easily made. Two major transitions onto land, those of vascular plants and tetrapods, seem to have taken the route through freshwater.

What are other problems associated with making this transition (Table 13.1)? A plant living in water is buoyed by the water, which is a relatively dense medium compared to air. (Is it easier to run in air or under water?) Under water, a plant does not need to support itself fully to be erect. However, in the less dense medium of air, a plant or animal needs some kind of stiff, supportive tissue. Imagine yourself on a beach with a long piece of seaweed and a long branch of a buckeye tree. You can stick the branch into the sand and it will remain erect. However, if you try to put the seaweed upright, as it was when it was in the water, it will flop over and become covered with sand. The problem of support had to be solved in animals by their skeletal system, which, with modification, could perform this function. Plants solved the problem of support, along with fluid circulation, as described below.

Water-dwelling plants and animals do not need any special way to obtain water for photosynthesis or to maintain water balance in their tissues; they live in "100-percent humidity" and can take water directly from their immediate environment. For organisms living in air, the water problem can be severe. In most environments air does not contain enough moisture to be the sole water source for plant survival. The most reliable source of moisture for a plant is from water trapped in soil. Thus, a land plant needs some kind of special structure to take water from the ground. Because rootlike structures were present in aquatic algae to anchor the algae to the ocean bottom, this was a likely candidate to develop into a water-uptake system. For animals that are mobile, getting to water is less of an issue, but animals typically cannot wander too far from a water source.

Table 13.1 MAJOR DIFFICULTIES TO OVERCOME IN THE TRANSITION FROM AN AQUATIC TO A TERRESTRIAL EXISTENCE. IF TRANSITION WAS DIRECTLY FROM THE OCEANS, THE OSMOTIC BALANCE FROM SALT WATER TO FRESH WATER MUST ALSO BE CONSIDERED.

Problem	Solution	
	Early Vascular Plants	Amphibians
Structural support	Vascular tissue	Changes to skeletal system already present
Circulation	Vascular tissue	(Not applicable)
Obtaining water	Vascular tissue in roots	(Not applicable, but needed to stay near a water source)
Desiccation	Cuticle with stomata	Not solved by amphibians[1]
Reproduction	Not fully solved by rhyniophytes[2]	Not solved by amphibians[3]
Breathing	Stomata	Lungs of rhipidistians

[1]Scales on reptiles solved the desiccation problem faced by amphibians.
[2]Not until the development of the seed did plants become more completely independent of moist habitats for reproduction.
[3]Reproduction on land first solved in tetrapods by the amniote egg of reptiles.

Once water is obtained, it must be distributed throughout the body. Again, this was not an issue for animals that had a circulatory system and internal organs, but modifications were required for plants. In plants, the above-ground parts, which are exposed to sun, are obviously those parts of the plant involved in photosynthesis. Thus, water taken into the plant at its roots is needed for photosynthesis at the upper extremities of the plant. Some sort of circulatory system was necessary to move water and dissolved minerals and salts throughout the plant and to distribute the manufactured carbohydrates of photosynthesis within the plant. Land plants evolved a special system of conductive tissues, called **vascular tissue**, to perform the functions of both support and fluid circulation. This consists of many very narrow, elongated, tubular cells through which water and food circulate within the plant. All land plants that have vascular tissue are called vascular plants. The walls of the vascular tissues contain organic compounds called **lignin** and **cellulose**, that provide structural rigidity. The tubular shape of this tissue serves both functions ideally because not only does a tube provide a conduit for fluid movement, but this also optimizes the use of materials for support. For a given volume of material, a hollow tube is more rigid than a solid rod—think of tent poles.

The vascular tissues are called **xylem** and **phloem**. Xylem cells are hollow and dead and provide for upward movement of water and dissolved nutrients from the roots. Phloem cells are living and transport manufactured food.

Obtaining water is only half the problem because desiccation is also a major issue for land organisms. Air is relatively dry, and a land organism must also have some way of preventing water in its tissues from being lost to air (Table 13.1). This is typically provided for by an outer coating that is impervious to water. For plants, an outer layer of waxy **cuticle** was developed to seal the inner plant from the dry air. However, a plant entirely encased by this outer coating would suffocate. Consequently, a series of specialized cells, **stomata**, typically located on the under surface of leaves, open and close to allow a plant to breathe. The first vertebrate animals, amphibians, did not solve the desiccation problem; but, as discussed later, reptiles did.

A final adaptation crucial to life on land was reproduction (Table 13.1). Green algae, organisms from which vascular plants arose, live in water. One mode of algal reproduction includes production of **spores**, which are distributed by water currents. In order to reproduce effectively, early land plants had to develop spores that could be transported by wind as well as by water and, most importantly, be able to resist drying out—a situation that is not a problem for a water plant. Reproduction for the vast majority of animals in the oceans occurs by shedding gametes into the water. Fertilization occurs externally, and this begins a planktonic larval life stage prior to reestablishment of a benthonic or nektonic habit in the adult. Reproduction had to be recast for fully terrestrial life, both for plants and animals, and as we will discuss in the ensuing chapters, innovation in reproductive success on land has been one of the key factors leading to advancement of terrestrial organisms.

THE FOSSIL RECORD

How did the earliest fossil plants conform to this series of requirements for success on land? The oldest known vascular plants are probably Late Silurian in age. However, plant microfossils (spores) and preserved tissues suggest that the earliest vascular plants may be as old as the Late Ordovician. Certainly various kinds of algae, lichens, and bryophyte-like plants lived in low-lying areas or near the water's edge long before vascular plants emerged. Perhaps some of these organisms even moved into ponds and puddles. Also, terrestrial microbes as old as the Proterozoic have been reported. However, when we say land plants, we are referring predominantly to vascular plants. The critical factors for recognizing vascular land plants in the fossil record are 1) vascular tissue (xylem); 2) cuticle; 3) stomata; and 4) spores with trilete scars. Trilete (or monolete) scars are the result of the meiotic division of the reproductive cells of land plants. This produces a cluster of four spores. Because these spores have thickened walls that prevent desiccation, each spore bears a scar marking where it was attached to the other three. Trilete scars also exist on the spores of bryophytes, in addition to those of vascular plants.

Trilete spores are known from Late Ordovician sediments, but these first spores probably represent bryophytes because other evidence for vascular plants is lacking. The earliest vascular plants belong to the plant division Rhyniophyta. Several fossils are known between the first occurrence of trilete spores from Late Ordovician strata and the first Early Devonian examples of *Cooksonia* that possess all four definitive attributes of vascular plants. *Cooksonia* is known from rocks as old as the Middle Silurian, but these plants do not have all the diagnostic features of vascular plants preserved. The oldest examples of cuticle are from the Late Ordovician, the oldest preserved simple vascular tissues are at least Late Silurian, and the oldest stomata are Early Devonian.

The transition of marine primary producers to land occurred over tens of millions of years, and apparently vascular plants acquired the essential features for a terrestrial existence gradually, probably as early land plants became less and less dependent on an aquatic habitat. Probably by at least the Late Silurian, plants that we would readily consider vascular plants existed in marginal stream and lake habitats, even though these plants lacked all four attributes.

Plants of Late Silurian to Early Devonian floras were all quite small. They were all less than 1 meter (3 feet) high, but commonly only a few centimeters. The main axis or stem of the plant had a central, solid core of vascular tissue, the xylem, forming the center of the stem and surrounded by a cylinder of phloem tissue. There were no leaves for photosynthesis, but the entire outer wall of the stem contained photosynthetic pigments. The entire above-ground plant was green. These early plants had an anchorage system called **rhizomes** that grew underground, but this was not a true root system that provides both support and uptake of water and nutrients from the soil. The spore-producing structures, the **sporangia**, were situated at the terminal tips of the plant branches, which, compared to more advanced plants, is judged to be a primitive feature. The plants had **dichotomous branching**; that is, each branch had two equal halves, like a tuning fork (Figure 13.1). Again, this is thought to be a primitive feature inherited from algal ancestors. A more advanced mode of branching is **monopodial branching**, which is characteristic of Middle Devonian and younger vascular plants. These have a central, large main axis, like the trunk of a tree, with smaller, more slender side branches.

The sediments in which the earliest vascular plants occur indicate that they lived in low, wet, marshy freshwater bogs. Some undoubtedly still had their "feet in the water," so to speak, but plants were becoming independent from the aquatic habitat.

As previously mentioned, a few plants that live on land are not vascular. These are primarily bryophytes, liverworts, hornworts, and mosses. These plants are all quite small and live mainly in low, moist places because they have not solved all the prob-

Figure 13.1   Model of the Devonian rhyniophyte *Rhynia*. Note the lack of leaves and true roots; also note the dichotomous branching, terminal sporangia, and small size; about 20 cm (8 in) high. (Courtesy of Field Museum of Natural History, Chicago.)

lems required for living on land. They do not have an adequate support or root structure to allow them to become larger and move into drier habitats. Mosses may well have preceded vascular plants onto land, but if so, we have no fossil record of them except perhaps spores with trilete scars.

An interesting question is, What was the landscape like during the early Paleozoic, the Cambrian and the Ordovician? We have no record of land plants during this time, so was the dry land only bare rock? Does it mean that there were primitive vascular plants for which we have no fossil record? Or, could Earth have been clothed in mosses, lichen, and other nonvascular plants? The first vascular plants were so simple, and the fossil record indicates rapid evolution after their first appearance, so it seems unlikely that they had occupied this habitat very long before they first appeared in the fossil record. We are left with two possibilities. Land areas may have had only bare, weathered rocks. In the absence of land plants, there would have been no soil, or humus, in the modern sense, although rock fragments may have been broken down by bacteria, fungi, and chemical weathering. Alternatively, nonvascular plants, if they existed, would have been confined to low, wet areas. In either event, there were no plants away from moist areas, and the landscape certainly would have been stark and very different from that of today.

Where did the vascular plants come from? We have noted that they probably arose from some kind of freshwater algae. The most likely candidates are from among the green algae. This is because these water-dwelling plants have a variety of photosynthetic pigments, including particular kinds of chlorophyll that are identical to those in vascular plants. In addition, green algae include many types that have relatively large and multicellular plant bodies. Other groups of algae are mainly small and single-celled, confined largely to marine water, or have a complement of pigments that is different from that of land plants.

CHANGES IN THE REPRODUCTIVE CYCLE

Our discussion of the evolution of vascular plants will focus on leaves, roots, and other important structures in their early evolutionary history (Chapter 14). However, before turning to that, we must consider one of the most important aspects of early plant evolution that has to do with the reproductive cycle. Many of the major events in plant evolution, such as changes from spore-producing to seed-producing plants and eventually the evolution of flowering plants, involved changes in reproduction.

The earliest vascular plants were spore producers. The commonly large, conspicuous plant that produces spores is called a **sporophyte**. The tetrads with trilete spores are produced in this phase. The sporophyte plant is **diploid** ($2N$ number of chromosomes), but the spores that it produces are **haploid** (N number of chromosomes). Spores are dispersed; when one lands in a suitable habitat, it germinates and begins to grow by a series of cell divisions. However, it does not grow into a new plant body like the one that produced it. Instead, it becomes a new and different plant body that remains very small; it is confined to low, moist habitats and has no vascular tissues. This new plant body may even be microscopic and subterranean, living underground. It is called a **gametophyte** because it produces gametes for sexual reproduction. Male and female organs develop on the haploid gametophyte and produce sperm and eggs. When released, these unite to form a **zygote**, which then grows into a new sporophyte plant body. Thus, in these primitive, spore-bearing vascular plants, two separate and distinct individual plants take part in a single reproductive cycle, and this is commonly referred to as "alternation of generations." Such a life cycle is typical of the earliest plants, as well as many of their aquatic algal ancestors. It is also present in many of the more advanced land plants, such as ferns. Details of plant reproduction are discussed in more detail in Chapter 14.

The reproductive life cycle of vascular plants contrasts sharply with the ordinary reproductive cycle of higher animals. Only one animal body exists; it produces gametes that after fertilization grow into a similar body. What are the advantages and disadvantages of the sporophyte-gametophyte alternation of generations life cycle of plants? One obvious answer is that a great many spores can be produced and released for dispersal of the plant. If they find a suitable habitat, they can germinate into a new individual without the need that animals have for close contact with another of its species. Thus, wide distribution is possible without dependence on chance encounters. Because plants are stationary and cannot move about to find a sexual partner, some dispersal system, such as that described, is necessary. In the sexual phase, the gametophyte lives in a wet habitat or microhabitat, mainly because moisture is necessary in order for the sperm to swim to an egg for fertilization. In a sense, the gametophyte is still almost an aquatic plant. The sporophyte phase is able to survive in more arid conditions, but a wet habitat is required for reproduction. Thus, comparable to the first vertebrate land animals described below, amphibians, the first land plants acquired a variety of critical characters that allowed them to live out of water, but they did not obtain independence from water with respect to reproduction (Table 13.1).

ORIGIN AND EARLY EVOLUTION OF ANIMALS ON LAND

Unlike vascular plants that probably evolved onto land a single time, animals evolved onto land many times, from worms to amphibians. Probably because these animals are mobile, fully aquatic animals could make brief excursions onto dry land much more readily than plants, thus leading to a greater number of lineages with fully terrestrial representatives. Furthermore, it is clear that some animals evolved directly from freshwater habitats, whereas others evolved from the ocean. The primary emphasis on terrestrial animals in this book is on vertebrates. However, invertebrates became fully terrestrial before amphibians, and at least six phyla of invertebrates (including seven groups of arthropods) made the transition. Invertebrates are briefly considered first.

TERRESTRIAL INVERTEBRATES

The oldest known land animals are arthropods, which have jointed legs and are related to trilobites, crabs, shrimps, and other arthropods of the sea. These oldest terrestrial arthropod fossils are from Upper Silurian rocks and consist of an arachnid and a centipede. Associated scorpions were once thought to be terrestrial but are now considered aquatic. During the Early Devonian, mites appeared and one reported arachnid is as long as 14 centimeters (5.5 inches) (Figure 13.2). By the Late Devonian, many kinds of arachnids, a probable bristletail insect, several mites, pseudoscorpians, myriapods, and an arthropleurid had appeared. These are all arthropods, and they formed a small ecosystem of herbivores and predators in and among the earliest land plants.

These fossil insects and spiders exhibit many primitive features. Early insects were wingless and related to living silverfish. Hundreds of fossil insect species date back to the Pennsylvanian Period. When wings evolved, they could not be folded back along the body but, instead, projected out to the sides like the wings of modern dragonflies. Various cockroaches, beetles, and relatives of crickets and grasshoppers were especially common. One must remember that the fossil record of insects is very sparse and that some groups may have evolved long before they appear as fossils.

During the Devonian Period freshwater aquatic invertebrates evolved. Freshwater bivalves are known from the Old Red Sandstone in Great Britain and from similar types of rocks in North America. Freshwater aquatic and land gastropods appear in the fossil record during the Pennsylvanian Period along with two additional groups of bivalves, worms with limy tubes, and freshwater arthropods with bivalved shells called ostracods. This freshwater and terrestrial community persisted through the remainder of the Paleozoic Era and continued into the Mesozoic Era when additional species evolved, increasing the diversity of terrestrial invertebrate life.

TERRESTRIAL VERTEBRATES

One of the most significant evolutionary events that occurred on Earth was the transition from water-dwelling fish to terrestrial tetrapods. As discussed, fish probably originated in the oceans, and our first records of them are in marine rocks. By the Devonian, however, they had radiated into almost all available aquatic habitats, including freshwater settings. One of the groups that is especially common in rocks deposited in fresh water is the lobe-finned fish. As previously discussed, the only surviving lobefin, *Latimeria*, lives in the ocean (see Chapter 7).

The freshwater Devonian fish of interest here are the **rhipidistian** crossopterygian lobe-fin. These fish lived in river channels and lakes on large deltas. The deltaic

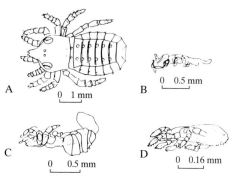

Figure 13.2   Examples of Devonian terrestrial arthropods from the Rhynie Chert in England (A) primitive arthropod. (B) Insect. (C) Questionable spider. (D) Mite. (Modified from M.S. Gordon and E.C. Olson *Invasions of the Land* Copyright © 1995 Columbia University Press)

rocks in which these fossils are found are commonly red, due to oxidized iron minerals, indicating that the deltas formed in a climate that had alternate wet and dry periods. If there were periods of drought, any adaptations allowing the fish to survive the dry conditions would have been advantageous—would have made them better fish. In these rhipidistians several such adaptations existed. First, we know that they had lungs as well as gills for breathing (Table 13.1). Cross sections cut through some of the fossils reveal that the mud filling the interior of the carcass was different in consistency and texture than in other parts of the fish. These differences reveal a saclike cavity below the front end of the gut that can only be interpreted as a lung. Gills were undoubtedly the main source of oxygen for these fish, but the lungs served as an auxiliary breathing device for gulping air when the water became oxygen depleted, such as during extended periods of drought. So, these fish had already evolved one of the prime requisites for living on land, that is, the ability to use air instead of water as a source of oxygen. A second adaptation of these fish was in the structure of the lobe fins (Figure 13.3). The fins were thick, fleshy, and quite sturdy, with a median axis of bone down the center (Figure 13.4). They could have been used as feeble locomotor devices on land, perhaps good enough to allow a fish to flop its way from one pool of water that was almost dry or anoxic to an adjacent pond that had enough water and oxygen for survival. These fins eventually changed into short, stubby legs. The bones of the fins of a Devonian rhipidistian exactly match in number and position the limb bones of the earliest known tetrapods, the amphibians. It should be emphasized that the evolution of lungs and limbs was in no sense an anticipation of future life on land. These adaptations helped these fish remain successful fish. It was serendipitous that these developments were also, later, useful for life on land.

What ecological pressures might have caused fishes to gradually abandon their watery habitat and become increasingly land-dwelling creatures? Changes in climate during the Devonian may have had something to do with this if freshwater areas became progressively more restricted. Another impetus may have been new sources of food. The edges of ponds and streams surely had scattered dead fish and other water-dwelling creatures. In addition, plants had emerged into terrestrial habitats immediately adjacent to streams and ponds, and various arthropods were also members of this earliest terrestrial community. Thus, by the Devonian the land habitat marginal to freshwater was probably a rich source of protein that could be exploited by an animal that could easily climb out of water. Evidence from teeth suggests that these earliest

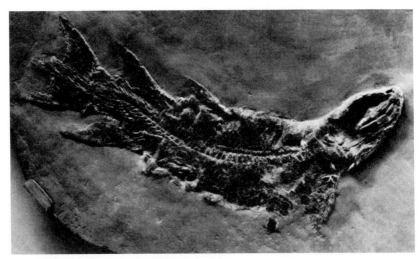

Figure 13.3   A specimen of the Devonian lobe-finned fish *Eusthenopteron*. Notice the thick, fleshy fin at about the middle of the body and the peculiar tail with the prominent middle section. Considerably reduced. (Courtesy of Field Museum of Natural History, Chicago.)

tetrapods did not utilize land plants as food; they were presumably all carnivorous and had not developed the ability to feed on plants.

How did the first amphibians make the transition to a terrestrial habitat? Similar to rhyniophytes, they only made a partial transition; they were still quite tied to water. First, many problems that faced plants were not applicable to the first tetrapods. The ancestors of these animals already had a circulation system, and they were mobile so that they could move to water to drink. Furthermore, they already had lungs, which rhipidistians presumably used for auxiliary breathing. The principal changes for the earliest amphibians was in the skeletal system—changes in the bones of the fins, the vertebral column, pelvic girdle, and pectoral girdle.

The earliest known land vertebrates are **amphibians**. They are related to the living amphibians—toads, frogs, and salamanders. The first definite land vertebrates are from the very Late Devonian rocks of Greenland. They belong to genera like ***Ichthyostega***, a name that refers to their fishlike skulls (Figure 13.5). Actually, many of the skeletal parts

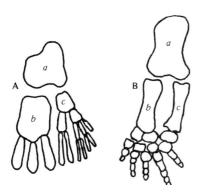

Figure 13.4   (A) Idealized bone arrangement in the front (pectoral) fin of a primitive crossopterygian fish. (B) The same basic arrangement in the front limb of an amphibian. Bones labeled *a* (humerus), *b* (radius), *c* (ulna) are the same in each instance.

of ichthyostegids, especially the skull, were very similar to their fishy rhipidistian ancestors. The tail is long and fishlike, and the limbs were short, stubby, and sprawled to the
side of the body, barely long enough to keep the belly from dragging in the mud. The
internal bony supports for the limbs and the girdles (pelvic girdle behind and pectoral
girdle in front) were weak and small. The backbone was also not strongly constructed;
however, the entire skeletal system was sufficiently strong for walking on land. The skull,
especially in the back, had many bones that were identical to the bones of rhipidistians.
The exterior had a bony armor of scales, inherited from fish, that probably aided in
retaining body moisture so that the animal would not dry out too quickly on land.
However, most amphibians lacked any means by which to prevent desiccation. Although
amphibians are terrestrial organisms, they are not completely adapted to land. All
amphibians have to return to the water to lay their eggs; the eggs lack a hard outer shell
to keep them from drying out on land. It is probable that these first amphibians spent a
large part of their life in the water, and they certainly returned to water during breeding
season, as do living amphibians. They were surely slow and awkward moving about on
land. So despite the successful adaptations that brought amphibians out of water, they
were and still remain dependent on moist habitats and water for reproduction.

   Another feature these amphibians exhibited has to do with their new environment.
Fish sense water movements largely through a special series of fluid-filled tubes in the skin
on the sides of the body and on the head. These constitute the lateral line system. Water is
incompressible, and these tubes sense water movement next to the fish. Because this system was not useful on land and because air is highly compressible, amphibians developed
an ear to sense changes in air pressure. The ear was situated low on the back of the skull
where a tympanic membrane, or eardrum, stretched across the **otic notch**, or ear notch in
the bone at the back of the skull (Figures 13.5, 13.6). Movements of the eardrum had to be
transmitted into the interior of the skull where they could be recorded by nerves. This transition was effected by a single earbone, called the **stapes**. The bone supporting the gill just
behind the jaws is called the **hyomandibular arch**. It moved forward during the evolution of
the fish and came to support and strengthen the hinge area of the jaws and to prop the
upper jaw against the braincase (see Chapter 12). The gill opening between the jaw arch
and the hyomandibular arch was reduced in size, migrated upward, and became a respira-
tory tube (spiracle) in more advanced fish. In amphibians, the old fish hyomandibular bone

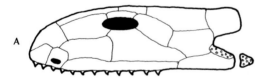

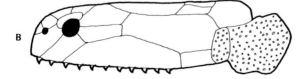

Figure 13.5   Side views of skulls.
(A) The earliest amphibian
Ichthyostega. (B) Its ancestor, a
rhipidistian fish. Note the otic or
ear notch at the back of the
amphibian skull and the reduction
in size of the opercular bones
(stippled).

Figure 13.6   A mounted skeleton of the Permian amphibian *Eryops* from Texas. The otic notches in the back of the skull are on either side of the neck, the skull is low and broad, and the short limbs sprawl to the side. (Courtesy of Field Museum of Natural History, Chicago.)

became the earbone or stapes. It was present in the area where the ear developed and was converted to this new function. The spiracle became the eustachian tube of amphibians.

During their early stages of evolution, amphibians underwent a number of evolutionary trends that improved their fitness for living on land. The bones of the pectoral and pelvic girdles, which supported the limbs, became larger, stronger, and more firmly attached to the backbone for support. Bones at the back of the skull became reduced in size and some were lost. A major change occurred in the nature and construction of the bones comprising the vertebral column. In fish these are weak and disc-shaped, much of the weight of the fish being buoyed by water. In amphibians these bones become larger, with bony crests for muscle attachment and with processes that articulate and support one bone in relation to another, front to back. Several different lineages of amphibians evolved from ichthyostegid amphibians, each of which had one diagnostic character in the construction of the backbone (Figure 13.7). Thus, one series

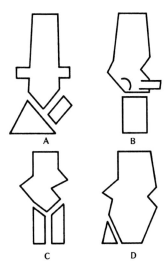

Figure 13.7   Side views of various bone arrangements in single vertebra of fossil amphibians. The two or three bony elements in each vertebra have a different arrangement and prominence in each of the four major groups of fossil amphibians represented.

of early evolutionary innovations and experimentation among amphibians was for a stronger backbone for survival on land. Some of these early amphibians are interpreted as probably spending much of their lives in water, because they had relatively weak limbs. Others developed very massive, strengthened legs; these animals probably returned to the water only to breed.

KEY TERMS

amphibian	*Ichthyostega*	spore
cellulose	lignin	sporophyte
cuticle	monopodial branching	stapes
dichotomous branching	otic notch	stomata
diploid	phloem	vascular tissue
gametophyte	rhipidistian	xylem
haploid	rhizome	zygote
hyomandibular arch	sporangia	

READINGS

Alexander, R.M. 1975. *The Chordates*. Cambridge University Press. 480 pages. The most comprehensive book yet written on the biology of living vertebrates, including minor discussion of fossils.

Banks, H.P. 1970. *Evolution and Plants of the Past*. Fundamentals of Botany Series. Wadsworth. 166 pages. An excellent short paperback that emphasizes the early history of land plants by one of the authorities on Devonian plants.

Benton, M.J., ed. 1988. *The Phylogeny and Classification of the Tetrapods*. Systematics Association Special Volume No. 35, 2 vols. Clarendon Press. 377 pages, 329 pages. Detailed discussion by a series of authors on the evolution and classification of all land-dwelling vertebrates.

Carroll, R.L. 1988. *Vertebrate Paleontology and Evolution*. Freeman. 698 pages. The modern-day version of the classic Romer textbook on vertebrate fossils. Detailed discussion of all fossil vertebrates with a classification listing all valid fossil genera.

Gensel, P.G. 1986. "Diversification of Land Plants in the Early and Middle Devonian." In R.A. Gastaldo, ed., *Land Plants: Notes for a Short Course*. Studies in Geology 15, University of Tennessee, Department of Geological Sciences. Pages 64–80. A good up-to-date discussion of early land-plant evolution and radiation.

Gordon, M.S., and Olson, E.C. 1995. *Invasions of the Land, The Transitions of Organisms from the Aquatic to Terrestrial Life*. Columbia University Press. 312 pages. A detailed but readable consideration of the invasion of life onto land.

Knoll, A.H., and others. 1986. "The early evolution of land plants." In R.A. Gastaldo, ed., *Land Plants: Notes for a Short Course*. Studies in Geology 15, University of Tennessee, Department of Geological Sciences. Pages 45–62. A technical discussion of physiological aspects of early land-plant evolution.

Terrestrial Primary Producers: The Land Plants

PHANEROZOIC FLORAS

Terrestrial habitats have been populated by a series of major floras through time (Figure 14.1). Each marks basic morphological innovations that led to diversification and dominance. Rhyniophytes and other early vascular plants dominated into the Middle Devonian, after which lycopods, sphenopsids, and ferns dominated. These more advanced spore-bearing plants formed the vegetation of the vast Pennsylvanian coal swamps; but during the Permian, the seed-bearing gymnosperms—seed ferns, conifers, cycads, and ginkgos—dominated in many habitats. Gymnosperms continued to dominate plant communities until the evolution of the angiosperms, the flowering plants, that have dominated in most terrestrial habitats since the Cretaceous.

PLANT REPRODUCTION

EVOLUTION OF REPRODUCTIVE STRUCTURES

The most complex and important evolutionary changes that took place in land plants was with reproductive structures—their formation, complexity, and position on the plant. One can almost say that the fossil history of vascular plants is principally a history of plant reproduction. In order to understand this aspect of plants, we must first briefly summarize the generalized alternation of generations life cycle of plants.

We will begin with the adult plant body that is typically large and conspicuous; this is what one usually thinks of as the plant. As introduced in Chapter 13, this is the **sporophyte**, and the cells of this plant body are all diploid (2*N*) (Figure 14.2). The sporophyte has specialized reproductive structures, called sporangia, where spores are produced. This takes place by reduction division (**meiosis**), so that each spore produced is haploid (*N*). In early land plants the sporangia were small and simple, but in later ones they became larger and arranged in clusters. Upon release, a small spore is carried by wind or water to a new living site, where it germinates and grows into a new, different, and usually quite small plant body, the **gametophyte**. The gametophyte produces sex cells or gametes. Notice that the spores do not undergo combination or fertilization before they produce the gametophyte. Each gametophyte has all haploid cells from these spores. It develops male and female sex organs in which the sex cells are produced by **mitosis**, or ordinary cell division. The male and female gametes that are produced unite through fertilization to form a new diploid (2*N*) cell, the zygote, that develops into a new diploid sporophyte plant body.

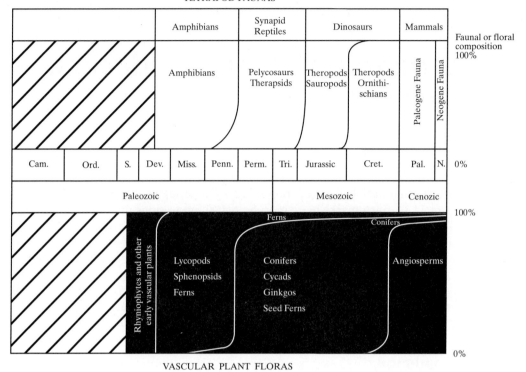

Figure 14.1   Evolutionary floras through time are highlighted on this chart of the history of terrestrial flora and fauna.

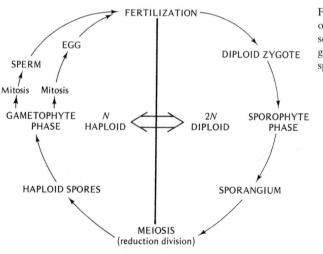

Figure 14.2   Generalized life cycle of a vascular plant with two separate individuals—the gametophyte (haploid) and the sporophyte (diploid).

This alternation of generations life cycle of land plants differs essentially from that of most higher animals in that an additional adult body is interposed between meiotic division, which takes place in spore formation and the union of haploid cells in fertilization to form a diploid adult. The interposed adult phase is, of course, the gametophyte. In all the vascular plants discussed here, the sporophyte is the dominant phase. In mosses and liverworts (not important as fossils), the gametophyte is dominant, and the sporophyte is inconspicuous. In lower vascular plants, such as ferns, the gametophyte is a separate body, although it may be microscopic or subterranean. In the more advanced seed-bearing plants, the female gametophyte is retained on the sporophyte plant. Sex cells are produced. The male gametophyte is a **pollen** grain, which is dispersed through wind, water, or insects. Pollen grains typically come to rest on reproductive structures of another plant, although self-fertilization is possible. A pollen grain produces sperm and fertilizes the egg produced by the female gametophyte. The embryo is encased in a protective coat and a food supply. This is the **seed** that contains the developing sporophyte of the next generation.

SILURIAN THROUGH MIDDLE DEVONIAN FLORA

Land surfaces were devoid of vascular plants during the Precambrian and earliest part of the Paleozoic Era. If primitive plants such as small lichens and mosses existed, they may have only occurred at the water's edge, and they left no fossil record. Certainly, most of the continents were bare rock; Earth was not green, as it is today. The scarcity of plants would have had a profound effect on various weathering processes, both chemical and physical, and on soil-forming processes, especially because humic acids produced by land plants, which help form soil, would have been lacking. Even the weather and climate would have been affected because land plants play an important role in recycling water from soil back to the air. On a summer day, an apple orchard with 40 trees can move 16 tons of water from the soil back into the atmosphere.

The oldest vascular plants, the **rhyniophytes** (Division Rhyniophyta), appeared at least by the Late Silurian (Chapter 13). Once established on land, plants diversified very rapidly. Rhyniophytes are quite small, rarely more than a few centimeters high. Many may have lived in bogs and other wet places so that even though they were true vascular plants, they were still very closely associated with water.

One of the simplest and oldest of these early land plants is typified by the Late Silurian to Early Devonian genus *Cooksonia*. These plants were small, less than 50 centimeters (1.5 feet) high; lacked woody tissue; had small, simple sporangia on the branch tips; and a stem that branched dichotomously (Figure 14.3A). *Cooksonia* lacked roots but was anchored by a rhizome, an extension of the stem. It also lacked leaves, so the entire above-ground part of the plants was probably green, contained chlorophyll, and carried on photosynthesis. These rhyniophytes seem to have lived in permanently wet habitats, such as shallow ponds and bogs.

Additional, somewhat more advanced, and larger vascular plants first occur in Lower Devonian rocks. They still lack roots but have scales on the stem that may have been green and precursors to leaves. The sporangia are globose- to kidney-shaped and arranged in rows on the stem. These plants are typified by *Zosterophyllum* (Figure 14.3B).

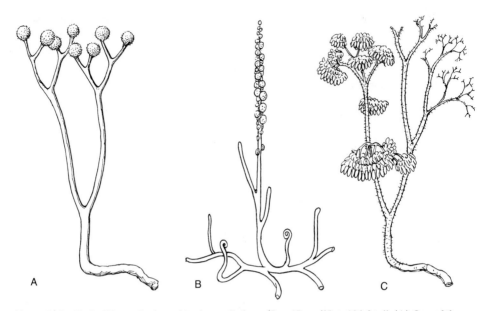

Figure 14.3 Typical Devonian vascular plants, all about 30 to 45 cm (12 to 18 in) tall. (A) One of the earliest known vascular plants, *Cooksonia*, with dichotomous branching and simple terminal sporangia. (B) *Zosterophyllum*, a vascular plant with smooth stem and rhizome, dichotomous branching, and sporangia arranged along sides of a fertile branch. (C) *Psilophyton*, with clusters of terminal sporangia and hairlike spines on the stem. (Modified from Gensel, 1986.)

Psilophyton is representative of a third group of Early to Middle Devonian primitive vascular plants that again are quite small, although some species exceeded 60 centimeters (2 feet) in height (Figure 14.3C). They have very complex branching patterns that also may be a precursor to leaf development. Some species have hairy or spiny stems, and the sporangia are arranged in drooping clusters.

LATE DEVONIAN THROUGH PENNSYLVANIAN FLORA

EVOLUTION OF LEAVES

Early in the history of vascular plants, two quite different kinds of leaf structures evolved. Leaves are the sites of photosynthesis in more advanced plants and are distinguished because they are thin flaps of the plant that contain vascular tissue. One leaf type is small and simple, issuing directly from the stem and with a simple midvein (Figure 14.4A). Leaves of this type are thought to have evolved from small spines on early plant stems. Several of the leafless early vascular plants have abundant, hairlike spines preserved on the stems. In lycopods, these spines are thought to have enlarged and, eventually, to have had a branch of vascular tissue develop to conduct water to the leaves (Figure 14.5).

Other early plants had leaves that were quite large and much divided, similar to the leaves of ferns (Figure 14.4B). They are not thought to have evolved from spines,

Figure 14.4 Two contrasting types of primitive leaves. (A) The leaf of a lycopod—long, slender, and with a midvein. (B) A leaflet of much-branched, fernlike foliage with a complex network of veins. Each leaf is several centimeters long. (Courtesy of Field Museum of Natural History, Chicago.)

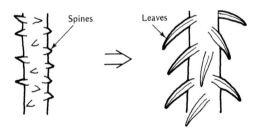

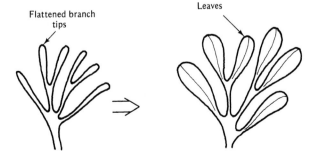

Figure 14.5 The two modes of evolution of leaves in vascular plants. Small, simple leaves (above) evolved from spines on the stem; large complex leaves (below) evolved from flattened tips of branches.

but from the flattened and increasingly subdivided tips of branches. That is, the ends of the branches became flattened and lateral tissues developed along the branches (Figure 14.5). Chlorophyll became concentrated in these modified branch tips, which contained vascular tissue. This kind of foliage may have developed in the progymnosperms and early ferns, which are also distinctive in having the reproductive sporangia on the undersides of the leaves.

DEVELOPMENT OF WOODY TISSUES

The xylem and phloem tissues that together constitute the vascular system of a land plant initially formed a solid cylinder in the center of a primitive plant stem. The **xylem** was situated in the center of the stem surrounded by the **phloem**. As long as plants were small, as with the early rhyniophytes, this structural arrangement was satisfactory. But there was an early trend, beginning during the Devonian, for plants to become increasingly larger. This put more stress on the stems from external forces, such as wind. Plants evolved two features of the vascular tissue that allowed the plant body to increase in size. First, the primary vascular tissues evolved from a solid central core to a hollow cylinder with the center filled with soft pith tissue. As discussed for vascular tissue, a hollow cylinder has considerably more strength to resist bending and breaking than does a solid cylinder; thus, this arrangement of vascular tissues helped increase the strength of the stem. Second, the primary vascular tissues are not woody. Woody tissue is stiff and resistant and evolved from the xylem tissues. The softer, conductive xylem is called primary xylem. It produces secondary xylem, or **wood**, that is still conductive but also supports the stem and branches of the plant. Woody tissues are known from Devonian plants and evolved in all major advanced groups of vascular plants. Wood is among the most commonly found fossil plant material because of its resistance to decay and destruction.

DOMINANT PLANTS OF THE FLORA

The late Early to Middle Devonian Period was a time of very rapid vascular-plant evolution (Figure 14.1). By the end of the Devonian, most of the major Paleozoic vascular plant groups had evolved, and trees graced the landscape.

The more advanced spore-bearing plants that evolved during the Devonian include the lycopods, sphenopsids, ferns, and progymnosperms (forerunners of gymnosperms). These groups developed true leaves—some large and complex, others small and simple—and true roots. Many developed woody tissues in the stem and, hence, had a much stronger stem and could grow much taller. Ultimately, at least one group of these advanced spore-bearing plants developed seeds.

The **lycopods** (Division Lycophyta) are represented today by very few living genera. Most are small plants that seldom grow more than a few inches, but some may be as tall as 2 meters (6 feet). In contrast to the rhyniophytes, lycopods have true leaves and true roots, which means that vascular tissue penetrates both, and they also have more complex reproductive structures. The common names for living lycopods are club moss and ground pine. Their extinct relatives included large trees, some of which reached heights greater than 36 meters (117 feet) tall. The lycopods are characterized by long, slender, very simple leaves with a single midvein that issue directly from the trunk and that are arranged in spirals (Figure 14.6). In many lycopods, the bark of the trunk has distinctive leaf pad scars that are diamond shaped and reflect the spiral leaf pattern (Figure 14.7). Lycopods have more advanced reproduction than any of the early vascular plants discussed above. From small, simple terminal sporangia of rhyniophytes and other early plants, lycopod sporangia shifted to a position where they are protected in the angle between a leaf and a stem. Two different things then happened to the sporangia, not necessarily both in the same plant. First, they arranged themselves in clusters, forming a large conspicuous structure called a **cone** (Figure 14.8). Second, two different kinds of sporangia evolved, one that produces a female megaspore and another that produces a male microspore (Figure 14.9). However, rather than having both sexes on one gametophyte body, there are two separate gametophytes.

The *Baragwanathia* flora of Australia (Early Devonian) has a variety of primitive plants, including early lycopods (Figure 14.10). By far the best known fossil lycopod is the coal swamp tree *Lepidodendron* (Figures 14.7 and 14.11).

Sphenopsids (Division Sphenophyta) are characterized by a hollow stem that is jointed (Figure 14.12). There is a single living genus, *Equisetum*, which is quite common in roadside ditches and other wet habitats, such as along streams. Common names include scouring rush and horsetail, and this genus is divided into approximately 20 living species. In sphenopsids, whorls of leaves and sporangia occur at nodes on the stem where two joints meet. In *Equisetum* the leaves are reduced to tiny leaf traces, but fossil forms have larger leaves with star-shaped leaflet patterns (Figure 14.13). Although different in detail from the lycopods, sphenopsids also have repro-

Figure 14.6   Portion of a restoration of the Pennsylvanian scale tree *Lepidodendron*. Long, simple, straplike leaves are arranged in spirals that clothe the stem. A female cone is also visible. (Courtesy of Field Museum of Natural History, Chicago.)

Figure 14.7 Impression of the surface of the trunk of *Lepidodendron*, showing the diamond-shaped leaf base scars; a simple leaf was attached to each scar. Specimen is of Pennsylvanian age from Illinois; each leaf scar is approximately 1.5 cm (0.5 in) across. (Courtesy of Field Museum of Natural History, Chicago.)

Figure 14.8 A large compression cone, *Lepidostrobus*, of a Pennsylvanian lycopod from Illinois; several centimeters in length. Notice the modified leaves (called sporophylls) that form the cone preserved around the edges and base of the cone. (Courtesy of Field Museum of Natural History, Chicago.)

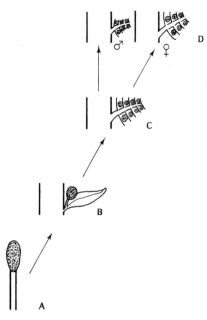

Figure 14.9 Generalized pattern of evolution in the reproductive structures of vascular plants. (A) Terminal small sporangia. (B) Single sporangium in angle between leaf and stem. (C) Cluster of sporangia (all with the same size spores) and associated leaves, forming a cone. (D) Separation of male cones with small spores and female cones with large spores.

Figure 14.10   Stem and leaves of *Baragwanathia*, a primitive Early Devonian lycopod from Australia. Long, slender, simple leaves cover the stem. Slightly reduced. (Courtesy of F.M. Hueber, National Museum of Natural History.)

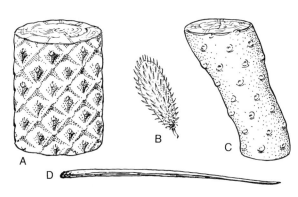

Figure 14.11   Typical Pennsylvanian coal swamp lycopods, all parts of *Lepidodendron*. (A) Trunk or stem with bark pattern showing diamond-shaped leaf base scars (see Figure 14.7). (B). Female cone. (C) Root, with large pith cavity and scars where rootlets attached. D. Long slender leaf with simple midvein (see Figure 14.6).

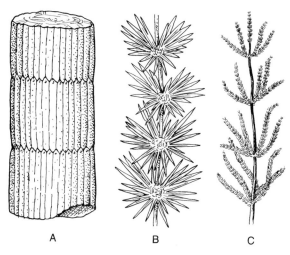

Figure 14.12   Pennsylvanian coal swamp sphenopsids. (A) Jointed stem of *Calamites* with large central pith. (B) Vinelike whorled leaves of *Annularia* (see Figure 14.13). (C) Stem and small leaves of *Calamostachys*.

Figure 14.13   A branch of a sphenopsid, *Annularia*, preserved in an ironstone concretion of Pennsylvanian age from Illinois. The stem is divided into segments, with a whorl of leaves present at the junctures of the segments. (Courtesy of Field Museum of Natural History, Chicago.)

ductive structures organized into cones. The cones of *Equisetum* are terminal, and generally only a few individuals are fertile in a given stand of plants.

The name scouring rush comes from the fact that the stem tissues of this living form are full of microscopic crystals of the mineral silica, creating a very hard and harsh tissue. American pioneers used bundles of these stems to scour skillets and pots. However, the silica crystals cause the stems to be poisonous to livestock. Fossil sphenopsids may be large trees, like the coal swamp *Calamites* or nonwoody vines.

The third group of advanced spore-bearers are the **ferns** (Division Pteridophyta), which have large, complex leaves. Ferns lack cones. Instead, sporangia are commonly in rows of small, dotlike structures on the undersides of leaflets. Most ferns are small- to medium-sized plants, but other living and fossil ferns grew to tree size and are called tree ferns. In contrast to other tree-sized plants, tree ferns do not produce woody tissue. Instead, the outside of the stem is armored in rigid roots. The origin of the ferns is poorly understood. They were very common late Paleozoic plants and are common fossils during the Mesozoic. They are the most common and diverse spore-bearing vascular plants today, with more than 10,000 species.

Finally, there is a small group of spore bearers that are the direct ancestors of the seed-bearing **gymnosperms** that represent one spore-to-seed evolutionary transition. These are called progymnosperms. Many of these plants have leaves that look superficially like those of ferns, however the leaves are spirally arranged on branches. Also, the woody stems have vascular tissues that are essentially similar to some conifers. They may be either homosporous (all spores the same size) or heterosporous (large female spores and small male spores). The progymnosperms are confined to the Devonian and Mississippian. During the Late Devonian, the first gymnosperms appeared. These were the seed ferns or pteridosperms (Figure 14.14). Other Paleozoic plants also developed seeds; seed ferns and these other early seed plants were also important in Pennsylvanian coal swamps.

Figure 14.14   Two large seeds from Pennsylvanian-age seed ferns. Each seed was contained within an ironstone concretion that has been split in half so that two views of each seed can be seen. Approximately two-thirds natural size. (Courtesy of Field Museum of Natural History, Chicago.)

COAL-SWAMP PLANTS

Rocks of the Pennsylvanian age are a major source of the world's coal supply. Coal beds are composed of the altered and degraded remnants of plants that grew in very large, forested swamps and mires. Pennsylvanian coal swamps were mostly coastal, bordering the sea, because marine rocks and fossils are commonly found in the strata above individual coal beds, deposited when the sea flooded over the swamp and killed all the terrestrial life that built the swamp. Today the areas most like these swamps can be found in southeastern Asia, such as the peat swamps of Indonesia and Malaysia.

These coal swamps had a rich and varied flora and fauna. The plants that dominated the swamps were diverse and highly structured in their community relationships. Large trees that stood more than 30 meters (98 feet) tall and were a meter or more across at the base were present. These largest plants were mostly lycopods, represented especially by *Lepidodendron* (Figures 14.6, 14.7, and 14.11). The sphenopsids are represented by *Calamites* (Figure 14.12). These were probably the two most common plants with large, woody stems. Another common tree was *Cordaites*, which belongs to an extinct group of early conifer plants, the Cordaitales (Figure 14.15). These had long, narrow, straplike leaves and fruiting bodies of loosely constructed cones, or "catkins," with the female catkins producing large, rounded, flattened seeds. These three trees, each belonging to a different group of plants, towered over the remainder of the coal-swamp flora. A middle story was formed by tree ferns (true ferns) and seed ferns (gymnosperms) that produced an abundance of fern-type foliage (Figures 14.16 and 14.17). Maximum heights were as much as 15 meters (50 feet) for tree ferns and 10 meters (32 feet) for seed ferns. The understory of these forests consisted predominantly of smaller ferns and seed ferns, of which there was a wide variety. Ferns are identified as fossils mainly on the basis of the shape of the reproductive structures and other morphology. Each leaf can be immense, as much as several meters long, with numerous subdivisions, each of which has many small pinnules.

Some of our best evidence for the nature of Pennsylvanian swamp plants come from **coal balls**. These are rounded, mineralized concretions preserved in coal soon

Figure 14.15   Restoration of a Pennsylvanian *Cordaites* tree, a typical coal-swamp gymnosperm. The long, slender leaves are quite unlike the small, needlelike leaves of most living gymnosperms. (Courtesy of Field Museum of Natural History, Chicago.)

after burial, before the plant material was compressed and coalified. Normally coalification, a type of carbonization, obliterates all fine detail of the plant material. However, coal ball fossils have cellular detail preserved. Plants preserved in coal balls display wide variation in terms of their abundance and diversity. At some localities, virtually every coal ball will contain remnants of *Lepidodendron* but no other plants. At other localities, an amazing diversity of plants is present. This seems to indicate that coal swamps were not uniform in their floral content. Imagine one edge of a swamp near the margin of a sea, with the landward p lants becoming progressively less and less influenced by the ocean. It seems reasonable that different plants would respond differently according to whether or not the water was fresh, brackish, or salty and to

Figure 14.16   Pinnules from a seed fern, *Neuropteris*, of Pennsylvanian age from Illinois, preserved in an ironstone concretion. (Courtesy of Field Museum of Natural History, Chicago.)

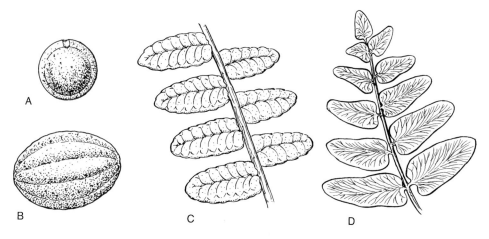

Figure 14.17   Pennsylvanian coal- swamp fern and seedfern fossils. (A, B) Typical seeds. (C, D) Portions of two different large leaves, showing shape of leaflets or pinnules. All slightly enlarged.

the extent of wetting and drying. There is some evidence that *Cordaites* tended to dominate around the landward edges of the swamp and that *Lepidodendron* occupied the seaward edges. It is quite clear that there were immense pure stands of plants such as *Lepidodendron* and broad clearings occupied by low-growing ferns. In this respect these ancient forests may have resembled modern coastal swamps.

PERMIAN THROUGH JURASSIC FLORAS

PERMIAN PLANTS

The close of the Paleozoic Era was a time of tremendous change in terrestrial communities. Coal swamps virtually disappeared in the northern hemisphere, although they still persisted in colder climates of the southern hemisphere where seed ferns, *Glossopteris* and *Gangomopteris*, and other seed-bearing plants formed thick layers of coal. With increasing continentality during the Permian, broad alluvial plains with low relief formed across large areas of the southwestern United States. These alluvial areas spread from Kansas to Oklahoma and to northern Texas, across into New Mexico, Arizona, and Colorado. In this area we have one of the best fossil records of terrestrial life, especially during the early part of Permian time.

The plants of the Permian reflect climatic change. The dominant lycopods and sphenopsids of the Pennsylvanian coal swamps were still present, but they were much less abundant. *Cordaites* (Figure 14.15) and its relatives, which occupied edges of Pennsylvanian swamps, were some of the most conspicuous plants during the Permian, living in drier habitats. These and other early conifers had small needlelike leaves that clothed the branches. These are a sharp contrast to the fern foliage of the seed ferns and the long, straplike leaves of the cordaiteans. **Conifers** (Division Coniferophyta) became one of the more conspicuous groups during the Permian. Seed ferns were present in northern areas, but they were declining in number. True ferns were still conspicuous. The overall aspect of Permian floras is one of plants growing in flood plains and drier upland areas, with various groups of gymnosperms dominating the floras.

The evolution of seeds has been clearly demonstrated as a feature of great adaptive significance. In seed plants, the female megaspores are retained within the female reproductive structure, pollen is produced by the male reproductive structure, and seeds ultimately develop after fertilization in the female plant. This is the familiar condition in pines and most other seed-producing conifers. The male pollen-producing structures are much smaller and inconspicuous. Male- and female-producing reproductive structures are not necessarily on the same plant. Even if both sexes are present on the same plant, they may require the pollen from another individual for reproduction. With seeds, plants are no longer tied to moist habitats for reproduction so that a separate gametophyte plant can germinate and grow. Instead, fertilization takes place on the female sporophyte, and zygote development takes place within the seed. The seed then produces another sporophyte. A seed is able to persist in the soil and germinate only when growing conditions, such as a rainy season, are suitable.

TRIASSIC AND JURASSIC PLANTS

Although large sphenopsids and lycopods were still present during the Mesozoic, they were less common. Lycopod trees became extinct by the close of the Triassic, and large sphenopsids survived only a little longer. Four groups of plants clearly dominated Triassic and Jurassic landscapes. The dominant understory plants were still ferns, which included a variety of foliage types. A middle story of plants was quite diverse, including tree ferns, seed ferns, and a group of gymnosperms including **cycads** and cycadeoids (Division Cycadophyta) (Figure 14.18). Cycads and especially cycadeoids typically dominated Mesozoic landscapes. A few cycads are still living. Cycads have a stem or trunk that commonly looks like a large pineapple and is composed of the coalesced bases of large leaves. The leaves break off as the plant grows, leaving a cluster of sturdy bases surrounding the stem (Figures 14.19 and 14.20A). The leaves are large and palm-like, growing in a cluster at the tip of the stem (Figure 14.20B). The cones of fossil cycads project from the stem, whereas in living cycads the cones are produced at the top in the center of the leaf whorl (Figure 14.20D).

The upper story of Triassic forests was formed by a variety of conifers. These had distinctive patterns of cells in the wood that permit us to ally them with primitive living conifers called **araucarians** (Figures 14.20C and 14.21). Today, these conifers are restricted to small areas mainly in the southern hemisphere, including Australia, New Zealand, and South America. The Norfolk Island pine is an example that is commonly grown as a houseplant or outdoor specimen on lawns. The best known fossil occurrence of these conifers is the Petrified Forest National Park in northeastern Arizona (Figure 14.21). Here, immense logs of araucarians from Upper Triassic rocks of the Chinle Formation are exposed due to weathering. This conifer forest covered a large area of the southwestern United States, probably growing on broad, flat flood plains and in coastal areas.

Another new group of gymnosperms, the **ginkgos** (Division Ginkgophyta), first appeared during the Permian. These gymnosperms are typically small to large and are slow-growing trees. Each individual is either male or female, bearing small repro-

Figure 14.18   Compound leaf of a fossil cycad from Mexico. The large, much-divided leaves issued directly from the top of a short, barrel-shaped trunk. (Courtesy of Field Museum of Natural History, Chicago.)

Figure 14.19   The short, stubby trunk of a fossil cycad of Cretaceous age from Maryland. The trunk is approximately 45 cm (1.5 ft) high and is composed of the coalesced bases of large, much-divided leaves that issued from the top of this trunk. The triangular pattern represents leaf base scars. (Courtesy of National Museum of Natural History.)

Figure 14.20 Mesozoic plants. (A) Reconstruction of a fossil cycad, about 2 m (6 ft) tall. (B) Enlargement of a small part of a cycad leaf showing several leaflets. (C) Twig of an araucarian pine. (D) Cone of a fir conifer. (E) Ginkgo leaf. (F) Twig of *Metasequoia*.

ductive structures of one sex or the other. The leaves are quite distinctive, having a fan shape with parallel veins and the outer marign split or entire (Figure 7.12, Figure 14.20E). The group was quite common during the Mesozoic all over the world. By the Cenozoic, however, it had dwindled in abundance and had virtually disappeared from the fossil record. The group is represented by a single living species, *Ginkgo biloba* (the maidenhair tree), which survives only in domestication, as discussed in Chapter 7.

This Mesozoic flora was the vegetation eaten by dinosaurs, other reptiles, and mammal herbivores during the Jurassic. Various kinds of conifers, including araucarian pines, were the principal large trees of the Jurassic. The other most common plants were various types of ferns, both tree ferns and herbaceous ferns and gymnospermous cycadeoids, which formed a small tree or middle story to the vegetation.

CRETACEOUS THROUGH HOLOCENE FLORA

EVOLUTION OF FLOWERS

The major evolutionary innovation of plant communities during the Mesozoic was the appearance and rapid adaptive radiation of the flowering plants, the **angiosperms**.

Figure 14.21   Petrified log of an araucarian conifer from the Petrified Forest National Park, Arizona. Maximum log diameter approximately 1 m (3 ft). (Photograph by W.I. Ausich.)

How did these most advanced plants, which make up more than 90 percent of the land plants alive today, evolve? What were the traits that made them so successful? First, we consider the record.

Plants with some but not all diagnostic angiosperm characters are known from the Triassic and Jurassic. However, the oldest definite flowering plants are first known from the Early Cretaceous with the oldest flowers and fruits known from later during the Early Cretaceous. Initially, they were rare and not very diverse. By the beginning of the Upper Cretaceous, angiosperms were abundant on a worldwide basis (Figure 14.22) and dominated most habitats by the end of the Cretaceous. Once they evolved, they rapidly expanded into virtually all terrestrial habitats, even those with the most severe climatic contrasts of hot and cold, wet and dry. Today, conifers principally dominate only in cold and dry climatic zones. This extraordinary domination over groups of plants that held sway for millions of years—the conifers, cycads, ginkgos, and ferns—has several explanations. No single morphological change was responsible for the success of the flowering plants; rather, it probably involved a combination of several factors.

Some people erroneously consider seed plants and flowering plants to be synonymous. This is not the case at all, as we have seen. The seed habit—encasement of the developing embryo in a protective coat with a food supply—originated with seed ferns as early as the Devonian. All gymnosperms, the dominant plants of the early and middle Mesozoic, are also seed plants. In order to understand the evolution of the **flower**, we begin with the familiar pine cone. Many primitive land plants have the reproductive structures grouped together into a cone. This is the case in some lycopods, sphenopsids, and other gymnosperms. Each segment of the cone is a modified branch

Figure 14.22 A sandstone slab containing several angiosperm leaves of Late Cretaceous age from Idaho. These plants all belonged to species that are now extinct. The leaf types indicate a mild climate in the northwestern United States at the close of the Cretaceous. (Courtesy of National Museum of Natural History.)

and leaf, bearing either male or female reproductive structures on the inner side. In a pine, the seeds are exposed; if you break off segments of a ripe female cone, you can see the large seeds.

Imagine the following modifications of a cone at the tip of a branch. The leaves, which are sporophylls, or spore-leaves, at the base of the cone become sterile, without reproductive organs. The sporophylls are arranged in whorls around the cone. Two whorls of sporophylls become sterile and encase and protect the immature, budding cone. In flowering plants, these become an outer whorl of sepals and an inner whorl of petals (Figure 14.23). The next series of sporophylls along the cone bear pollen-producing male reproductive structures. These become modified into long, tubular stalks that bear pollen-producing sporangia. Each modified sporophyll is called a **stamen**. The terminal sporophylls of the cone modify into egg-producing structures called **pistils**. The pistil is composed of a stigma, style, and ovary. In flowers with a compound pistil, each separate pistil is termed a carpel. Thus, a flower is a collection of highly modified leaves, some of which bear reproductive structures. Rather than having all similar sporophylls, the different whorls of sporophylls are highly modified and differentiated, resulting in a quite complex structure.

Most gymnosperms have separate male and female cones, so the change to flowers with both male and female structures was a significant step. Early angiosperms have a diversity of flower types that includes both unisexual and bisexual forms. The origination of angiosperms and their early evolution is interpreted as morphological innovations to improve insect pollination and increase cross-fertilization.

Why were these modifications and the evolution of the flower structure so important to the success of the angiosperms? One of the most conspicuous differences between angiosperms and gymnosperms has to do with the usual mode of achieving pollination. Living gymnosperms are generally wind pollinated; there is no reason to believe that their fossil relatives were not also wind pollinated. Wind pollination is a random process. In order to be successful, gymnosperms must produce a tremendous quantity of pollen that is easily liberated and that is small and light enough to be car-

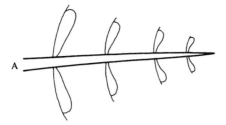

Figure 14.23 Diagrammatic cross sections of (A) a pine cone and (B) an angiosperm flower, showing four whorls of reproductive structures. These are all alike in the gymnosperm above and highly modified in the angiosperm flower.

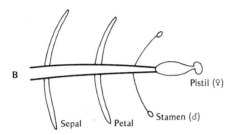

ried considerable distances by the wind. The number of female receptors must also be great to ensure adequate fertilization.

Many but by no means all angiosperms have departed from this pollination method and have evolved the consortium with insects as pollinators. Imagine that initially during the Cretaceous, or perhaps the Jurassic, beetles or other insects that fed on gymnosperm cones had pollen from male cones accidentally attach to their bodies. So when they fed on female cones, some of the female gametophytes were fertilized in the process. Natural selection may have favored those plants that were increasingly attractive to insects or made themselves more readily accessible to insect cropping in order to achieve this method of cross-fertilization. Evolutionary trends such as closure of the carpel on the stigma and reduction and evolution of the stigma prevented self-fertilization. The evolution of petals or distinctive smells and nectar to attract insects can be viewed as part of the selective process that led to the evolution of more complex flowers (Figure 14.24). Also, during early angiosperm evolution, the positions of flowers shifted to be more visible, thus attracting pollinators and improving insect pollination. Concurrent with these changes, there was no longer a need to produce such large quantities of pollen or of pollen receptors; thus, the number of stamens and pistils could be progressively reduced, as in more advanced kinds of flowers.

Further evolutionary trends among flowers included development of bilaterally symmetrical flowers that aid in pollinator recognition. Also, the pollen could be larger, heavier, and more ornamented—the better to cling to insects or other animals. This important change in effecting fertilization had a tremendous impact on plant life. After all, plants are mostly stationary organisms; they cannot move together for reproduction as do many animals. Evolution of fruit and seed types that could be dispersed by animals or the wind also developed. Any adaptation that would increase their repro-

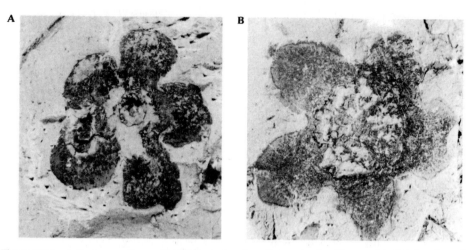

Figure 14.24   A Cretaceous flower. (A) Blossom showing a five-part calyx (five sepals), slightly enlarged. (B) The same kind of flower, about twice natural size, showing remains of fruit in the center. From Rose Creek, Dakota Formation, central Kansas. (Courtesy of David Dilcher, University of Florida.)

ductive potential should have been, and was, exploited to the fullest. Flowering plants have evolved a bewildering variety of forms, especially with respect to flowers.

The heightened relationship between flowering plants and insects also had a great effect on the insects themselves. Many became dependent on flowers as a source of food. The advent of flowering plants during the Cretaceous undoubtedly led directly to the tremendous increase in abundance and variety of living insects (e.g., butterflies, bees). We live in a terrestrial world that is primarily a flowering plant-insect world. Mammals (including humans), birds, worms, and other animals play a much smaller role in shaping our natural environment than do insects and flowering plants. It is very doubtful that insects were as diverse at the species level prior to the evolution of flowering plants during the Cretaceous. Many species of flies, beetles, bees, and other insects have life cycles that are predominantly controlled by and dependent on specific kinds of flowering plants.

A second aspect of the success of flowering plants has to do with their successful dominance of several habitats of terrestrial communities. The various kinds of gymnosperms—the conifers, seed ferns, cycads, and ginkgos—are mostly woody plants in the form of either trees or woody shrubs. In Jurassic terrestrial plant communities, what seems to be lacking is an understory of nonwoody, small, soft, herbaceous or weedy plants. This role was presumably filled by ferns and by the few vestiges of small lycopods and sphenopsids that survived into the Mesozoic. One of the conspicuous differences between flowering plants and gymnosperms is that the former very successfully invaded the understory habitat, producing a great variety of small, low, nonwoody plants. A great part of the diversity of flowering plants is contained within this category. Proportionally, there is a much smaller number of flowering plants that are woody trees or shrubs. And, it is exactly in the understory group of angiosperms where the greatest emphasis on insect pollination occurs. Many angiosperm trees, such as willows or birches, are still wind pollinated. Some angiosperm herbaceous plants, such as ragweed, are also pollinated by wind, but the great majority depend on insects. In summary, the dominance of flowering plants in the modern world can be largely understood on the basis of two radical innovations that evolved in these plants: a change in the mode of achieving fertilization and a successful invasion and exploitation of many understory habitats.

PLANT COMMUNITIES DURING THE CENOZOIC

We have already noted that by the close of the Cretaceous Period flowering plants had already become the dominant floral type in terrestrial communities. They were widespread and diverse. There are several aspects of Cenozoic land plant distribution that are important. These include evidence for progressive increase in diversity of angiosperm floras, especially during the early part of the Cenozoic; the gradual modernization of the floras through the Cenozoic, so that they came to look more and more like modern floras; and the biogeographical distribution of the plants, which reflected the shifting of ancient climates.

Western North America has yielded many collecting sites for Cenozoic plants. Many floras from individual localities are quite diverse and may include more than a hundred species. For this reason, Cenozoic paleobotany for North America is very well known and consequently, our discussion will focus primarily on these floras.

Beginning during the Paleocene, several localities in California, Oregon, and Washington yielded extensive floras that consist primarily of tree leaves. These include magnolias, figs, a persimmon related to Asian forms, a custard apple, and others (Figure 14.25). The relationships of these plants point to two important conclusions. First, during the Paleocene and Eocene, the northwestern United States had a subtropical climate (Figure 14.26), with rainfall estimated in the range of 170 to 200 centimeters (70 to 80 inches) a year and a mean annual frost-freeze temperature of 20 degrees centigrade (68 degrees Fahrenheit). Second, these floras show a relationship to plants that are now widely scattered in distribution. Some of the fossils closely resemble trees that still live along the Pacific coast and that have apparently lived in the area for many millions of years. Others represent plants that are now confined to Asia, whereas still others are related to living plants that are confined to subtropical and tropical parts of the western hemisphere, especially southern Mexico and Central America.

Proceeding farther north along the Pacific coast, several Eocene floras have been found in Alaska. These have an aspect that is quite different from that of the Oregon and Washington fossils. The Alaskan floras include a preponderance of plants that are considered to be temperate or north temperate in present climatic distribution. Some of these are walnuts, chestnuts, elms, oaks, pines, cypresses, and *Sequoia*. There is also a less common element of warmer-climate plants, including relatives of the breadfruit tree and the avocado.

Taking these two areas together, we can conclude that Eocene climates were considerably more moderate than they are today. The boundaries between tropical and subtropical zones and between subtropical and temperate zones were situated conspic-

Figure 14.25   A large fan-palm leaf from the Eocene of Wyoming. The presence of palm trees this far north during the Eocene is a clear indication that climates in this area were much milder at that time than they are today. (Courtesy of National Museum of Natural History.)

Figure 14.26   Distribution of early Cenozoic (Paleocene and Eocene) floras in North America. Note the northern extension of subtropical floras.

uously farther north in North America than at the present. Accompanying this benign climate, plant communities were seemingly much more cosmopolitan and widespread than they are today. Floras very similar to those of Alaska have also been found in Greenland and Spitsbergen, well within the Arctic Circle. These high-latitude temperate forests had a circumpolar distribution during the Eocene.

In Oligocene rocks of Oregon, fossil floras record a distinct climatic cooling from that of Eocene time. All the Oligocene plants have temperate affinities. They include redwoods, hawthorns, beeches, alders, and oaks.

During the Miocene, the cooling trend of the Oligocene continued (Figure 14.27). Not only are the fossil plants indicative of temperate climatic conditions, but now they begin to show evidence of decreasing rainfall and somewhat more arid conditions (Figure 14.28). Grasslands also began to expand. Fossils continue to show a more cosmopolitan distribution of certain plants than there is today. For instance, these Miocene floras include several plants that are today found as natives only in Asia, such as the ginkgo, tree-of-heaven, and a water chestnut. Also, there are trees that are now native to eastern North America, having been gradually excluded from western areas. These include elms, sweet gums, magnolias, and figs.

During the Pliocene, several western floras indicate exclusion of Asian and eastern North American elements. Virtually all the fossils have a close relationship with plants still growing in the general western area. Several of the earlier Miocene floras indicate rainfall less than 75 centimeters (30 inches) a year, and the Pliocene plants record a continuation of this tendency toward aridity.

This sequence of fossil plants from the Eocene into the Pliocene records a climate that very gradually became cooler and drier. These climatic trends culminated during the Pleistocene with the onset of the first continental ice sheets covering much of Canada and the north-central part of the United States (Figure 14.29). Obviously, the distribution of plants was profoundly affected by glaciation. At the maximum limit of glaciation, the coniferous forests of the north pushed far south into the central United States. Concurrently, temperate hardwood forests shifted into the southern United States and Mexico with similar restriction of subtropical and tropical forests in southern Mexico and Central America. During interglacial times (periods when the ice sheets had melted), these various plant communities migrated north again to positions close to those of today (Figure 14.30). This extensive migration of forests north and south occurred four times during the Pleistocene, with the waxing and waning of major ice sheets. The result was an intensification of the process of fragmentation of plant distribution, resulting in sharply different plant communities in different parts of the continent.

Accompanying these changes in climate were two other effects that had a profound influence on the nature and distribution of modern floras in North America. We have already seen that there is clear evidence for increasing aridity in the western United States during the later Cenozoic. This culminated in the onset of truly desert conditions in the southwestern United States and northern Mexico. A host of plants evolved that were especially adapted to these severe conditions, including many kinds of cacti, shrubs, and small trees that could survive very low rainfall.

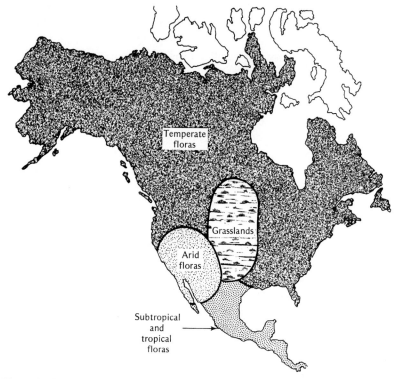

Figure 14.27 Distribution of Cenozoic floras during the middle of the Miocene in North America. Note that temperate floras now extend much farther south than they did during the early Cenozoic and that grassland prairies and desert floras are now evident as a result of climatic changes.

Figure 14.28 Two angiosperm leaves of Miocene age. (A) This specimen from Idaho is related to the grape family. (B) An oak leaf from Oregon. (Courtesy of National Museum of Natural History.)

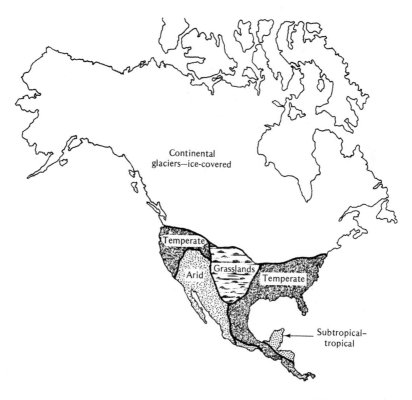

Figure 14.29   Distribution of floras in North America during the Pleistocene at a time of maximum glacial advance. Temperate floras are pushed far to the south, as are subtropical plants.

A second major development involved the area of the Great Plains of the United States and Canada. This area stands in the rain shadow of the Rocky Mountains. As the area became progressively drier, starting during the Miocene, larger plants, especially trees, were gradually excluded or confined to the banks of major streams. A variety of grasses became increasingly prominent, leading to the establishment of the extensive grassland prairies that still persist over much of this area. These extensive prairies offered large areas of new habitat for mammals. The fossil record of these prairies has been well preserved in the Great Plains where Miocene and Pliocene deposits contain the very resistant siliceous seed husks of grasses.

By far, the great majority of Cenozoic plant fossils that are known consist of leaves of trees, wood, and pollen, the latter two being very decay resistant. Fruits, seeds, flowers, or the remains of small herbaceous plants are much less common. Tree and shrub leaves tend to be rather tough and commonly do not decay as rapidly as those of herbaceous plants. Furthermore, trees live for years and produce millions of leaves over their lifetime. On the other hand, herbaceous plants have soft leaves and the entire plant tends to die at once. Consequently, we have a much poorer fossil record for

Figure 14.30   Distribution of floras in North America during the Pleistocene at a time of maximum glacial retreat. Grasslands and temperate forests extend far to the north; low rainfall at such times resulted in the spread of arid and semiarid plants.

small, soft angiosperms than we do for the longer-lived trees. Furthermore, most of the former plants are either annuals or biannuals; they live for only one or two seasons, set their seed, and then die. By contrast, trees and shrubs, the woody angiosperms, are all perennials; they live for many years. This is one of the very marked differences between the angiosperms and earlier dominant groups of plants. Among gymnosperms, there are virtually no annual or biannual forms known. The conifers, cycads, ginkgos, and others are all perennials. The development of an annual habit is one strategy for avoiding adverse, harsh conditions of either extreme cold or aridity. By germinating, growing, and setting viable seed within a short growing season, the annual plant can ignore winters and droughts. This way of life has been exploited to the fullest by angiosperms but not, as far as the fossil record indicates, by other major groups of plants. It appears that the number and diversity of herbaceous plants increased during the Cenozoic as climatic conditions became cooler and drier.

Another distinctive feature of Cenozoic angiosperms is the progressive development of the deciduous habit, in which all leaves are shed each year. This is in contrast to the evergreen habit, typical of modern conifers, in which only some leaves are lost during the course of a year. There are a few deciduous gymnosperms, such as the conifer *Metasequoia* (Figure 14.20F), ginkgo, and the bald cypress, the latter deriving its common name from the fact that it loses its leaves each autumn. The great majority of gymnosperms retain their leaves for more than one year, losing a few and growing some new ones each year, so that the total leaf complement is replaced only over a period of years. Angiosperms, on the other hand, include a great variety of deciduous trees and shrubs. The deciduous habit, like the annual habit, is an adaptation that is especially suited for survival under adverse conditions. After a deciduous tree drops its leaves, it goes into a period of dormancy; in this state, cold weather or drought cannot affect it nearly to the extent it could if it were evergreen. Deciduousness was common during the past in other plant groups, such as the southern hemisphere Permian seed ferns (e.g., *Glossopteris*). In angiosperms, it seems likely that the deciduous habit spread during the course of the Cenozoic as the climate deteriorated. Many angiosperms in tropical and subtropical climates are evergreen; the proportion of deciduous species increases conspicuously in higher latitudes, where the deciduous habit is advantageous.

KEY TERMS

angiosperm	gametophyte	pollen
araucarian	ginkgo	rhyniophytes
coal ball	gymnosperm	seed
cone	lycopod	sphenopsids
conifer	meiosis	sporophyte
cycads	mitosis	stamen
ferns	phloem	wood
flower	pistil	xylem

READINGS

Beck, C.B., ed. 1988. *Origin and Evolution of Gymnosperms*. Columbia University Press. 504 pages. Current ideas on gymnosperm paleobotany written by a group of authorities. Suitable for readers with a background in botany and paleobotany.

Cronquist, A. 1988. *The Evolution and Classification of Flowering Plants*. New York Botanical Garden. 555 pages. A revision of an important book dealing with family level and higher evolution and classification.

Dilcher, D.L. 1995. "Plant Reproductive Strategies: Using the Fossil Record to Unravel Current Issues in Plant Reproduction." *Monographs in Systematic Botany from the Missouri Botanical Garden* 53:17–198. Discussion of early angiosperm macrofossils and the key processes responsible for their evolution.

Friis, E.M., and others. 1987. *The Origins of Angiosperms and Their Biological Consequences*. Cambridge University Press. 358 pages.

Gastaldo, R.A. 1986. *Land Plants: Notes for a Short Course*. Studies in Geology 15. University of Tennessee, Department of Geological Sciences. 226 pages. This set of notes includes several papers on various groups of fossil land plants, including gymnosperms and angiosperms.

Niklas, K.J. 1997. *The Evolutionary Biology of Plants*. Chicago University Press. 449 pages. Textbook on the evolutionary history of morphology in plants.

Taylor, T.N., and Taylor, E.L. 1993. *The Biology and Evolution of Fossil Plants*. Prentice-Hall. 982 pages. Detailed modern textbook on paleobotany.

Thomas, B. 1981. *The Evolution of Plants and Flowers*. Peter Lowe Press. 116 pages. A short, easy-to-read, illustrated book on the evolution of plants. Emphasis is on the angiosperms.

Tidwell, W.D. 1975. *Common Fossil Plants of Western North America*. Brigham Young University Press. 173 pages. An excellent handbook for the identification of fossil plants. Very well illustrated and useful for any area, although emphasis is on western fossil floras.

White, M.E. 1996. *The Greening of Gondwana*. Reed Books. 256 pages. Very well illustrated book with color and black-and-white photographs of Australian fossil plants, with an emphasis on the Gondwana flora.

Early Consumers on Land: Reptiles

PALEOZOIC AND MESOZOIC TERRESTRIAL FAUNAS

Tetrapod faunas through the Paleozoic and Mesozoic are divisible into five faunas (Figure 15.1). Either morphological innovations or extinctions brought about these faunal turnovers, and, of course, the dinosaur era came to a close due to the end-Cretaceous mass extinctions.

Amphibians were predominant through the Mississippian and, as far as we know, were the only tetrapods present in coal swamps of the Early Pennsylvanian. Through the Middle and Late Pennsylvanian, synapsid reptiles came to dominate, first pelycosaurs during the Pennsylvanian and Permian and later the therapsids during the Triassic. Dinosaurs dominated through the Jurassic and Cretaceous with different faunas in each of these periods (Figure 15.1).

LATE PALEOZOIC AND TRIASSIC REPTILIAN FAUNA

EVOLUTIONARY TRANSITION TO REPTILES

By Early Pennsylvanian time, the first reptiles had evolved representing a great advance over the amphibians because reptiles were not closely linked to water. Scales on their bodies prevented desiccation, but, more important, they no longer needed water for reproduction. This major evolutionary step to a fully terrestrial existence was accomplished primarily due to innovation in reproduction. As one of their diagnostic features, reptiles have an **amniote egg**, a reproductive character that eventually allowed them to dominate many available land habits (Figure 15.2). The amniote egg is, in effect, a space capsule for the reptilian embryo. The amniote egg is covered by a hard but porous shell. Inside, the embryo is surrounded by tissues that nourish it and take care of wastes; respiration occurs through the shell. This egg can be laid on dry land without desiccating, unlike the soft amphibian egg that must be laid in water. Reptiles also forego the intermediate larval or tadpole stage of the amphibians; reptiles have direct development from embryo to juvenile.

The evolution of amphibians to reptiles was seemingly gradual, with fairly common intermediate forms that combined a blend of typical amphibian and reptilian characters (Figure 15.3). In general, early reptiles stabilized a particular style of backbone construction. Another diagnostic feature is the lack of an otic notch in reptiles, the ear being situated at the rear of the skull. Bones of the back part of the reptilian skull are reduced in number and size, a continuation of the trend present from rhipidis-

TETRAPOD FAUNAS

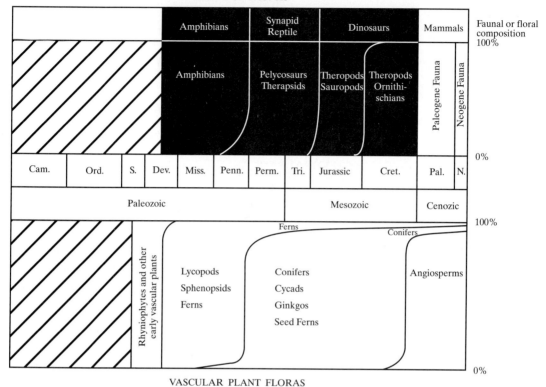

Figure 15.1   Terrestrial faunas and floras through the Phanerozoic, Paleozoic, and Mesozoic faunas highlighted.

Figure 15.2   Fossil dinosaur eggs from the Cretaceous of Mongolia. Although fossil amniote eggs are rare, they have been found in rocks as old as Permian. (Courtesy of Field Museum of Natural History, Chicago.)

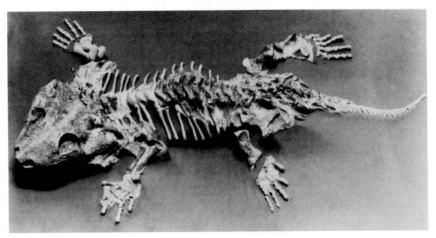

Figure 15.3   A mounted skeleton of the Permian amphibian *Seymouria* from Central Texas. This small animal, approximately 60 cm (2 ft) long, shows a unique blend of amphibian and reptilian characters but is much too young to have been the direct ancestor of reptiles. Notice that an otic notch is still present in the back of the skull. Other anatomical features resemble those of reptiles. (Courtesy of National Museum of Natural History.)

tian fish to amphibians. The reptilian skull tends to be somewhat narrower and higher than that of amphibians. The bones of the pelvic and shoulder girdles are enlarged and have wider areas of support with the backbone. The limbs still spraddle to the sides, but the bones tend to be somewhat longer and more slender. Wrist and ankle bones are reduced in number, and the finger and toe bones are stabilized into a consistent pattern of 2-3-4-5-3. In this system, each number indicates the number of bones per digit, the first number representing the inside digit (big toe) and the last, the outside digit (little toe).

Most important for distinguishing one type of early reptile from another is the structure of the bones in the temple region of the skull, behind the eye. Reptiles had either one, two, or no openings in the skull in this location; openings presumably accommodated bulging jaw muscles. The nature and arrangement of these openings, called **temporal openings**, provide data that is used in subdividing all major reptile groups (Figure 15.4). The group with the skull that is solidly encased in bone, with no opening at the rear of the side, is considered the most primitive (Figure 15.5). The reptiles of this group are referred to as stem reptiles or **anapsids** because they are the ones from which the other, more advanced reptiles are thought to have evolved. The only living anapsids are the turtles and tortoises. Another group, the **synapsids** or **mammal-like reptiles**, has a single temporal opening low on the side of the skull, beneath the postorbital and squamosal bones. They are extinct but very important because mammals evolved from this group of reptiles. The third group has two openings, one above the other, separated by a bony connection between the postorbital and squamosal bones (Figure 15.6). These are the **diapsids** or ruling reptiles. They include the dinosaurs of the Mesozoic Era, as well as most living reptiles—the crocodiles, alligators, snakes, and lizards. The final group, the **euryapsids**, has a single opening high on the skull, above the postorbital and squamosal bones, a condition derived from their diapsid ancestor. Euryapsids

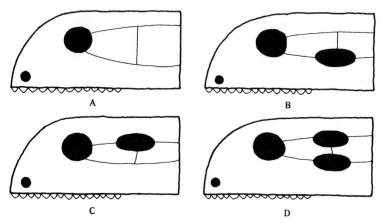

Figure 15.4   The basic types of reptilian skulls. (A) Anapsid, with solid bone and no temporal openings. (B) Synapsid, with the postorbital and squamosal bones meeting above the opening in the temporal region. (C) Euryapsid (aquatic and semiaquatic reptiles), with the temporal opening above the two bones. (D) Diapsid, with two openings and a bony bar between.

include an assorted lot of reptiles, including the ichthyosaurs and plesiosaurs. Most euryapsids had an aquatic or semiaquatic way of life.

LATE PALEOZOIC PELYCOSAURS AND THERAPSIDS

Throughout the Early Pennsylvanian, when reptiles first appeared, the fossils are relatively rare and without very much variety, but by Permian time, reptiles appeared in much greater abundance and variety and are clearly predominant over amphibians. This ascendancy of the reptiles was due, at least in part, to changes in climate. The coal swamp forests of the Pennsylvanian had dwindled by the Permian, giving way to seasonal rainfall and perhaps to more extremes in temperature. The various groups of

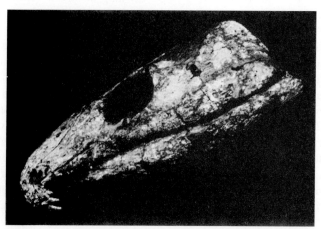

Figure 15.5   Side view of the skull of an anapsid, *Labidosaurus*, from the Permian of Texas. The solid bony temporal region is behind the large orbit for the eye. (Courtesy of Field Museum of Natural History, Chicago.)

Figure 15.6   Skull and lower jaw of a hadrosaur dinosaur, *Lambeosaurus*. Distinctive features include the flattened beak that lacks teeth, the long row of sturdy grinding teeth that denote a plant diet, a peculiar bony crest on the forehead, and the two large openings on the temporal region of the skull (diapsid condition). Skull is approximately 80 cm (32 in) long. (Courtesy of Field Museum of Natural History, Chicago.)

amphibians that had typified the Pennsylvanian were still present during the Permian but with a somewhat different aspect and lesser importance.

Amphibians evolved along two quite distinct lines with respect to habitat. Some amphibians became increasingly land-dwelling animals. These included some that were relatively large, up to 2 meters long, with massive short limbs and a large, flat, aligatorlike skull. The majority of amphibians took a different route. They had given up life on land and returned to a dominantly aquatic life, living all or most of their time in freshwater habitats. In some of these amphibians, the limbs are so reduced in length and size that the animals could not have supported themselves on land. In these types, the backbone became smaller and weaker. The tail was a swimming structure, and the skull was typically very low, flat, and broad. Reptiles, with somewhat larger brains, were also providing more intense competition. Due to their ability to lay eggs on land, the reptiles could flourish under drier conditions. Reptiles underwent a conspicuous radiation during the Permian and evolved a wide variety of herbivores and carnivores. How can we determine the diet of these extinct animals? One clue is body shape. Carnivores tend to be slender, whereas herbivores typically have a barrel-shaped trunk. Both body types are present in Permian reptiles. Some of the small, slender reptiles with sharp teeth are thought to have been insect eaters.

Two groups of reptiles were dominant during the Permian Period. One was the anapsids (Figure 15.7). These primitive reptiles were probably the ancestors of the synapsids, the mammal-like reptiles. The diapsids, which were barely represented during the Permian but which dominated the Mesozoic scene, led to the euryapsids. The synapsids flourished during the Permian Period and were represented by an archaic group, the pelycosaurs (Figures 15.1 and 15.8). These reptiles evolved directly into mammals. The pelycosaurs included *Dimetrodon*, the sail-back reptile. There has been controversy concerning the function of the conspicuous sail along the spine of this reptile (Figure 15.9). It has been suggested that the sail served primarily as a secondary sexual character; as a device for defense, making the animal look larger and fiercer than it really was; and also as a thermoregulatory device. This last idea has recently gained general acceptance. It is postulated that the animal, when it was cold, would turn itself broadside to the sun's rays. The sail, being richly supplied with blood vessels

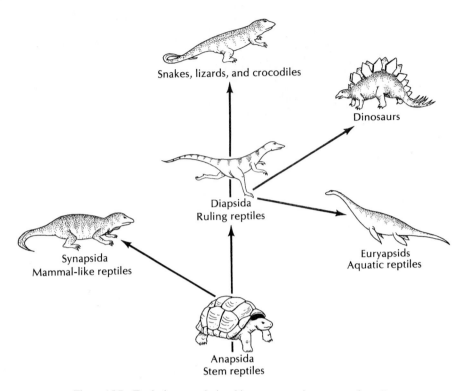

Figure 15.7 Evolutionary relationships among major groups of reptiles.

as evidenced from the preserved bones, served to warm this cold-blooded animal. If *Dimetrodon* became too hot, it simply turned itself 90 degrees until the sail was parallel to the rays of sunshine. Mammal-like reptiles of the Lower Permian bear little resemblance to mammals, but by Late Permian time they show a series of features clearly indicative of their evolution along lines that would ultimately culminate in mammals. These late Permian forms are quite rare in North America, but they are well known from a sequence of Late Permian and Triassic terrestrial rocks from South Africa known as the Karroo Group and in coeval rocks in Antarctica.

Mammal-like reptiles continued to flourish during the Triassic. The dominant Triassic forms were the **therapsids**. They had evolved considerably from their Permian ancestors. A variety of features in therapsids indicates a transition toward mammals. Their limbs became longer and more slightly built than those of other reptiles. Also, their limbs became tucked beneath their body instead of spraddled to the side. This new posture allowed for more efficient support of the body, and the entire body motion for running changed. In reptiles with the legs spread out laterally, movement takes place with the entire body flexing from side to side in a motion analogous to a swimming fish. In therapsids, reptiles and other animals with the legs directly beneath the body, the body flexes dorso-ventrally. This is a much more efficient movement that allows for much more rapid running. Also, the therapsid braincase became progressively larger in proportion to the rest of the skull, and the number of bones in the toes of the feet was reduced in number to a 2-3-3-3-3 formula.

Figure 15.8   A mounted skeleton of a primitive pelycosaurian synapsid from the Permian of New Mexico. The obscure, single, large temporal opening is behind and below the orbit for the eye. (Courtesy of Field Museum of Natural History, Chicago.)

The therapsids also developed a secondary, hard, bony palate. This consisted of a bony extension at the front of the mouth from bones of the upper or primary palate. Presence of a secondary plate allows an animal to eat and breathe at the same time. This is especially useful, if not necessary, for maintenance of body temperature in a warm-blooded animal. It has been suggested that the presence of a bony secondary palate in these reptiles is a clue that they may have already become warm-blooded— one of the important characteristics of mammals that is nearly impossible to detect directly in the fossil record.

The teeth of mammal-like reptiles also underwent fundamental changes. The number of teeth was reduced and confined to the bones of the jaw edges; they were not scattered over the roof of the mouth as in many other reptiles. Pelycosaurs began to develop a differentiated dentition (Figure 15.8), but it was the therapsids that devel-

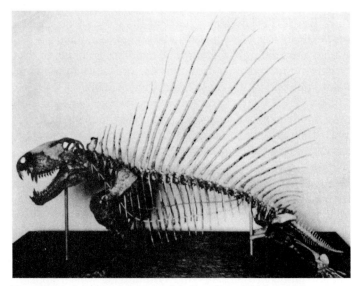

Figure 15.9   A mounted skeleton of the Permian sail-back reptile *Dimetrodon*, approximately 3.5 m (11 ft) long. Notice the large single opening in the rear of the skull behind the orbit of the eye. The highly elongate bones rising from the vertebrae supported a fleshy sail in life. The animal was probably an aggressive carnivore, as witnessed by the impressive teeth in the jaws. (Courtesy of National Museum of Natural History.)

oped dentitions that were like mammals. Teeth became differentiated into nipping incisors in front; stabbing canines next; followed by shearing, cutting, and grinding premolars and molars at the back of the jaw. Most other reptiles have simple, cone-shaped teeth that are basically all alike. This differentiation of the tooth row is another feature in which the mammal-like reptiles resemble mammals.

One final feature must also be mentioned. The defining characteristic to differentiate a fossil therapsid reptile from a true mammal is the number of bones in the lower jaw. Mammals are characterized by the presence of a single bone in the lower jaw, the **dentary**. In reptiles the dentary articulates with a bone of the upper jaw, the **squamosal**. All reptiles have more than one bone in the lower jaw, and one of these, the **articular**, articulates with the **quadrate** bone of the skull. Advanced mammal-like reptiles, the therapsids, approach the mammalian condition in that the dentary is much enlarged and other bones of the lower jaw are reduced in number and size. The articular and quadrate bones become quite small. The variety of features mentioned above, i.e., differentiated dentition, posture of the legs, etc., are typical mammalian characters, and the mammal-like reptiles gradually developed these features. The transition to mammals was truly gradual. The evolutionary transition was slow with many intermediate forms between a typical mammal and a typical reptile. However, our classification system dictates that we place organisms in one group or another. The number of jaw bones was chosen as the diagnostic character. Therefore, if the jaw articulation is effected by the dentary-squamosal, the specimen is a mammal; if by the articular-quadrate, it is a reptile. The first mammals evolved during the Late Triassic.

Another major group of land animals from the Mesozoic was the diapsid, or ruling reptiles. One important diapsid group was the **thecodonts**, which gave rise to the groups known as **dinosaurs** before the close of the Triassic. The thecodonts first appeared during the Early Triassic and became extinct at the close of the period. Whereas some thecodonts ran on all four legs (**quadrupedal**), others exhibited a new **bipedal**, or two-legged stance. They used the front feet for food handling, grabbing, and so on. Most thecodonts were relatively small reptiles, from the size of a chicken to about 2 meters long. Their bones were thin and lightly constructed; some had hollow leg bones, like birds. The skull was thin-boned with large vacuities, or areas where bone was not developed. The hind limbs were quite elongate and powerful for swift movement. The front limbs were short and weak. They give the impression of having been fast runners for their time, in some respects resembling large, ground-dwelling birds, such as ostriches. Some may have been insectivorous, others perhaps caught other small prey. However, they had feeding habits that required speed. Their swiftness may also have been advantageous in escaping larger, heavier predators, such as some of the amphibians and mammal-like reptiles. By the close of the Triassic, thecodonts had given rise to both groups of dinosaurs, the Saurischia and Ornithischia. The **Saurischia** comes from *sauria*, meaning reptile, and *ischia*, referring to the ischium bone of the pelvis (Figure 15.10). These dinosaurs had a pelvis built like that of many other reptiles, hence the name lizard-hipped dinosaurs. From them evolved the other major group of dinosaurs, the **Ornithischia**, which appeared at the end of the Triassic. This group had a birdlike pelvis.

The saurischians were relatively common by the Late Triassic. These earliest saurischians were bipedal, small, and lightly constructed, with insectivorous or carnivorous food habits.

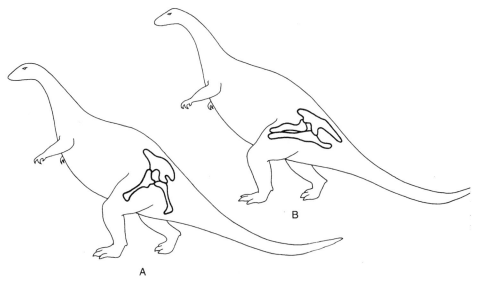

Figure 15.10   Pelvic girdles in bipedal dinosaurs. (A) In saurischians the girdle of three bones (ilium, ischium, and pubis) is triradiate. (B) In ornithischians the pubis bone extends both fore and aft.

JURASSIC AND CRETACEOUS DINOSAUR REIGN

By the close of the Triassic, there were many kinds of vertebrates populating terrestrial ecosystems, including snakes, lizards, therapsids, dinosaurs, and mammals; but it was the dinosaurs that emerged during the Jurassic as the dominant land animals. Dinosaurs remained in this dominant role for nearly 100 million years, until their extinction at the close of the Cretaceous.

Mesozoic ecosystems are divisible into the Jurassic and Cretaceous based on both the dominant dinosaur groups and on floras (Figures 15.1 and 15.11). The remarkable evolution of angiosperms during the Early Cretaceous (Chapter 14) undoubtedly had a major impact on the herbivorous dinosaurs.

JURASSIC ECOSYSTEM

Gymnosperms, cycads, ginkgos, and conifers were the dominant forms of Jurassic vegetation. By comparison to angiosperms, these gymnosperms seem like poor fodder, but this harsher vegetation supported a diverse array of herbivores, including the largest terrestrial animals to ever live, the sauropods.

During the Jurassic, the major group of large and small carnivores were the **theropods**. These were saurischians, and all were bipedal animals. Many of them were quite small, 6 feet or less in length. By the close of the Jurassic, others had become very large beasts, up to 30 feet long, with heavy bodies and hind legs (Figure 15.12). The front legs became progressively smaller and weaker, and the teeth evolved into long, spearlike, stabbing teeth. These animals must have been formidable predators. The theropods were the only carnivorous dinosaurs. Jurassic examples of the theropods include *Allosaurus* and *Ceratosaurus*.

Period	Ornithischian Herbivores	Saurischian Herbivores	Saurischian Carnivores	Plants
Cretaceous	ORNITHOPODS *Corythosaurus* *Lambeosaurus* *Parasaurolophus* CERATOPSIANS *Protoceratops* *Triceratops* ANKYLOSAURS *Ankylosaurus* *Euoplocephalus*	only a few sauropods	THEROPODS *Albertosaurus* *Tyrannosaurus*	Angiosperms
Jurassic	STEGOSAURS *Stegosaurus*	SAUROPODS *Apatosaurus* *Brachiosaurus* *Brontosaurus* *Diplodocus* *Seismosaurus*	THEROPODS *Allosaurus* *Ceratosaurus*	Conifers Cycads Ginkgos

Figure 15.11   Common dinosaurs and plants present during the Jurassic and Cretaceous; dinosaurs given by trophic type.

Figure 15.12   A large carnivorous dinosaur from the Mesozoic Era. Note the bipedal stance, the triradiate structure of the pelvic girdle, the reduced forelimbs that were used for grasping prey, and the large teeth. (Courtesy of National Museum of Natural History.)

In addition to the carnivorous dinosaurs, the saurischians include the major group of Jurassic herbivores, the enormous **sauropods**. These were strictly herbivorous, quadrupedal animals. Again, some were small, but others became very large, indeed. Sauropods include such animals as *Apatosaurus, Brachiosaurus, Brontosaurus, Camarasaurus, Diplodocus* (Figure 15.13), *Seismosaurus*, and *Suprasaurus*. The heaviest ones are estimated to have weighed from 45 to 65 tons (*Brachiosaurus*), and the longest ones were nearly 42 meters (137 feet) long (*Suprasaurus*). The legs were pillarlike, constructed similar to those of an elephant in order to support the tremendous weight of the body.

The legs of most sauropods show clearly that these animals evolved from a bipedal ancestor. All sauropods, except *Brachiosaurus* had front legs shorter and less massive than the hind limbs. The neck and tail were commonly elongate and the head was quite small. The teeth were small and peglike. Exactly what plants formed the bulk diet of sauropods is not certain, but during the Jurassic it was certainly gymnosperms, such as cycads and pines, which is puzzling because this certainly is not "soft, lush vegetation" that is easy to digest. Sauropods possessed giant gizzard stones in a crop that surely helped to break this tough vegetable material. At one time there was controversy concerning whether these animals were truly land dwelling, or whether they spent most of their lives in freshwater pools and lakes, where the water could help buoy up their tremendous weight. However, it is now believed that their legs were built massively enough to support their enormous weight, and footprints have been found deeply impressed into what was once soft sediment. In addition, careful paleoenvironmental work on the rocks that contain sauropod remains indicate that they did not live in water habitats; in fact, many lived in arid or at least monsoonal-type (very dry during most of the year) climates. Indeed, the most recent reconstructions of sauropods depict them as high-level grazers, standing on their two hind

Figure 15.13   The giant sauropod *Diplodocus* of Late Jurassic age from Utah. Note the arch shape of the spinal column and the massive legs, designed to support many tons of weight. This dinosaur is the longest known, close to 28 m (90 ft) in length, but it was not one of the heaviest, having a somewhat lighter build than other large sauropods had. (Courtesy of National Museum of Natural History.)

legs, feeding from the leaves of trees. Their large size was probably effective at protecting them from all but the largest Jurassic predators, such as *Allosaurus*, which was 9 meters long. It is interesting to speculate whether several of these large predators ganged up on one of the gigantic herbivores in pack fashion, or if they preyed on the old, sick, or young.

The other major group of dinosaurs, the ornithischians was much less diverse and less common than the saurischians during the Jurassic. An exception to this was the Jurassic **stegosaurs**, the best-known example of which is *Stegosaurus* (Figure 15.14). These ornithischians were quadrupedal but, again, show clearly that they evolved from a bipedal ancestor. The front limbs are typically shorter and less heavily constructed than are the hind legs. The stegosaurs reached lengths of about 7 meters. They had a very small head and brain cavity for their size and an enlarged nerve center in the pelvic region of the spinal column.

The outstanding feature of the stegosaurs is the distinctive row of heavy, triangular bony plates, or scutes, arranged along the back from the head to nearly the end of the tail. These plates have traditionally been viewed as some type of defensive protection from Jurassic predators, along with the spines at the end of the tail that would have been a formidable defensive weapon. However, more recently it has been suggested that these plates may have functioned in temperature regulation, much the same as the sail of *Dimetrodon*, because compared to normal bone, these plates have an extremely high density of blood vessels.

CRETACEOUS ECOSYSTEM

Dramatic changes occurred in terrestrial ecosystems at the onset of the Cretaceous (Figures 15.1 and 15.11)—the dominant plants changed to angiosperms (Chapter 14), and the dominant herbivorous dinosaurs changed. The only constant was the predators. The carnivorous theropods remained the dominant dinosaur predators during the Cretaceous. Examples are *Albertosaurus* (Figure 15.15) and *Tyrannosaurus*. These

Figure 15.14   The ornithischian dinosaur *Stegosaurus*, of Jurassic age. Stegosaura were approximately 6 m (20 ft) long and characterized by two rows of large bony plates along the backbone. Notice the long bony spikes on the end of the tail. These dinosaurs were herbivores. (Courtesy of National Museum of Natural History.)

Figure 15.15 The skull and lower jaw of one of the large carnivorous theropod dinosaurs, *Albertosaurus*. The specimen is about 1 m (3 ft) long. (Courtesy of Field Museum of Natural History, Chicago.)

were the largest and probably the most ferocious of the dinosaurs. One of the largest and most popular dinosaurs is *Tyrannosaurus rex*, the "king tyrant lizard," known from Upper Cretaceous rocks. This reptile is surely one of the largest predators known; it was 13 meters long and weighted up to 8 tons. Its spearlike teeth were up to 15 centimeters long.

Among herbivores, the giant sauropods of the Jurassic dwindled during the Cretaceous. They were exceedingly rare in North America during the Cretaceous but occur in somewhat greater numbers in India and Mongolia. In North America there is some indication that they may have populated higher elevation habitats, which, if true, may indicate that they followed gymnosperms into this habitat.

Orthithischians were the dominant Cretaceous herbivores. Three groups were especially important, ornithopods, ankylosaurs, and ceratopsians (Figure 15.11). These, along with the Jurassic stegosaurs and others, constitute the bird-hipped (ornithischian) dinosaurs that were all herbivores and either bipedal or quadrupedal. The ornithischians evolved from a bipedal saurischian ancestor.

Several groups of **ornithopods** were important during the Cretaceous, including the hypsilophodonts, iguanodontids, and hadrosaurs. **Hadrosaurs**, or duck-billed dinosaurs (Figures 15.6, 15.16–15.18), were especially common. They were bipedal with reduced front legs. Some of the earliest Jurassic forms had relatively large front legs that could have been used for walking, but the Cretaceous forms were clearly bipedal. Examples include *Corythosaurus, Lambeosaurus*, and *Parasaurolophus*. A characteristic feature of hadrosaurs is that the tooth row did not reach the front of the mouth; it is probable that these animals had a horny beak. The herbivorous duck-billed dinosaurs were abundant and diverse during the Cretaceous. Many evolved elongate nasal passages that served as sounding tubes.

Another group of ornithischians, the **ankylosaurs**, were large armored, tanklike animals that somewhat resembled living armadillos. The back and sides of ankylosaurs were completely encased by columns of close-fitting, thick bony scutes. In addition, there were long bony spikes at the shoulder region and on the tails of some. They had very feeble dentition; some lacked teeth altogether.

Figure 15.16 A mounted skeleton of the hadrosaur (ornithopod) dinosaur *Anatosaurus*. This dinosaur lived during the Late Cretaceous. Note the typically ornithischian structure of the pelvic girdle bones. The jaw lacked teeth in the front of the mouth, where the teeth were replaced by a horny beak. Hence the name duck-bill dinosaur is commonly applied to this group of dinosaurs. (Courtesy of National Museum of Natural History.)

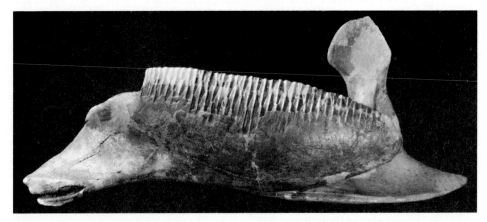

Figure 15.17 Inner view of the lower jaw of a hadrosaur (ornithopod) dinosaur, possibly *Kritosaurus*, showing the battery of large grinding teeth for shredding plant debris. Although quite different in origin, the tooth battery somewhat resembles the single molar tooth of modern elephants. The total length is 90 cm (nearly 3 ft). (Negative No. 120324 courtesy of Department Library Services, American Museum of Natural History. Photograph by J. Kirschner.)

Figure 15.18   Enlargement of the preserved skin from the rib area of a mummified hadrosaur (ornithopod) dinosaur, *Edmontosaurus annectens*, from the Kirtland Formation, San Juan County, New Mexico, of Cretaceous age. The skin surface is covered with large and small polygonal to oval bony tubercles. Actual width of specimen is 50 cm (1.5 ft). (Negative No. 35608 courtesy of Department Library Services, American Museum of Natural History. Photograph by Anderson.)

The final group of ornithischians that, like the ankylosaurs, was confined to the Cretaceous is the **ceratopsians**, or horned dinosaurs. The most famous of this group is *Triceratops* (Figure 15.19). These dinosaurs were mostly of moderate size, 5.4 to 6 meters long. They were typified by having on the back of the skull a wide, flat bony frill that extended over the neck region and surely protected that vital and vulnerable area from attack. Another interesting feature of these animals is that they gradually evolved horns on the nasal region of the face above the eyes. The oldest and smallest ceratopsians from the Early Cretaceous lacked horns.

Typical Cretaceous dinosaurs were the large theropods, such as *Tyrannosaurus*, and three groups of herbivorous ornithischians: the ornithopods, ankylosaurs, and the

Figure 15.19   A mounted skeleton of *Triceratops*, the youngest and most advanced of the herbivorous ceratopsian dinosaurs. There was a horny beak at the front of the mouth and three horns on the skull. A large bony frill extended back from the skull, protecting the neck region. The specimen is Late Cretaceous in age and approximately 6 m (20 ft) in length. (Courtesy of National Museum of Natural History.)

horned dinosaurs (Figure 15.11). All became extinct by the close of the Cretaceous. It is not clear whether all groups disappeared more or less simultaneously or whether the extinction took place gradually over several millions of years. The very youngest dinosaur remains known in the United States consist of scraps of ceratopsian bone from beds that are latest Cretaceous or, perhaps, youngest Cenozoic in age. Thus, it is possible that *Triceratops* outlived other groups of dinosaurs and was the last one of these large reptiles to become extinct.

ADDITIONAL DINOSAUR FACTS

WARM-BLOODED DINOSAURS

For the last several years a broad discussion has ensued concerning the physiology of dinosaurs—were they cold-blooded like living reptiles or warm-blooded like living birds and mammals? Living cold-blooded animals, or **ectotherms**, cannot maintain constant internal body temperature, so they must rely on the external environment for internal body warmth. This is why lizards and snakes are sluggish on a cool morning. Alternatively, warm-blooded animals, or **endotherms**, maintain a nearly constant body temperature through metabolic heat. Being warm-blooded exacts a high price on animals because they must eat large quantities of food to maintain a constant metabolism for heat generation and they die if their body temperature varies too much. Both metabolic extremes have adaptive value for certain habitats. However, warm-blooded organisms are much more active animals. We now know that the classical definitions of ectotherms and endotherms are end members of a spectrum and that some animals have intermediate metabolisms.

No one can obtain a direct measure of a dinosaur's body temperature, so a variety of indirect means have been devised to examine this question. Nearly all these methods indicate evidence from dinosaurs that is consistent with endothermy, but in every case questions are raised: a) data are consistent only with some dinosaurs and not with all, b) the method has untestable potential flaws, or c) explanations other than endothermy could explain the data. Examples of attempts to determine dinosaur metabolism include, among others, posture and gait, predator-prey biomass ratios, paleogeographic distribution, bone histology, and body size.

Dinosaurs have limbs directly beneath their bodies, which is a posture present in living mammals and birds. This characteristic of living warm-blooded animals affords a much more active life style than the posture of cold-blooded organisms with legs spraddled to the side. Therefore, dinosaur posture is consistent with the active life of warm-blooded mammals and birds, but where this may record the general activity of dinosaurs, it does not necessarily record their metabolism. The predator-prey biomass ratio has been calculated for numerous fossil faunas. This is an estimate of the ecosystem biomass of the predator to that of the prey species. A warm-blooded predator must eat more often than its cold-blooded counterpart, so the predator-prey biomass ratio would be smaller. Again, the ratios calculated for dinosaurs are comparable to mammals rather than to cold-blooded reptiles. Although this method has a basic appeal, critics question the validity of the data due to the vagaries of fossil preservation and claim that, at best, this method only provides information on the predators.

Mesozoic dinosaurs lived at polar extremes, the north slope of Alaska in the north and Antarctica in the south. Even with adjustments of plate positions due to plate tectonics, these dinosaurs lived inside the arctic and antarctic circles and would have experienced winter months with 24-hour darkness. Although this would be impossible for living ectotherms, did these dinosaurs only migrate to these polar extremes during the summer months or was the Mesozoic climate uniformly warmer?

When first announced, evidence from the histology (fine detailed structure) of bones was convincing. Living ectotherms have compact, dense bones with few blood vessels and growth rings, indicating periodic growth. In contrast, living endotherms have porous bone with numerous blood vessels and no growth rings. This structure of the bone is regarded as a direct measure of the metabolism of an animal, and dinosaur bones were nearly identical to those of mammals. With subsequent, more comprehensive work in this field, the evidence is not as clear. Finally, body size has been considered, which suggests that enormous dinosaurs were more likely faced with problems of venting excess body heat rather than a lack of body heat.

When all these various types of information are analyzed, the metabolism for dinosaurs is still unclear. Probably, dinosaurs as a whole exhibited a variety of metabolisms. The most consistent evidence for endothermy is for certain theropods and ornithopods. Smaller dinosaurs tend to have more consistently convincing evidence, and it is clear that small theropods were ancestral to the endothermic birds. The largest dinosaurs may have been **gigantotherms**. Simply because of their enormous size, they could maintain a nearly constant body temperature. Dinosaurs of intermediate size may have been ectotherms or partial ectotherms.

THE SOCIAL RELATIONSHIPS OF DINOSAURS

In the past few years there has been a great advance in our understanding of how dinosaurs lived and what kinds of social interactions they may have had. Much of this new information must be credited to Jack Horner, a dinosaur paleontologist at Montana State University and the Museum of the Rockies in Bozeman, Montana. He and his fellow workers have discovered the greatest concentration of dinosaur nests, their contained eggs, and baby dinosaur skeletons that has ever been found. These discoveries permit us to make inferences about dinosaur behavior. The nests occur in two different layers in the Hell Creek Formation of the Late Cretaceous, and they were built by two different ornithopods, adults of which have been found associated with the nests.

The nests are quite large and were excavated into the soil by the dinosaurs (Figure 15.20). The nests may be up to 2 meters (6 feet) in diameter and are shaped like large bushel baskets. The eggs are very carefully arranged in the nests and have a definite spiral pattern around a central egg. The eggs are elongate and stand on end, indicating a precise behavioral pattern either in laying the eggs or manipulating them after they were laid. The nests are filled with vegetable material, now plant fossils. These may have helped to incubate the eggs or served as camouflage to help prevent egg eaters, such as other small reptiles or perhaps mammals, from preying on the eggs. Decomposing plant material generates heat, as in a compost pile, and if heat was needed, this may be additional evidence that the dinosaurs were warm-blooded. The nests are in large clusters and have a definite spacing, about 25 feet apart. This spacing indicates that a large cluster of dinosaurs shared the nesting territory at the same time

Figure 15.20 Dinosaur egg clutches of a hypsilophodontid (ornithopod), reduced in size. From the Late Cretaceous of central Montana. (Courtesy of the Museum of the

and that they may have returned to the same nesting site for more than one year. Thus, nests provide clear evidence of social behavior.

The great majority of the hadrosaur nests contain only broken egg fragments, indicating that most of the baby dinosaurs hatched successfully. However, a few nests contain whole eggs, and studying these by X-ray and CAT-scan techniques reveals that they contain the bones of embryonic dinosaurs (Figure 15.21). These are the most numerous and best preserved fossils of unhatched dinosaurs ever discovered. There is one very important difference between the contents of the nests in the two layers at the Hell Creek Formation. In the older layer a few skeletons of recently hatched baby dinosaurs have been found. These are only slightly larger than the eggs and indicate that the dinosaurs died shortly after hatching. In the other nests, however, juvenile dinosaurs have been found that are up to 4 feet long. These clearly inhabited the nests long after they hatched and underwent a conspicuous growth interval before leaving the nest. Thus, they must have been fed by one or both parents. Fossil evidence of this kind of parental care is virtually unique to these sites in Montana.

However, another indication of social behavior by dinosaurs has been developed by the study of fossil footprints. In Cretaceous rocks in the western United States, dinosaur footprints are abundant in some layers of rock, especially in western Canada, Montana, Wyoming, Utah, and New Mexico. The rocks were deposited on a coastal plain that bordered the western edge of an extensive inland sea that spread from Texas to northern Alaska during Cretaceous time. This was the last major flooding of the North American continent by shallow seaways. The dinosaur tracks indicate several things. First, exactly the same kinds of tracks have been found over a very wide latitudinal range, from New Mexico to western Canada. This indicates that the dinosaurs

Figure 15.21   Egg of the theropod dinosaur *Troodon*, showing evidence of the embryo skeleton within, from Upper Cretaceous rocks of central Montana. (Courtesy of the Museum of the Rockies, Bozeman, MT.)

may have migrated, perhaps on a seasonal basis. Second, the same kinds of tracks, especially those of some herbivorous dinosaurs, occur in dense concentrations, thus indicating group or herd behavior. Furthermore, in the dense, herd trackways, juveniles tend to be in the center, which is herd behavior for protection of the young. This, coupled with the nest information, is further strong indication of the social instincts of dinosaurs. Jack Horner has called some of the hadrosaurs the "cows" of the Cretaceous, living in herds, migrating with the seasons, and taking care of their young.

HOW TO MEASURE A DINOSAUR

If you pick up any book about dinosaurs, you can find many vital statistics on these animals—how much they weighed, how long or tall they were, how fast they could run, and so on. In a few cases, specimens have been found that are complete enough to be used as a basis for estimating length or height. In other cases fossil footprints, identified as to the kind of dinosaur that made them, can be used to estimate weight, based on cross sections of leg bones. But if you look into the matter of what actual bones have been found for a specific dinosaur, you may find that a single complete skeleton has never been uncovered. Rather, a partial skull, a couple of bones, one or two vertebrae, and a couple of ribs may be all that is known for a particular species. How, then, do paleontologists come up with seemingly complete and accurate measurements from such incomplete remains?

Using footprints to obtain an estimate of animal length does not provide an overall length, but rather the trunk length, from the hipbone socket to the forelimb socket. The length of the tail and neck cannot be directly measured from footprint data. First, the front and hind footprints must be identified. These may differ slightly in size, shape, number of toe impressions, and so on. The front footprints may occur slightly in front of the hind footprints or on either side of them, or the two kinds of prints may overlap, with the hind foot stepping directly or partially into the mark left by the front foot. Second, the pattern of footprints is determined by considering the relative length of the trunk and the legs. Four successive prints must be identified. On a drawing of these, the center of the left-side front footprint is connected with the center of the right-side front print. This will be an oblique line crossing the center of the trackway. The same thing is done for two successive hind prints, left and right. Next, the center points of the two oblique lines along the center of the trackway are connected. This distance is called the stride and provides a direct measure of the trunk length. Finally, estimates of tail and neck length are added to this to obtain the overall length of the dinosaur.

The weight of an animal is related to the total volume of bone and the total volume of tissues that form the animal. Actually, we are using weight here inaccurately; what we are really calculating is the mass of the animal not its weight (which is the mass multiplied by the force of gravity). First, we can determine the mass of various living animals, large and small. We express this mass, for large animals, in tonnes. A tonne is not the same as the English ton, but it is a metric ton or 2,205 pounds. Masses of living animals range from the blue whale, at 90 to 100 tonnes (about 200,000 pounds), to humans, at 0.05 to 0.07 tonnes (110 to 150 pounds), to an amoeba, at 0.0000003 tonnes (0.01 ounce). To calculate mass, it is necessary to have an estimate of the volume of the animal and the average density of the stuff forming the animal—bone, muscles, gut, and so on. It turns out that most animals are about the same as the density of water. Some animals, such as crocodiles, which have heavy bone scutes, are slightly denser than water—about 8 percent more dense. However, most animals are slightly less dense than water; they can float in water. The volume of large animals such as dinosaurs has been determined by building scale models of the animal, determining the volume of the model, and then multiplying that by the scaling factor. Such calculations of volume for a large sauropod, such as Brachiosaurus, yield a volume of between 45 and 50 cubic meters. If the dinosaur were the same density as water, its mass would be between 45 and 50 tonnes, approximately one half that of a living blue whale and nine times heavier than a modern elephant.

Another way to estimate mass is by measuring the circumference of the upper leg bones (the femur in front or the humerus behind). This has been done for different sizes of living animals. A nearly straight-line relationship exists, so by extrapolation with living animals, dinosaur mass can be determined. The result of these analyses may result in estimated masses that are somewhat less than those from scale-model analysis. For instance, by the second method, *Brachiosaurus* weighs 32 tonnes. A different formula must be used for bipedal than for quadrupedal dinosaurs.

We can also estimate how fast the animals could run by studying the footprints. We determine the animal's stride, just as we did above to find the length of the animal. Speeds of various animals have been determined relative to their strides. The speed, adjusted to account for gravity, is called the dimensionless speed. Using this and leg length, it is possible to calculate the speed of the animal. The greatest speeds for dinosaurs obtained in this way are for ornithopods, with running speeds of approximately 25 to 27 miles per hour. This is faster than a human can run and slightly slower than a horse can gallop, which is quite respectable for an animal weighing 1,300 pounds. Tracks recording running have probably not been found for the largest sauropods; it is possible that the really big sauropods ambled along at a much slower pace.

Finally, it has been possible to measure the tones (sounds) that dinosaurs may have made. The crested hadrosaurs had long tubular nasal cavities in the head crest that connected with the roof of the mouth. These hollow cylindrical cavities may have been sound-producing chambers. If so, the pitch of the sounds that they made, assuming air passed over the openings to the cavity as in an organ pipe, would be directly related to the length of the chamber; long pipes produced bass sounds and the short pipes, treble sounds. The frequency of the sound in cycles per second can be calculated by dividing 170 by the length in meters of the pipe. Performing this calculation for one hadrosaur dinosaur resulted in a pitch of F natural. In any one species of hadrosaur some specimens have smaller crests than others. The crests are thought to be a secondary sexual characteristic, and the females are presumed to have had the smaller crests. If this is so, then females would have had made higher tones than males. Even among dinosaurs, the females apparently sang soprano.

THE EXTINCTION OF DINOSAURS

One of the most fascinating problems of paleontology, about which much has been written, is the cause or causes of the extinction of the dinosaurs. First, we should reemphasize that most dinosaurs became extinct long before the end of the Cretaceous. The stegosaurs lived only during the Jurassic, and the sauropods were already scarce during the Late Cretaceous. The most common dinosaurian remains in very young Cretaceous rocks, immediately prior to their final extinction, are ceratopsians, theropods, and ornithopods. Thus, the final dinosaur extinction involved these three groups. A host of different theories have been advanced to explain the disappearance of these large animals. It has been suggested that a) the gradually developing small mammals may have eaten all their eggs; b) they were bombarded by cosmic rays that rendered them sterile; c) mountain building somehow affected them; d) there were changes in climate, in vegetation, and in distribution of land and sea. Many other possible, and some outrageous, ideas have been suggested.

A recent theory of dinosaur extinction that has gained considerable support is that dinosaurs, ammonites, and other groups that became extinct at the end of the Cretaceous Period were forced to extinction when an enormous asteroid struck Earth. Evidence for this impact is a globally distributed clay layer greatly enriched in the rare element iridium at what is called the **K/T boundary** (K is the map symbol for Cretaceous and T is the map symbol for the Tertiary, an outdated term for much of the lower Cenozoic). This iridium-enriched layer is present at numerous localities in Europe, North America, and the deep sea. It is suggested that the dust from the impact was injected into the upper atmosphere and caused global darkness, restricting photosynthesis and creating imbalances in both the marine and the terrestrial food chains. The question of the end-Cretaceous extinctions has been a topic of recent active research and lively debate.

As discussed in Chapter 5, the end-Cretaceous extinctions are regarded as the third most devastating mass extinction to have occurred on Earth. However, unlike the two more severe events, the end-Permian and end-Ordovician extinctions that principally devastated shallow-water marine faunas, the end-Cretaceous mass extinctions were more selective. Certain dominant organisms in each ecosystem became extinct while others survived. In the oceans, planktonic foraminifera were devastated; nektonic ammonite cephalopods and reptile predators became extinct; benthonic rudistid bivalves became extinct; and benthonic corals, echinoids, bryozoans, and sponges suffered considerable extinctions. On land, of course, dinosaurs and pterosaurs (flying reptiles) became extinct. Therefore, the cause of this extinction must explain, for example, why echinoids and bryozoans suffered severe extinctions and bivalves and gastropods did not, and why dinosaurs became extinct but many other reptiles and mammals were much less or, in some cases, largely unaffected.

In order to come to a firm concensus on the cause or causes of this important juncture in Earth history, it is essential to 1) understand which organisms became extinct and which survived; 2) document the pattern of decline of those groups that underwent significant extinctions; 3) develop a good understanding of the physical environment through the Late Cretaceous; and 4) document evidence for an asteroid impact, widespread volcanism, and other putative causes. As mentioned above, the extinction was selective. Zooplankton, nektonic predators, reef-forming rudistid bivalves, and others were greatly affected in the oceans. On land the dinosaur herbivores and carnivores became extinct, and plants were also impacted.

Although we can determine the ultimate success or non-success of organisms across the K/T boundary, it may never be possible to develop a robust understanding of the pattern of extinction leading to the boundary. Due largely to the problems of preservation and sampling, it may not be possible to discern whether the extinctions were gradual or abrupt. For example, the fossil record of angiosperm leaves indicates that at low latitudes in North America, species extinction was approximately 80 percent at the end-Cretaceous; however the spore and pollen record in the same area was as low as 30 percent. Which is correct? Even if both were the same, are the western North American data, the only comprehensive well-studied records, representative of the whole globe?

We do know that Earth conditions were changing during the Late Cretaceous and that both substantial volcanism and a large asteroid impact occurred approximately 65 mya. The coincidence of these factors, rather than any one in isolation, may explain the end-Cretaceous mass extinctions. Climates and environments were changing during the Late Cretaceous. The generally warm Cretaceous climate was cooling, and a global regression of the seas was occurring. Thus, coastal habitats, where dinosaurs lived, were changing rapidly, as were shallow oceanic habitats. In addition, extensive volcanic activity in the Indian subcontinent occurred during this time. Vast basalt lava flows from this volcanism are called the Deccan Traps. Eruption of these flood basalts certainly introduced particulates into the atmosphere, which would have contributed to climatic cooling. Finally, geochemical and geophysical evidence strongly supports the impact of an extraordinarily large asteroid at approximately 65 mya. The most convincing evidence for this are 1) unusually high concentrations of the element iridium at the boundary in many places throughout the globe; 2) shocked quartz; and 3) a large crater the correct age. Iridium lacks terrestrial sources sufficient to explain its distribution at one horizon globally. It does occur in extraterrestrial bodies, so if significant quantities of the ejecta from an enormous asteroid became incorporated into the upper atmosphere, it could be distributed globally. Shocked quartz is formed by extra-intense pressure. The known occurrences of shocked quartz include ejecta from meteorite impacts. Grains associated with the boundary sediment of the K/T extinction, which also have a high concentration of iridium, contain shocked quartz. Finally, a large crater, the Chixculub crater (perhaps as large as 300 kilometers in diameter), on the Yucatán Peninsula of Mexico is the correct age and of an appropriate size that it could be the scar left by such an asteroid.

Earth's biosphere was under great stress 65 mya. Dinosaurs were one casualty. Some scientists are now arguing that multiple coincident causes may be responsible rather than a single factor. This seems like a plausible solution. Future research will reveal additional evidence to support this notion or suggest otherwise.

THE FOSSIL RECORD OF BIRDS

Except for dinosaurs, probably the most famous fossils are those of ***Archaeopteryx***, the oldest bird (Figure 15.22). The first specimen, discovered in 1861, consisted of a single feather. Recognition of this feather was possible because of its exceptional preservation in a very fine-grained lithographic limestone, the Solenhofen Limestone. This lithographic limestone is quarried commerically for making printed lithographs, including geological maps. Late that same year a complete specimen was found that was bought and sold several times before it was purchased by the British Museum of Natural History. The British Museum paid 700 pounds (about $1,400 today), a very large sum of money at the time, for this and an additional 1,700 specimens. Since 1861, four additional specimens and isolated feathers have been extracted from the lithographic limestone quarries, each of which was treated as a valuable commercial object and sold at least once before deposition in a scientific institution. The second specimen went to the Berlin Museum of Natural History. This specimen, sold in 1881 for 20,000 German marks (about $19,000 today), was discovered in 1877 approximately 10 miles away from the site of the first complete specimen. A third specimen was not discovered until 1956 and presently remains in private hands, where it is not available to the scientific community for study. The fourth specimen, labeled as a pterosaur, a flying Mesozoic reptile, was collected in 1855 and stored in a museum drawer until 1970 when it was recognized as an *Archaeopteryx*. Finally, a fifth specimen, originally collected in 1951, was recognized in 1973, again in a museum, where it had been classified as a small dinosaur.

Archaeopteryx differs from modern birds in several ways. Teeth are still present in both jaws. The tail is long and has a series of vertebrae down its midline. The front limbs, or wings, are short and stubby, with the bones not fused. There are still three clawed fingers at the wing tips, the wrist bones occupy the midlength of the wing. The sternum, or breastbone, is small and not expanded to accomodate large flight muscles

Figure 15.22 *Archaeopteryx*, the oldest bird known, from Jurassic rocks in Germany. In life this bird was about the size of a crow. Note the impressions of the feathers along the wings and tail, the clawed toes at the leading edge of the wing, and the long bony extensions of the backbone into the tail. (Courtesy of National Museum of Natural History.)

as in modern birds. Other features are more birdlike. The braincase is enlarged and solidly encased by bone. The orbits for the eyes are large, and the undoubted bird feature, feathers, are preserved. The small breastbone of *Archaeopteryx* has led to speculation that the animal may not have had large enough chest muscles for sustained, powered flight. On the other hand, the primary wing feathers of *Archaeopteryx* are clearly asymmetrical, with unequal-length small barbs on either side of the central shaft. This pattern is typical of all living flying birds, whereas flightless birds have barbs of equal length on either side of the shaft (Figure 15.23). Thus, this feature would seem to indicate that the oldest known bird could indeed fly, although it was probably not as powerful or adroit a flyer as living birds.

Birds are generally considered to have evolved from a small, bipedal, lightweight coelurosaurian dinosaur like *Compsognathus* (a small Jurassic theropod). Indeed, if *Archaeopteryx* did not have preserved feathers, it would be classified with this group of dinosaurs. There are two quite different theories concerning the origin of flight from these dinosaurs to birds. One idea, the arboreal theory, is that the bird's ability to fly originated from a climbing and gliding ancestor. In this view, bird ancestors climbed trees and eventually learned to parachute to the ground, then to glide from one tree to another, and finally to fly. This trait would perhaps have originated as an escape from predators or in the search for food such as leaves, seeds, and fruits. Climbing into trees and nesting in trees for safety may also have preceded gliding and flying. Several different kinds of animals can glide in addition to the well-known flying squirrels of North America. There are gliding marsupials in Australia, as well as gliding lizards, snakes, and frogs. It has been suggested that the small clawed digits at the midwing of *Archaeopteryx* allowed the animal to haul itself up into trees. An analogous feature is found in the chicks of a single living bird, the hoatzin of the Amazon Basin. This bird nests in trees overhanging water. The young swim to the base of the tree and then use two small functional claws at the middle of each wing to clamber back up into the nest. It loses these claws as an adult. However, the fingers of juvenile hoatzin are assuredly not a primitive feature derived from *Archaeopteryx* but, rather, a secondary derived character.

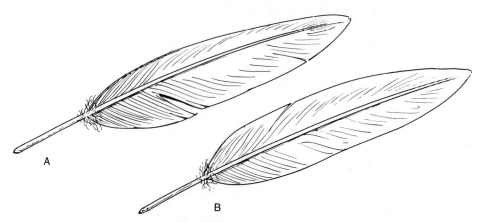

Figure 15.23   Primary wing feathers of birds. (A) An asymmetrical feather of a bird that can fly. (B) A symmetrical wing feather of a flightless bird. The oldest bird, *Archaeopteryx*, had asymmetrical primary wing feathers and was, therefore, presumed to have been capable of flight.

The second hypothesis about flight, the cursorial theory, is that birds evolved from ground-dwelling ancestors. In this view, feathers first evolved as insulation for temperature control because the reptilian ancestor was becoming or was warm-blooded. These small reptiles ran along the ground and captured insects and other small animal prey. The feathers of the forelimbs became elongated and were, perhaps, used to sweep up and capture insects, much as a butterfly net would be used. From this stage, the animals acquired the ability to jump into the air to capture prey, and the developing wings helped to stabilize the body and let it down gently to the ground. Finally, they used the forelimbs as wings for flight. It is also possible that leaving the ground could have been an antipredator adaptation. This latter idea, the cursorial theory, is generally favored because the skeletal morphology of *Archaeopteryx* is much more similar to ground-dwelling reptiles or birds than it is to a reptile that crawls up into trees.

In 1990 another astonishing discovery of fossil birds was announced. A small, sparrow-sized bird from Jurassic rocks was found in China. This bird was about 10 million years younger than *Archaeopteryx*. It had a considerably larger breastbone and, thus, was more clearly a good flyer.

After the Jurassic, a series of Cretaceous birds are known, most notably from the marine Niobrara Chalk of western Kansas. These two birds, *Hesperornis* and *Ichthyornis*, were both water birds; one was small and ternlike and the other, a large diving bird with small wings. Bones of these birds were discovered during the nineteenth century, and for many years they were the only two Cretaceous birds known. In recent years, at least four additional bird genera have been found in the Niobrara Chalk, and other Cretaceous birds have been found in several other areas, including England, Texas, Alabama, Montana, Canada, Chile, and South Dakota.

During the Cenozoic Era birds are rare as fossils. Their aerial habitat, generally small size, lack of teeth, and thin hollow bones all work against their preservation as fossils. Some of the most primitive and common fossil birds are the very large flightless birds called ratites. These include living ostriches, emus, cassowaries, and others, as well as the large extinct dodos, elephant birds, and moas. Birds were important predators in South American faunas during the Cenozoic, as described in Chapter 16. There are 27 orders of modern birds. The more advanced birds, excluding the ratites, can be divided into two main groups, the land birds and the water birds. Among the former, the most advanced forms are the small, perching songbirds, the passerines.

KEY TERMS

amniote egg	ectotherm	quadrupedal
anapsid	endotherm	Saurischia
ankylosaurs	euryapsid	sauropods
Archaeopteryx	gigantotherm	squamosal
articular	hadrosaurs	stegosaurs
bipedal	K/T boundary	synapsids
ceratopsians	mammal-like reptiles	temporal opening
dentary	Ornithischia	thecodonts
diapsid	ornithopod	therapsids
dinosaurs	quadrate	theropod

READINGS

The books listed below include several recent books on dinosaurs. The literature on dinosaurs, both technical and popular, is immense, and it is difficult to keep track of it all. The listing below represents a good, readable selection for an introduction to dinosaurs, reptiles, and birds.

Alexander, R.M. 1989. *Dynamics of Dinosaurs and Extinct Giants*. Columbia University Press. 167 pages. This book explains how the size, weight, running speeds, and strengths of dinosaurs can be estimated. It includes a section on the sounds that dinosaurs may have made. Other large animals including flying reptiles, marine reptiles, giant birds, and mammals are included with simple mathematical analysis.

Archibald, J.D. 1996. *Dinosaur Extinction and the End of an Era, What the Fossils Say*. Columbia University Press. 237 pages. A balanced treatment of evidence for and against rapid extinction at the K/T boundary. A readable account that supports multiple causes for the extinction.

The Asteroid and the Dinosaur (videotape). 1981. WGBA, Ambrose Video. 60 minutes. Treats the extinction of dinosaurs due to asteroid impact.

Barthel, K.W., ed. 1990. *Solnhofen: A Study in Mesozoic Paleontology*. Cambridge University Press. 236 pages. A modern study of this most important fossil area from which the oldest bird, *Archaeopteryx*, was collected. Includes many other exquisitely preserved fossils.

Benton, M.J. 1989. *On the Trail of Dinosaurs*. Crescent Books. 143 pages. A brief, simple book on dinosaurs.

Farlow, J.O. 1989. *Paleobiology of Dinosaurs*. Geological Society of America Special Paper 238. 100 pages. A series of articles that focus on the biology of dinosaurs and their nesting, social behavior, diet, and temperature regulation.

Farlow, J.O., and Brett-Surman, M.K., eds. 1997. *The Complete Dinosaur*. Indiana University Press. 752 pages. Collection of contributions from numerous experts on the history, study, biology, ecology, and media perception of dinosaurs. Detailed information is presented; it is well illustrated.

Feduccia, A. 1980. *The Age of Birds*. Harvard University Press. 208 pages. Excellent coverage of both fossil and modern birds, their ancestry, origin, and the evolution of flight.

Glut, D.F. 1982. *The New Dinosaur Dictionary*. Citadel Press. 286 pages. An illustrated definition of all genera of dinosaurs and the higher classification of dinosaurs.

Horner, J.R. 1988. *Digging Dinosaurs*. Workman. 210 pages. A personal narrative of important discoveries of dinosaurs and their nests and eggs in Montana.

Lucas, S.G. 1997. *Dinosaurs: The Textbook*, 2nd ed. W.C. Brown Publishers. 292 pages. A well-illustrated introductory textbook on dinosaurs.

McLoughlin, J.C. 1980. *Synapsida*. Viking. 148 pages. An advanced book on mammal-like reptiles.

Pabian, K., and Chure, D.J. 1989. *The Age of Dinosaurs*. The Paleontological Society, Short Course in Paleontology 2. 210 pages. A compilation of 16 articles by different authors, each an expert on some aspect of dinosaur paleontology. Especially designed as a sourcebook for college teachers.

Russell, D.A. 1989. *An Odyssey in Time: The Dinosaurs of North America*. Northwood Press. 239 pages. An oversize book replete with many color pictures, especially of Cretaceous dinosaurs and Canadian specimens. Speculation on what dinosaur descendants, had they existed, might have been like.

Weishampel, D.B., and others. 1990. *The Dinosauria*. University of California Press. 730 pages. The most comprehensive and thorough modern treatment of dinosaurs. Quite technical and organized mainly by major groups, but includes essential information on evolution and paleobiology.

Advanced Consumers on Land: The Mammals

MAMMALIAN FAUNAS

By the close of the Mesozoic, reptiles no longer dominated terrestrial communities. They had been the main herbivores and carnivores, both large and small, for more than 100 million years. However, as soon as the dinosaurs and other large Mesozoic reptiles became extinct, mammals took over. This is one of the major examples of large-scale ecological replacement. Mammals surely did not force out the reptiles or outcompete them. They played a waiting game, remaining in the background throughout much of the Mesozoic until the reptiles no longer dominated and most became extinct. Then, mammals diversified very rapidly; they underwent a remarkable adaptive radiation. Based on the fossil evidence, they became more abundant than reptiles had ever been during the Mesozoic.

Mammals are an extremely diverse group of animals. There are many different orders, suborders, infraorders, superfamilies, and families. In addition to their success on land, they, like Mesozoic reptiles before them, invaded the sea in the form of whales and dolphins; and like the pterosaurian reptiles, some mammals (bats) acquired the ability to fly. On land, we recognize two major groups of mammals: the **carnivores**, or flesh-eating mammals and the ungulates, or the hoofed, herbivorous mammals. Other mammals include many smaller, less diverse groups, such as the primitive insectivores, the primates (lemurs, monkeys, apes, and man), the subungulates (elephants and their relatives), the rodents (a very large group), the lagamorphs (rabbits and hares), the edentates (sloths, armadillos, and anteaters), and others.

It is convenient to divide mammals into three major faunas for discussion (Figure 16.1). The first fauna was the mammalian fauna that lived alongside dinosaurs during the Late Triassic through Cretaceous. During the early part of the Cenozoic, there was an **archaic** or **Paleogene mammalian fauna**. This fauna flourished during the Paleocene, Eocene, and Oligocene epochs of time. Most dominant groups of the archaic fauna have become extinct, although there are still some vestiges of this old fauna alive today. Beginning with the Miocene and continuing today, the **modern** or **Neogene mammalian fauna** has existed. It is similar in many respects to living mammalian fauna.

During the Cenozoic, the evolutionary history of the mammals is most fully expressed in the evolution of the molars and other teeth. No other parts of the hard anatomy show such widespread and diverse changes as do these feeding structures.

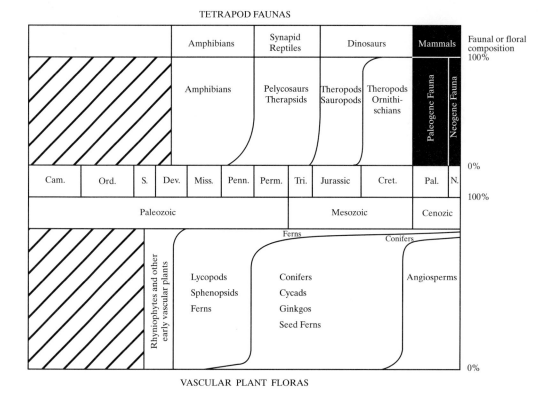

Figure 16.1   Terrestrial faunas and floras through the Phanerozoic with the mammalian faunas of the Cenozoic highlighted.

THE EARLIEST MAMMALS—MESOZOIC ERA

Mammals evolved from mammal-like reptiles that lived during the Triassic Period. As mentioned in Chapter 15, the defining character for mammals is the lower jaw that becomes composed of a single bone, the dentary. Another defining characteristic of mammals is a newly constructed ear. Reptiles have a single bone for sound transmission in their middle ear, the stapes, which is the old fish hyomandibular. Mammals have three bones in the middle ear. The two new bones are the old articular and quadrate bones of reptiles that were the jaw and skull articulation in mammal-like reptiles. These bones were shifted over to the ear, which is situated close to the jaw joint in mammal-like reptiles and early mammals. The bones, the **malleus** and **incus**, are the hammer and the anvil bones of the mammalian ear.

Our fossil record for these earliest mammals is quite skimpy, consisting mainly of tiny teeth and jaw fragments. Only a few reasonably complete skulls or postcranial skeletons have been found that allow us to make a reasonable reconstruction of what these earliest mammals were like (Figure 16.2). The earliest mammals were all quite small, generally about the size of a mouse, and were very primitive, with a small brain and short legs compared to their advanced descendants. The Jurassic mammalian record is also quite fragmentary; these fossils are known from only a few sites. Mammals were quite small, generally mouse- to rat-sized, during this time. Complete skeletons are

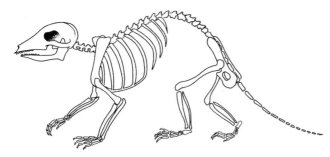

Figure 16.2   Reconstruction of an early Mesozoic mammal, *Megazostrodon*. The animal was approximately 8 cm (3 in) long.

unknown, isolated teeth and fragments of jaws being the most common fossils. Several different kinds of mammals had already evolved by the Jurassic; all the Jurassic groups are now extinct, so we do not know exactly what they looked like. Knowledge of these animals is based mostly on their molar teeth, which are definitely mammalian in character, but they are simpler and more primitive than those of most living mammals. The early mammals were clearly subordinate to reptiles. They were probably small carnivores, insectivores, herbivores, and omnivores. One group, the **multituberculates**, were rat- to beaver-sized, with skull and teeth reminiscent of some living rodents. Multituberculates were omnivores. The other groups, the **pantothere** and **triconodont** mammals, were more likely small predators. We can tell they were probably warm-blooded from features we have already discussed in their ancestors, the mammal-like reptiles. They also probably had hair, an insulating feature that may have first appeared in mammal-like reptiles. Hair was a structurally new feature for vertebrates, unlike scales of reptiles or feathers of birds and not derived from either.

An interesting problem concerns the mode of reproduction of these primitive mammals. Three grades of mammals are alive today in which the method of reproduction is quite different. The **monotremes** are quite primitive; they lay eggs but have primitive mammary glands. Only two species are alive: the spiny anteater (echidna) and the duck-billed platypus of Australia. Whether they are relicts of primitive Mesozoic mammals, is very difficult to know. Monotremes have several characteristic skeletal features by which they can be identified, but skeletons of Jurassic mammals do not show monotreme affinities. The critical feature would be the molar teeth of living monotremes because we know what these structures look like in fossils. Unfortunately, the molar teeth of monotremes are degenerate, little more than featureless pads of enamel. Thus, identification of monotremes in the fossil record is virtually impossible.

Another mammalian group is the pouched mammals, or **marsupials**. In marsupials, the infant is, for all practical purposes, born as an embryo that crawls from the birth canal to the pouch. They spend a long period of time attached to a teat in the pouch before they are sufficiently developed to emerge and live independently. Marsupials can be recognized as fossils, but they do not appear until the Late Cretaceous. Advanced mammals, the **placentals,** give direct birth, and they also first appeared during the Late Cretaceous. In placentals, which constitute most of the living mammals, there is considerable development of the young prior to birth. Placentals have well-developed mammary glands and many engage in extensive parental care of the young. Furthermore, the relative brain size of marsupials is smaller than that of the placentals. Thus, we have little evidence for the relationships of Jurassic mammals.

MARSUPIALS AND PLACENTAL MAMMALS

By Cretaceous time, mammals had undergone considerable evolution and diversification. Triconodonts survived into the Late Cretaceous but were rare. Pantotheres were present during the Early Cretaceous. The multituberculates continued to flourish. The most common Cretaceous mammals, however, were the marsupials. These were related to the living opossums of North America and to the extensive marsupial fauna of Australia and South America—the kangaroos, wombats, koalas, and many others. Most of the Cretaceous marsupials were small; some bore an amazing resemblance to the living opossum. In addition, a few small primitive placental mammal fossils have been found.

The oldest placentals, Upper Cretaceous, are of a group called the **insectivores**. As the name implies, the living representatives of this group mainly eat insects. Moles, shrews, and European hedgehogs are the best known living insectivores, and they are among the most primitive living placentals. All the Cretaceous mammals were also still quite small, none being larger than a fox or beaver, and were herbivores, insectivores, and carnivores. Fossils of these mammals are rare, and molar teeth are the most important anatomical parts for their study. By careful examination of the arrangements of the cusps, ridges, and other features on the molars, it is possible to readily distinguish a marsupial molar from that of a primitive placental molar.

THE PALEOGENE MAMMALIAN FAUNA

At the beginning of the Cenozoic, turtles, amphibians, birds, lizards, and mammals populated terrestrial ecosystems. All were relatively small; all had held subordinate roles to dinosaurs in Cretaceous habitats. Of these, mammals and birds diversified to dominate. Mammals underwent a remarkable adaptive radiation during the Paleocene and Eocene. The first fauna was composed of forms that would be unfamiliar to us. This is called the archaic or Paleogene mammalian fauna (Figure 16.1).

The groups of Paleogene mammals are far from being distinct. The difference between carnivores and herbivores is most striking in characters of the molar teeth. However, many of the early mammals have little specialization of these teeth. As mentioned when dicussing Cretaceous mammals, the oldest and most primitive placental mammals are the insectivores, ancient relatives of shrews and moles. During the Paleocene, insectivores were still common. Most of these and other mammals were relatively small during the Paleocene, the largest being about the size of a sheep. One of the most conspicuous groups consists of the early carnivores called **creodonts**. The earliest ones were still very similar to their insectivore ancestors. They were relatively small animals with short legs and were probably not very swift runners. During the Paleogene, these carnivores had a variety of specializations. Their front teeth (incisors) became bladelike for nipping and tearing flesh, and they retained their stabbing canines. Most important, a pair of upper and lower molars became increasingly larger and higher and developed a bladelike edge. As these teeth sheared past each other, they could tear flesh and cut through sinew and bone. These specialized molars, called **carnassials**, became increasingly prominent in the jaws of carnivores during the Eocene and Oligocene.

Although the creodonts were undoubtedly the dominant carnivores during the Early Cretaceous, more advanced carnivores also appeared during the early and middle Paleocene that would eventually replace them. These are the **miacidids**. The earliest creodonts were flat-footed; they walked around on the entire soles of their feet. On the other hand, the miacidids developed somewhat longer legs, and gradually the bearing surface of the feet was reduced to the surfaces of the toes. Miacidids also had the last premolar and molar specialized for shearing, but other features were still primitive, such as the area around the inner ear that was not calcified as it was in later forms. By the Oligocene epoch, the creodonts had dwindled to a few surviving stocks. Their place as the major predators on land was taken over by the miacidids. Two main groups of advanced carnivores evolved from miacidids during the Oligocene and continued to the present day. One group includes dogs, bears, raccoons, weasels, and their relatives; the other consists of cats, hyaenas, and the Old World civet cats.

By far the most abundant and diversified mammals during the early part of the Cenozoic were the **ungulates**, the hoofed, herbivorous mammals. Like the carnivores, the earliest of these mammals shows little departure from their presumed insectivore ancestors. They were mostly small in size, the body was relatively long and slender, and the legs were short and little specialized (Figure 16.3). Ungulates do show a somewhat more advanced foot structure than the earliest carnivores because they did not walk on the flats of their feet. Instead, their feet were slightly raised, so that only the lower surfaces of the toes were in contact with the ground. Their teeth were quite generalized initially but soon began to show adaptations for plant eating. The molar teeth became square-shaped, with a flat grinding surface for shredding plants.

The most common group of primitive herbivores is called the **condylarths**. They mostly fit the general picture of early herbivores as described in the previous paragraph. They evolved during the Lower Cretaceous and may have been the dominant plant eaters during the Paleocene and Eocene. However, condylarths died out at the close of the Eocene.

Other archaic ungulates included several groups that attained a rather large size early during the Cenozoic. One such group was the **pantodonts**; they were about the size of a sheep during the Paleocene but became cow-sized, approximately 2.5 meters (8 feet) long, during the Eocene. Some of these had strong claws instead of hooves and long, heavy upper canine teeth. These are thought to have been specializations for a root-grubbing style of food gathering. Another group was the **uintatheres**, known mainly from the Eocene rocks of the western United States and Asia (Figure 16.4). The name comes from the Uinta Mountains of Utah. Some of these animals were as large as a rhinoceros and are the largest archaic mammals known, apart from early toothed whales that were 25 meters (81 feet) long. They had peculiar bony swellings on the top and face of the skull, presumably for defense, and they had short, stubby legs to support their considerable weight. Males had long, stabbing upper canine teeth.

Two final groups of primitive ungulates go by the names **Notoungulata** and **Litopterna** (Figure 16.5). These animals were common during the very early Cenozoic. Their early remains are from Paleocene rocks in North America and Asia, but they soon became extinct in North America. Before that, notoungulates migrated to South America, which was in land communication with the other continents during this time. However, North and South America soon drifted far enough apart to prevent migra-

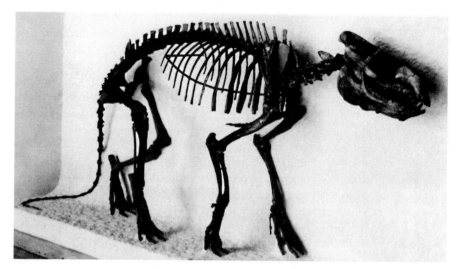

Figure 16.3 The mounted skeleton of an Oligocene artiodactyl, *Merycoidodon*. These primitive ungulates, called oreodonts, were relatively small and primitive in structure. The teeth were not very specialized. Notice that there are four functional toes on each foot, although the feet are partially raised off the ground. (Courtesy of National Museum of Natural History.)

Figure 16.4 The largest mammals of the Eocene were the uintatheres, one of which is shown here as a mounted skeleton. These were heavily built ungulates with short, massive legs, peculiar bony knobs on the skull, and long defensive canine teeth. The skull is approximately 75 cm (2.5 ft) long. (Courtesy of National Museum of Natural History.)

Figure 16.5   Side view of the skull and lower jaw of a South American notoungulate, *Homalodotherium*, of Miocene age (approximately 40 cm [15 in] long). The long battery of teeth indicates that this animal was a herbivore. (Courtesy of Field Museum of Natural History, Chicago.)

tion between these two continents, but notoungulates persisted in South America. The notoungulates, litopterns, other primitive mammals (especially marsupials), and large ground birds survived in South America in nearly complete isolation until fairly recent times. Primitive notoungulates and litopterns radiated into a host of different types. Some of them became quite large and were cattlelike in appearance. Others resembled horses and camels. Still others became rodentlike; some of these were very large, resembling a bear-sized beaver. This is a clear example of parallel evolution. South America had a set of habitats similar to those of North America, and the notoungulates and litopterns radiated into this great variety of South American habitats. In the process, they evolved various features of the skeleton (especially of the skull, legs, and teeth) that closely resembled, but are not related to, features that evolved in other groups of ungulate mammals in North America and elsewhere.

In addition to these various groups of archaic mammals that did not survive the Cenozoic, there were two other groups of hoofed mammals that originated quite early during the era and went on to become the dominant hoofed mammals alive today. These are the **Perissodactyla** (odd-toed) and the **Artiodactyla** (even-toed) hooved mammals. In the first group, the axis of the foot runs down the center, through the third digit (Figure 16.6). In the second group, the axis is between the third and fourth digits (your ring and middle fingers). Living representatives of the perissodactyls include the horse, tapir, and rhinoceros. The horse has a single (third digit) functional toe; the others have three toes. After the Eocene, artiodactyls were considerably more diverse than the odd-toed ungulates. Many had the number of functional toes reduced to two, the third and fourth, resulting in the split- or cloven-hoofed ungulates. Pigs, camels, sheep, goats, antelope, deer, and cattle are all examples of artiodactyls.

During the early Cenozoic, various groups of odd- and even-toed ungulates were already well differentiated and do not share a direct ancestor. One of the best known of these early mammals is *Hyracotherium*. This animal was about the size of a fox and was of slender build. It had three functional toes on each hind foot and four toes on each of the front feet. The teeth were low crowned and had a generalized plant-eating character. Although *Hyracotherium* is generally believed to be the ancestral stock for all later horses, it is sufficiently generalized that it could also be considered the ances-

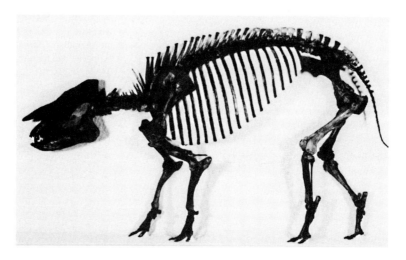

Figure 16.6 Skeleton of *Palaeosyops*, a primitive Eocene perissodactyl that was quite large compared to most Paleogene mammals. This ungulate belongs to an extinct group of odd-toed herbivores called titanotheres, some of which attained much larger size than the specimen shown here. (Courtesy of National Museum of Natural History.)

tor of all perissodactyls. It was surely a browser, living in the forest or on the forest edges, nibbling low leaves from trees and shrubs.

Although they are mainly a late Cenozoic assemblage, one final group of archaic mammals should be mentioned—the elephants and their relatives. These animals originated in Africa. The oldest and most primitive **proboscideans** (from proboscis, or trunk) are from Eocene rocks in Egypt. As adults, these were about the size of a baby elephant. They had four short tusks, each a few inches long, two each in the upper and lower jaws.

In summary, we can now describe the general character of the Paleogene mammalian communities. The primary consumers, or herbivores, included a wide variety of hoofed ungulates. Many of them were rather small by modern standards—generally smaller than a sheep—but a few, the pantodonts and uintatheres, reached the size of a cow, and brontotheres were rhino-sized. The most conspicuous of these were the condylarths. The notoungulates were largely confined to South America. Secondary and tertiary consumers were represented first by the creodonts and later by early members of the miacidids. All these carnivorous mammals gradually developed adaptations of the skeleton and teeth that made them progressively more effective predators during the Paleogene.

In addition to these main groups, most other kinds of mammals evolved during this time, and these others developed into the principal groups of herbivores and carnivores of today. These include insectivores, bats, primates, rodents, and edentates (sloths and their relatives). The oldest known whale fossils are also from marine rocks of Eocene age. Thus, there was a tremendous adaptive radiation and diversification during the early part of the Cenozoic. Some of the habitats they occupied had probably been left vacant by reptilian extinctions at the close of the Mesozoic, but other niches were undoubtedly new. The tree-dwelling habit, the fruit and flower diet of some primates, and the nut and hard seed diet of many rodents were adaptations that were not pursued by reptiles, as known. At least some of these new ways of life were directly related to the characteristics and expansion of the flowering plants during this same time interval.

THE NEOGENE MAMMALIAN FAUNA

During the Miocene Epoch, many of the mammal groups that had been conspicuous during the Paleocene, Eocene, and Oligocene had become either extinct or had dwindled to insignificant roles and the Neogene (or Modern) mammalian fauna arose (Figure 16.1). In the northern hemisphere, almost 75 percent of the mammal families known from the Miocene still persist. On a worldwide basis, this percentage is lower, about 50 percent, because of the larger number of endemic families of South America that later became extinct.

By the beginning of the Neogene, the miacidids were diminished in importance, as modern types of carnivores were undergoing a conspicuous adaptive radiation. There was an abundance of doglike or wolflike forms. Some of these were large, approaching small bears in length and weight. In addition to a variety of true cats, there were several different kinds of saber-toothed cats that were a conspicuous element of many faunas (Figure 16.7). Saber-toothed cats originated during the Eocene and continued into the Pleistocene Epoch when they became extinct, approximately only 10,000 years ago. They evolved long stabbing and slicing upper canines. The lower jaw was hinged so that it could be opened extra-wide. The molars were virtually reduced to very large, shearing carnassial teeth. The limbs of advanced saber-toothed

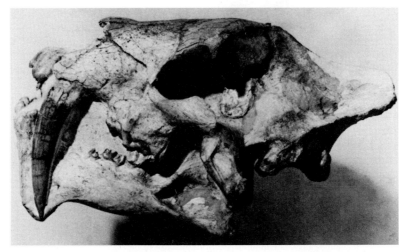

Figure 16.7   The skull and lower jaw of an Oligocene saber-toothed cat, *Hoplophoneus*, approximately 15 cm (6 in) long. Note the long, stabbing canine in the upper jaw, and the large, shearing carnassial teeth at the back of the jaw. (Courtesy of National Museum of Natural History.)

cats were massively constructed. These carnivores presumably preyed on huge mammals, perhaps mastodons and elephants. In addition to a variety of weasels, otters, and their relatives, bears became a conspicuous predator element during the Neogene. A final group of carnivores to first appear during the time was the marine predators—seals and walruses (Chapter 12).

Among the hoofed mammals, the two dominant groups were the perissodactyls and the artiodactyls, the latter of which has tended to be more diverse than the former. In both groups, a number of common evolutionary trends are evident (Figure 16.8). There was an overall tendency for the size of the animals to increase. This was accompanied by an increase in the length of the legs and neck and a larger brain size and skull. The feet changed their relationship to the ground so that only the lower surfaces of the digits had contact with the ground. The "palm" or "sole" of the foot was raised, serving to further increase leg length. In the odd-toed forms, this trend culminated in the advanced horses. Beginning in the Pliocene, only the third digit was functional in horses, and the feet were raised so that the animal ran on the tip of a single toe on each foot. In artiodactyls, this tendency resulted in only two functional toes on each foot (the third and the fourth digits) and, again, only the tips of the digits touched the ground. These evolutionary features can be viewed as adaptations for greater speed to escape predators and for a more arid environment with harder ground; both adaptations were partly a result of a major change in habitat from forest-dwelling browsers to plains-dwelling grazers.

Another trend in the hoofed mammals was for the jaws to become longer and the facial region of the skull to elongate. Accompanying this trend was progressive modifi-

Figure 16.8 A Miocene perissodactyl belonging to an extinct group called the chalicotheres. The animal, *Moropus*, was about the size of a small horse (approximately 2.75 m [9 ft] long) but was heavily built and probably slow moving. This extinct form is unusual in that there are claws on the feet, rather than hooves. These are probably related to a root-digging habit. (Courtesy of National Museum of Natural History.)

cation of the teeth. These became larger and more nearly square in outline. The premolars evolved to look like the molars, and there was an increasingly complex pattern of enamel on the grinding surface of the molars and premolars. Instead of being low, the teeth became high crowned, growing throughout the life of the animal. These changes resulted in a long battery of grinding teeth in both jaws that are very resistant to wear, especially from grasses that are high in silica content. Grasses comprised much of the plant life of the expanding prairies (Figure 16.9).

Not all the ungulates showed such advanced features. Many forms remained browsers and did not develop any of the advanced features that were just listed. Among the perissodactyls, both the tapirs and rhinos tended to be much more conservative than the horses with respect to evolutionary change (Figure 16.10). Even among the horses, not all of them became plains-dwelling grazers; some remained browsers with three toes.

Among the even-toed ungulates, deer, cattle, and camels had the greatest increase in diversity during the Neogene. These are the ruminants, or cud-chewing ungulates. The second stomach in these animals is an adaptation for eating the harsh grasses of extensive prairies that came into existence during this time. Swine and their relatives, the hippos, were important nonruminant groups of artiodactyls.

Elephants underwent a spectacular evolutionary history during the later Cenozoic. They increased tremendously in size, and their tusks became very long. Some had two lower tusks, some two upper tusks, and others had four tusks. The skull became high, and the neck became short to support the weight of the massive skull, trunk, and tusks (Figure 16.11). The legs became pillarlike to support the great weight of the animals. Two main groups can be recognized: 1) **mastodons** and 2) **mammoths** and elephants (see discussion below).

Figure 16.9   Bottom view of the skull of a Pliocene horse that was slightly smaller than a modern horse. Notice the long battery of molars and premolars in the jaws; these teeth are high crowned and have complex patterns of enamel. There is a conspicuous gap in the tooth row between these grinding teeth and the nipping incisors at the front of the jaw. (Courtesy of National Museum of Natural History.)

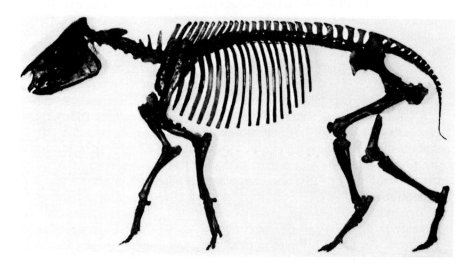

Figure 16.10 A primitive rhinoceros from the Eocene, *Hyrachyus*. The skull is approximately 30 cm (1 ft) long. Note the three-toed feet, typical of primitive perissodactyls, or odd-toed ungulates. (Courtesy of National Museum of Natural History.)

Figure 16.11 Skull and lower jaw of an advanced elephant, the mammoth of Pleistocene age. Notice the high back part of the skull that supported massive neck muscles. A single very large molar tooth is present in each half of the upper and lower jaws. The snout contains a large cavity where a tusk was situated in life, but the tusk is now missing. (Courtesy of National Museum of Natural History.)

SPECIAL ASPECTS OF CENOZOIC MAMMALS

Several aspects of mammalian paleontology during the Cenozoic are of special interest. These are 1) the existence of centers of evolution and migration for specific groups of mammals, 2) the isolation and eventual unification of South America, and 3) the extinction within the last few thousand years of many large mammals from the northern hemisphere.

CENTERS OF EVOLUTION

Each major lineage of mammals initially evolved in a restricted area and, as land bridges allowed, then migrated away from the center of evolution to appear as fossils in widely separated parts of the world. For certain groups of mammals, we have good documentation of the place of origin and times of migration.

Horses.   One of the best known of the Cenozoic mammalian fossil records is that for horses. The oldest horse, *Hyracotherium*, is from both North America and Europe. These occurrences provide evidence for each migration between these two continents during the Eocene. During the Oligocene, a succession of horse genera evolved in North America. Early horses were forest browsers, much like deer living in the forest today. *Hyracotherium* was a small animal, approximately 45 centimeters (18 inches) in length, and it had simple teeth and three toes on the hind feet and four on the forefeet (Figure 16.12). As discussed in Chapter 14, one of the revolutionary changes in North American floras during the Cenozoic was the development of prairie grasslands in the rain shadow of the Rocky Mountains. Horses were one of the mammals that evolved into this hew habitat. This evolutionary sequence is confined to western North America. There were no horses in Europe during the Oligocene. From the Eocene into the Pleistocene, there is an unbroken fossil record of North American horses, from which a fairly complete understanding can be developed. The evolutionary history of horses is quite complex because some horses stayed forest browsers, while others adapted to the plains.

The evolutionary story of the modern horse records the habitat shift of horses from forests to plains, or from being forest dwellers to plains grazers. Forest dwellers are typically timid organisms; camouflage and hiding are means of ensuring safety. They browse on leaves of bushes and small trees and run on the leaf litter of the forest floor. In contrast, the plains grazer is exposed, with no place to hide, and the grasslands developed in a semiarid setting, so the ground was much harder than it was in the forest. Furthermore, grass is a harsh vegetation that significantly wears teeth. Horses adapted to the open environment by becoming larger and running in herds for protection. As a result of being larger, their necks and faces became larger (Figure 16.13). Running on the hard prairies caused a progressive reduction in the number of toes to one functional toe (Figure 16.12). In fact, the modern horse not only has one functional toe, but it is standing up on the equivalent of its fingernails and toenails. Finally, the teeth changed in two ways. First, instead of being simple teeth, the horse's teeth became very elongate so that they could continue to grow for several years as they wore away from processing grasses. Second, the grinding surface of teeth changed from being small with cusps to large with a complex series of ridges, thus becoming a much more effective grinding surface.

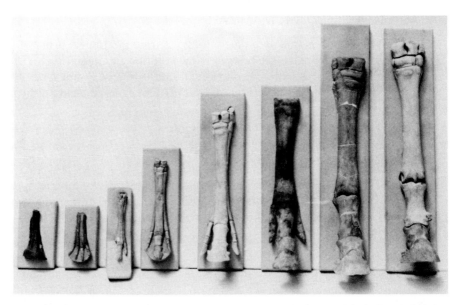

Figure 16.12 Evolution of the forefoot in horses. The two small, four-toed examples on the left are Eocene in age. The farthest left is *Hyracotherium*. The third from the left is Oligocene in age, the next two are Miocene, and the right-hand examples are Pliocene, Pleistocene, and Holocene (from the modern horse *Equus*). Notice the increase in size and length of the foot and the change from four to three to one functional toe. (Courtesy of National Museum of Natural History.)

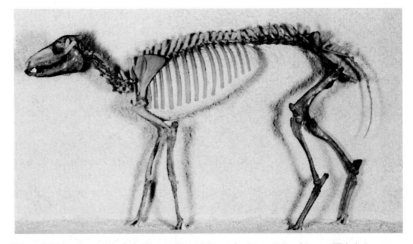

Figure 16.13 A mounted skeleton of the Oligocene horse *Mesohippus*. This horse was approximately 1 m (39 in) long, with a shorter skull and less specialized teeth. There were still three functional toes on each foot, and the legs were not yet highly elongate. (Courtesy of National Museum of Natural History.)

Through the Cenozoic, this basic progression of characters developed. Not all horses developed all characters—some early plains grazers evolved back into the forest. However, the modern horse *Equus* is the final product of these aggregate adaptations for life as a plains grazer.

One or more of these horses migrated from North America to Eurasia, probably via the Bering Strait, which was a major land bridge between the Americas and Eurasia during times of sea-level lows. European horses can be viewed as a series of migrants with intermittent, local evolutionary lineages. During most of the Cenozoic, North America was clearly the center of horse evolution until the close of the Pleistocene, only a few thousand years ago. Horses were still present in North America when humans first invaded the continent, also across the Bering Strait. Artifacts and fossil horse bones have been found together. Then, along with most large mammals, horses became extinct in North America, approximately 10,000 years ago. Horses continued to survive in Europe and Asia; in Africa they were represented by zebras. The wild horses of the western states in modern times are descended from horses that escaped from the Spaniards during the time of early exploration of North America.

Camels. A group of mammals with a pattern of evolution similar to that of the horses is the camels (Figure 16.14). They seemingly originated in North America, where we find them during the late Eocene. Again, there is a continuous sequence of fossil camels through the remainder of the Cenozoic in North America until the end of the Pleistocene. By this time, some of the North American camels had become very large, much larger than those living today. One group invaded South America during the Pliocene and survive today as the llama, but in North America camels became

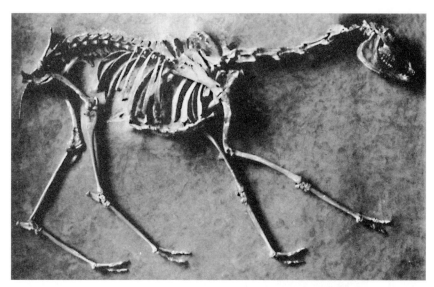

Figure 16.14   A Miocene camel, *Stenomylus*, from Nebraska. Note the elongated vertebrae in the neck and the elongate legs. Camels are artiodactyls, and this specimen shows two functional toes on each foot. (Courtesy of National Museum of Natural History.)

extinct at the close of the Pleistocene. They continue to survive in Eurasia as the modern Bactrian and dromedary camels.

Elephants and Other Mammals.   Several major groups of mammals seem to have had a center of origin in Africa, from where they migrated north into Europe, east into Asia, and finally into North America. One such group is the elephants and their relatives. Mastodons are first found in Africa during the Eocene; and by the close of the Miocene, they were in Europe and North America, where they persisted until their extinction at the close of the Pleistocene. Other groups that may have evolved in Africa, or at least in the Old World, and that never migrated to North America include the giraffes, true hippos, and many of the African and Asian antelopes, which are not related to the North American pronghorn.

The distribution of large elephantlike animals during the Pleistocene in North America forms an interesting pattern. First, there are two quite different kinds of these large animals, the mammoth and the mastodon. It is easy confuse these two early mammals. The mastodon belongs to a different family than the mammoth. The mammoth is in the Elephantidae and is closely related to the modern living elephants, which are placed in two genera, the African elephant, *Loxodonta*, and the Indian elephant, *Elephas*. Perhaps one reason for the confusion is the similarity of the scientific names of the two extinct animals; the mammoth name is *Mammuthus*, and the mastodon name is *Mammut*. The mastodon belongs in the family Mammutidae. The mastodon has relatively short, straight tusks, and each jaw contains two or three rather small molar teeth at any given time. These teeth have a few rounded cone-shaped crests and are suitable for browsing on rather soft vegetation. On the other hand, the mammoth has highly elongate and strongly curved tusks and a single, very large molar tooth in each half of each jaw (Figure 16.15). These molars were slowly replaced during life, but there was only a single functional molar at any one time. Each tooth has numerous cross ridges of enamel that provide a broad, hard grinding surface that is suitable for more harsh vegetable material, such as grasses.

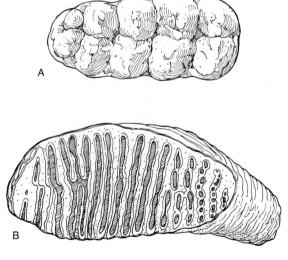

Figure 16.15   Comparison of the molar teeth of (A) a mastodon, and (B) a mammoth. The mammoth tooth is considerably larger than the mastodon tooth, has many parallel ridges of enamel, and occurs singly in the jaw. The mastodon tooth, on the other hand, occurs with one or two other molars, has only a few rounded cusps, and is smaller. (Modified from Osborn, 1940.)

Neither of these animals was present in North America until the very close of the Pliocene Epoch, when they are thought to have crossed over the Bering land bridge from Asia. The mastodon is represented by a single species, *M. americanus*. This large beast was a forest dweller and lived primarily in the enormous wooded areas in the more northern parts of eastern North America. Most of these fossils are found in peat bogs, old river sandbars, and other ice-age deposits in the eastern United States. Mammoth fossils are present but relatively rare in this area.

In contrast, the mammoths underwent considerable species evolution after they arrived in North America, although the precise taxonomy of these animals is still not known. A conservative view of their evolution would have three distinct species during the Pleistocene, rather than the sixteen that were proposed by some early paleontologists. The imperial mammoth was a very large animal that lived primarily in the southwestern and southeastern United States. It did not range very far north. The Columbian mammoth, on the other hand, was a temperate animal and is found throughout the central United States. Finally, there is the woolly mammoth, the northernmost species that ranged from Alaska and Canada down into the Great Lakes region. These species are differentiated in details of their molar teeth, tusks, and skull.

Mastodons and mammoths, as well as a variety of other animals, are found in the Rancho La Brea tar pits in downtown Los Angeles, one of the most famous fossil localities in the world. These pools of tar form by upward seeping of oil from buried sands. The oil loses light hydrocarbons that could be refined into gasoline, leaving behind heavy tarlike deposits that do not evaporate easily. Animals that stray onto the surface of the pools are caught in the sticky tar, which cannot support their weight, so they are trapped and eventually are engulfed in the tar. The tar is constantly churning as new material wells up from the depths, so there is no internal stratigraphy to the deposits. During the Pleistocene, thousands of animals were trapped and buried in these pools. Their bones are beautifully preserved. Mammals are the best known fossils from this locality, although birds, insects, reptiles, and other animals are also present. A single human (Amerindian) fossil is also known. Interestingly, more than 90 percent of the animals recovered, counted as individuals, are carnivores and only 10 percent are herbivores. The meat eaters were presumably lured onto the surface of the pool by other animals trapped in the tar. The predators in turn became trapped. The more common carnivores include the saber-toothed cats and a large extinct wolf, the dire wolf. Other carnivores that are less common include an extinct American lion, the mountain lion or cougar, bobcat, fox, coyote, weasel, badger, skunk, and short-faced bear. The herbivorous animals at La Brea include the mastodon, mammoth, camel, bison, peccary, antelope, horse, tapir, and two kinds of sloths.

SOUTH AMERICAN AND AUSTRALIAN ISOLATION

As already mentioned, the isolation of the South American land mass occurred during the Paleocene. At this time, the principal mammals in South America included a variety of marsupials and very primitive ungulate groups. None of the early carnivorous creodonts are known from South America. The endemic groups evolved and radiated into a variety of habitats. The hoofed mammals, especially notoungulates and litopterns became horselike, rhinolike, rodentlike, and bearlike in many of their skeletal adaptations. Yet, they also retained many very primitive features from their notoungulate ancestors. The marsupials diversified to fill most of the larger carnivorous niches in South America. One marsupial, *Thylacosmilus*, developed large, stabbing canines, like those of the placental saber-toothed cats of the rest of the world (Figure 16.16). *Thylacosmilus* is clearly a marsupial and represents another remarkable case of parallel evolution. Other marsupials evolved to resemble dogs and wolves in structure. Unique to South American communities were the large flightless birds that also developed as important predators. These evolved during the Oligocene and were a dominant predator until the late Pliocene. The largest of these birds was up to 3 meters (9 feet) tall, with a skull nearly .5 meters (1.5 feet) in length. With long legs for rapid running and a large, powerful beak, these were undoubtedly very effective predators.

The ecological balance of this South American fauna was changed abruptly toward the close of the Pliocene Epoch, approximately 3 mya, when the present-day Isthmus of Panama emerged due to plate tectonics. This produced a land migration route between North and South America for the advanced placentals of North America to move south and the endemic South American fauna to move north. The net result was the invasion of South America by advanced placentals. This was a filter bridge type of migration, with nearly a one-way migration south. Carnivores, especially large cats, such as cougars, dogs, and wolves, spelled the end for the various more primitive hoofed mammals that had successfully persisted in South America for so long. They all became extinct. Other groups, such as tapirs, the llama, and other advanced artiodactyls, probably competed directly with notoungulates for the same habitats. Large predatory birds became extinct. The marsupials, relatively slower animals with smaller brains by comparison to advanced placentals, also became extinct, except for a

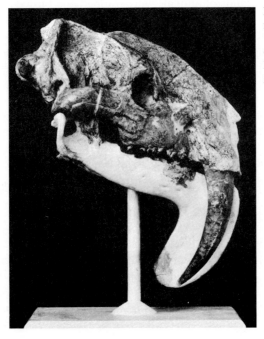

Figure 16.16 Mounted skull of a marsupial saber-toothed cat, *Thylacosmilus*, from Pliocene rocks of South America. (Courtesy of Field Museum of Natural History, Chicago.)

variety of the opossum. The opossum was able to survive and became a successful northerly migrant. Other endemic forms that were able to move north and compete successfully against members of the North American fauna included the extant armadillo and porcupine and the now extinct giant ground sloth, which survived in North America until very recently. In summary, this event devastated the South American fauna. Replacement of 100 percent of the native South American ungulates occurred; 40 percent of the families of living South American mammals were invaders from the Pliocene.

A similar sort of situation is taking place in Australia today. Here, the continent is isolated from mainland Asia, but it shared land connections with South America and Antarctica. The fauna included monotremes and marsupials. Only a few placental rodents (mice and rats) were able to migrate across the straits that separated Australia from the mainland of Asia, apparently by rafting. Other placentals included bats, which were not as seriously affected by water barriers. With the coming of humans, however, placental mammals have invaded Australia. The dog (dingo) was introduced by the aborigines, and Europeans brought more dogs, cats, rabbits, and additional rodents. These exotic animals have been competing very successfully with the endemic marsupial fauna, and some of the native species have already become extinct or are very, very rare. Without proper conservation and care, Australia may eventually become a second South America, with nearly total loss of its primitive fauna.

MAMMALIAN EXTINCTIONS

During the entire Cenozoic, mammals tended to evolve very rapidly. There are many genera and families of mammals that became extinct during the course of the last 60 million years. Yet one of the most significant waves of extinction occurred almost yesterday, at the close of the Pleistocene, approximately 10,000 years ago (Table 16.1). These **Pleistocene extinctions** are puzzling for two reasons. First, mostly large mammals became extinct. Smaller relatives of the same animals survive today. Second, all the mammals that became extinct at this time had survived through four major climatic shifts that each witnessed an advance and retreat of continental ice sheets during the Pleistocene. The last million years or so of Earth history has probably been as variable with respect to global climate as any other in Earth history. Immense ice sheets covered the northern part of North America, Europe, and parts of Asia. The Greenland and Antarctic ice caps are the dwindled remnants of the once enormous ice sheets. The final glaciers, called the Wisconsin glaciation in North America, retreated about 10,000 years ago. It was after this final retreat that the extinctions of large mammals occurred, especially in North America and Europe. On the North American continent, from 30,000 to 5,000 years ago, the mastodon and woolly mammoth became extinct, as did the imperial elephant, a camel, the horse, a giant beaver, large ground sloths, giant bison, the dire wolf, saber-tooth cats, large deer, and large moose. Ninety-five percent of the mammalian megafauna became extinct. The comparable situation in Europe witnessed extinction of large cave bears, woolly rhinos, and the Irish elk. In many respects it is fair to say that the living North American native mammalian fauna is depauperate—it is low in species diversity, especially of larger forms. These extinc-

Table 16.1 **EXTINCT PLEISTOCENE MAMMALS WORLDWIDE. (COMPILED FROM ANDERSON, 1984.)**

Groups	Total Known Genera	Extinct Genera	Extinct Families	Percentage of Extinct Genera
Marsupials	70	18	4	26
Insectivores	21	4	1	19
Edentates	40	32	4	80
Primates	27	11	0	41
Carnivores	63	12	0	19
Rodents	127	35	1	28
Lagomorphs	9	3	0	28
Litopterns*	2	2	1	100
Notoungulates*	3	3	2	100
Perissodactyls	12	6	1	50
Artiodactyls	103	45	0	44
Tubulidentates	2	1	0	50
Hyracoids	2	1	0	50
Sirenians	2	1	0	50
Proboscideans	8	6	3	75
Deinotherioids	1	1	1	100

*Endemic South American herbivore groups

tions did not seem to affect the African faunas nearly to the extent that they did those of the northern hemisphere. However, extinctions of very large marsupials also occurred at about the same time in Australia. The currently rich mammalian faunas of Africa may be indicative of what the North American fauna looked like before this latest wave of extinction.

It is difficult to ascribe these extinctions to climatic change. These animals lived through four major glaciations and three interglacial periods when the climate was at least as mild as it was 30,000 years ago. Some scientists have argued that early humans hastened the demise of these large animals by selectively preying on the large forms. It is not yet clear whether the timing of the extinctions and the arrival and spread of humans in North America is sufficiently close to allow this as a viable hypothesis. Certainly humans coexisted with this larger mammalian fauna in Eurasia long before the extinctions began to take place there. It is also difficult to imagine that the relatively small populations and crude hunting methods of the human beings who initially colonized North America would have placed sufficient hunting pressure on these animals to directly result in their extinction. Prior to the invasion of North America by Europeans, the Native Americans seemingly lived in balance with the mammalian fauna, without causing undue predatory pressures. Thus, this time of extinction, like that at the close of the Mesozoic and Paleozoic, raises many questions that cannot be fully answered. Both biological and physical factors may have been involved, including the crucial dependence of secondary consumers on specific primary consumers. The loss or depletion of just a few large herbivores may have triggered waves of loss throughout the larger carnivorous mammals.

COMPARISONS AND CONTRASTS

In summing up the record of mammals during the Cenozoic, it is worthwhile to make some comparisons and contrasts between the record of these land animals and that of the reptiles of the Mesozoic. First, a similar proportion of primary and secondary consumers occurred in the two faunas. Among the mammals, there are many more kinds of ungulates and other herbivorous types than there are carnivores. The same statement can be made about the dinosaurs. Only one group of dinosaurs was carnivorous—the theropods. All other saurischians and all ornithischians were herbivores. This is in accord with our ideas about food pyramids. Just as it takes a large amount of plant material to sustain one herbivore, it takes a number of herbivores to keep one carnivore alive. The basic aspects of community structure among Mesozoic reptilian communities were probably comparable to those of Cenozoic mammalian communities.

Mammals cannot be considered in isolation from other dominant groups of land life during the Cenozoic. If we could take a stroll through the woods today, it is clear that four groups make up the great majority of life that we see and hear. Except for the odd snake, turtle, worm, or frog, the dominant animals are mammals, birds, and insects; and of course, the dominant plants are the angiosperms, which had a major impact on the diversification of all these animals. Without the evolutionary diversity of flowering plants, it seems very unlikely that there would be anything like the great number of different kinds of mammals that are alive today or that thrived during the past. The increase in the number of mammalian primary consumers has probably also had a strong influence on the variety of mammalian secondary consumers or carnivores.

Among insects, many are especially adapted to feed on the leaves, flowers, seeds, wood, or other parts of flowering plants. Insects undoubtedly underwent an adaptive radiation, at least at the species level, as they began to exploit the increasing number of plants that evolved during the Cenozoic. Even the few insects that do not relate directly to plants, such as mosquitos and fleas, are intimately associated with warm-blooded mammals, which, in turn, are primarily or secondarily dependent on angiosperms.

The tremendous variety and abundance of birds is also closely related to the success of the angiosperms. Early birds of the Mesozoic, of which we have a very scanty fossil record, were probably all carnivorous. They fed on insects and other arthropods, such as spiders and crustaceans, fish, and small reptiles, amphibians, and mammals. They led a life comparable to that of shorebirds and waterbirds, hawks, and vultures of today. The great increase in variety of so-called songbirds, most of which have plant-related diets, is probably a direct consequence of the dominance of the flowering plants for their food, although a few do feed on conifer seeds and others, such as hummingbirds, feed directly from flower nectar. Even woodpeckers, which are primarily insect eaters, will eat seeds during times when conditions are adverse.

Thus, the successful adaptation and proliferation of the flowering plants during the gradually changing global climate conditions of the Cenozoic provided the prime impetus for a whole series of complex changes that ultimately led to the natural world that we live in today.

KEY TERMS

archaic mammalian fauna
Artiodactyla
carnassials
carnivore
condylarth
creodont
incus
insectivores
Litopterna
malleus

mammoth
marsupials
mastodon
miacidids
modern mammalian fauna
monotremes
multituberculates
Neogene mammalian fauna
Notoungulata
Paleogene mammalian fauna

pantodonts
pantotheres
Perissodactyla
placental
Pleistocene extinctions
proboscideans
triconodont
uintathere
ungulate

READINGS

Kurtén, B. 1986. *How to Deep-Freeze a Mammoth*. Columbia University Press. 121 pages. A collection of short essays on various aspects of fossil mammals and the Ice Age.

Kurtén, B. 1988. *On Evolution and Fossil Mammals*. Columbia University Press. 301 pages.

Lillegraven, J.A.; Kielan-Jaworowska, Z.; and Clemens, W.A., eds. 1979. *Mesozoic Mammals*. University of California Press. 311 pages. A technical book with chapters by various authors.

Lister, A., and Bahn, P. 1994. *Mammoths*. McMillan Co. 168 pages. Modern treatment of the proboscideans.

Savage, R.J.G. 1986. *Mammal Evolution: An Illustrated Guide*. British Museum of Natural History, Facts on File. 259 pages. An oversized book with many illustrations of bones and reconstructions, many in color. Little coverage of human evolution.

Sutcliffe, A.J. 1985. *On the Track of Ice Age Mammals*. British Museum of Natural History. 224 pages. An excellent, lavishly illustrated book on Pleistocene mammals worldwide.

Primate and Human Evolution

INTRODUCTION

Of all fossils, from tiny one-celled algae to giant dinosaurs, none engender as much fascination and controversy as fossils of human and human-related animals. These fossils are exceptionally rare, and they are the remains of our ancestors. The rarity of these fossils leads to many theories, and their importance leads to much emotion. Because of this importance, the amount of money, time, and effort that is needed to find these remains, with many from remote parts of Africa, far exceeds the expenditures for any other kind of fossils. The study of these fossils is a separate, distinct discipline—anthropology.

Our species name is *Homo sapiens*. The generic name *Homo* means man, and the species or trivial name *sapiens* means wise. We belong to an order of mammals named **primates**, meaning first, and to the family Hominidae. Where do we belong among the remainder of Earth's animals? How do we decide whether a specific fossil should be included in the genus *Homo*? If the fossil is included in *Homo*, then should it be included in our species, *H. sapiens*, or in another species?

In spite of the relative scarcity of **hominid** (members of the family Hominidae) fossils, intense collecting efforts have yielded many skeletons. Many taxa are known from only partial specimens, and population data are known for only a few species. Whereas this record is quite incomplete, it is a good framework with which to begin to piece together the 4.4-million-year hominid history. However, due to the incomplete nature of this record, few scientists agree on either the species names applied to the fossils or the phylogeny of the hominids. New finds and new species are reported every year, and no two publications agree on every detail. Because a consensus has not been reached, two alternatives are discussed here: 1) a conservative view with all known hominid fossils assigned to one of seven species and three genera; and 2) the interpretation with all currently debated names listed, totaling of 16 species in four genera (Figure 17.1). When a consensus is reached in the future, it will undoubtedly be intermediate of these extremes. It is important to remember that the fossil record of our evolution is well documented and that it is only the interpretation of the fossil specimens that differs among scientists.

The family Hominidae, which includes humans, is characterized by primates with an upright posture. The forelimbs are not used for locomotion and, hence, are usually shorter and less robust than the hind limbs. Freeing the forelimbs from a role in walking, running, or swinging though trees was an important, initial step in use of the arms

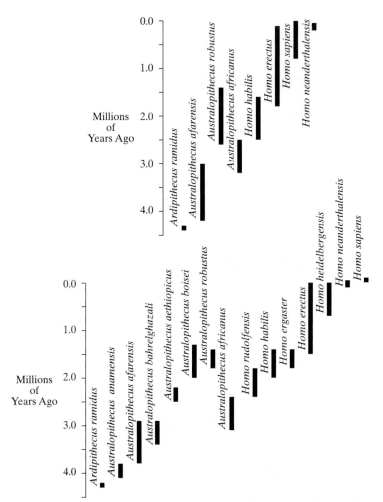

Figure 17.1　Comparison of hominid species names and durations. (A) Minimalist view of hominid species used here. (B) Maximum number of hominid species that could be recognized.

to gather food and to manipulate tools. It is now known that at least one living ape, the chimpanzee, uses twigs or grass stems as simple tools to accomplish specific tasks. At one time, the ancestors of humans surely went through comparable stages of tool development. Upright posture can be inferred even with fragmentary fossil bones. In particular, with bones from the shoulders and hips, it is possible to tell whether a specific animal had an upright stance. Also, the position on the skull where it attaches to the backbone reflects an animal's posture. For many years, it was assumed that development of an upright stance for walking and use of simple tools would have come after conspicuous enlargement of the brain. We now know that this is not the case. As discussed below, the small *Australopithecus afarensis* walked upright despite its seemingly small braincase.

The most common skeletal parts of fossil primates are teeth, just as in other fossil mammals. Primate teeth, especially molar teeth, are useful in classification of these fos-

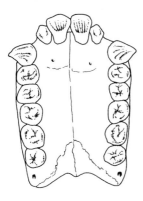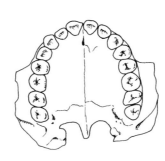

Figure 17.2   Comparison of the upper jaws of a human and an ape. Note the rounded, semicircular outline and the same-sized small teeth of the human jaw compared to the U-shaped jaw with large canine and front incisor teeth of the ape.

sils, including all hominids. As a rule, primates have quite generalized teeth that reveal whether they were insectivorous or fruit- or leaf-eating herbivores. In chimpanzees and humans, the molars reflect an omnivorous diet, including both plant and flesh foods. In the earliest hominids, the teeth are characteristically small and are all about the same size. This contrasts to the larger teeth of apes and monkeys with incisors and canines larger than other teeth (Figure 17.2). The teeth of apes and monkeys, when viewed from above or below, have a U-shaped outline rather than a rounded, arch-shaped outline as in humans. In lower primates the teeth of males, especially canines, are commonly larger than in females. These sexual differences have ceased to exist in humans.

The oldest preserved worked tools that primates used were smooth, rounded stream cobbles. These were used by hunters as hammers to break bones and extract bone marrow from animals that had been killed for food. Such cobbles can be recognized by typical wear patterns that are the result of repeated pounding. These tools have been found in sediments in Ethiopia approximately dated at 2.6 million years, slightly older than the oldest known specimens of *Homo*.

THE PRIMATES

The order of mammals to which humans belong, the primates, is one of the longest surviving and most generalized groups of mammals. The order first appeared during the Eocene and continues to the present day, represented by approximately 200 living species. The order is divided into three groups or suborders, with the two main suborders being the Prosimii and the Anthropoidea.

The first group of primates is the suborder Prosimii. These are the **prosimians** or premonkeys and include three groups: the living lemurs, the living tarsiers and lorises, and the extinct adapids from the Eocene. Lemurs and tarsiers are first known as early Eocene fossils. Living prosimians are all small, tree-dwelling animals of India, Madagascar, and southeastern Asia. The lorises and tarsiers are quite small insectivores (less than 1 pound), and the considerably larger lemurs (5 or more pounds) are fruit- and leaf-eating. Most of these have a partially opposable thumb and, hence, have grasping hands. The muzzle or snout is greatly shortened, resulting in a flattened face. This indicates a decrease in the sense of smell and an increased emphasis on sight. The eyes face forward rather than to the side, thus making stereoscopic or three-dimensional vision possible. Also, like a few other mammals, the primates see in color. Prosimians have either 38 or 36 teeth, differing from humans and apes with 32. The difference is that prosimians have three rather than two premolars in each side of the upper and lower jaws.

The second group of primates is the suborder Anthropoidea, which includes monkeys, apes, and humans. This group apparently evolved from a prosimian stock during the late Eocene or early Oligocene. Two major groups of monkeys exist, which are 1) the Old World monkeys of Africa and Asia that have 32 teeth; and 2) the New World monkeys of South and Central America that have 36 teeth and a prehensile tail. These and other important differences between these two groups indicate that the New World monkeys diverged early from the main line of anthropoid evolution and have existed in isolation for considerable time. Fossil records for both groups begin during the Oligocene.

The superfamily Hominoidea includes three families and many genera. In addition to humans, living members include the chimpanzee, gorilla, orangutan, and two genera of gibbons. Our family, the Hominidae contains three or four genera.

Modern studies of the biochemistry of living anthropoids, including comparisons of their proteins, indicate clearly that humans are much more closely related to the two African apes, especially the chimpanzee and to a lesser extent the gorilla, than they are to the Asian apes, the orangutan or the gibbons.

THE FOSSIL RECORD OF PRIMATES

The oldest confirmed primates are from Eocene rocks. The fairly common adapids were arboreal, insectivorous, and fruit-eating and were about the size of a mouse or woodchuck. The adapids include ancestors of later primates as well as many rodentlike forms. These include the ancestors of lemurs and lorises. Some of these were larger plant eaters and arboreal tree dwellers weighing up to 5 pounds. This group of primitive primates became extinct by the end of the Eocene. Little is known about the fossil record linking these early primates to their living descendants. Only rare fossil lorises are known from the Miocene to the Holocene.

At the base of the anthropoid radiation is the Egyptian Oligocene genus *Aegyptopithecus*. This animal weighed up to 10 pounds and was a fruit-eating tree dweller that exhibited distinct hominid features. It was clearly an ancestor of Miocene apes such as *Proconsul* (Figure 17.3) and, at the same time, was structurally similar to Old World monkeys. The features that separate monkeys and apes developed progressively with time as these two groups evolved during the Neogene. Primitive apes evolved during the Miocene and Pliocene, both in Africa and in Eurasia. As many as 16 ape and apelike genera occurred that are exclusively known as fossils. From this group during the late Cenozoic, the direct ancestors of humans ultimately evolved.

The fossil record of our family, the Hominidae, begins with East African fossil material from approximately 4.4 million years ago that is assigned to *Ardipithecus ramidus*. At this time, it is unclear whether this fossil is a direct ancestor to later hominids or a separate evolutionary branch. The next youngest hominid is the oldest species of **Australopithecus**, *Australopithecus afarensis*, from approximately 4.1 mya.

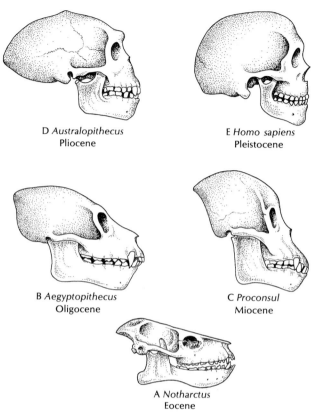

D *Australopithecus*
Pliocene

E *Homo sapiens*
Pleistocene

B *Aegyptopithecus*
Oligocene

C *Proconsul*
Miocene

A *Notharctus*
Eocene

Figure 17.3 Side views of skulls of various primates discussed in the text. (A) *Notharctus*, a primitive Eocene primate skull is 7.5 cm (3 in) long. (B) *Aegyptopithecus*, the oldest known ape, Oligocene in age. (C) *Proconsul*, the Miocene stem hominoid ape. (D) *Australopithecus*, the oldest hominoid. E. *Homo*, a modern human.

This and later primitive australopithecines can be grouped into one species or split into as many as three species. Other names applied to these earliest hominids are *Australopithecus anamenis* and *Australopithecus bahrelghazali*. Given the lack of numerous complete specimens to statistically assess population variability for each proposed species, it is difficult to determine how many of these names apply to truly different biological species versus different geographic or temporal variants of one or more species. Regardless of the taxonomic uncertainty, as a whole this group of primitive hominids is now quite well known.

Australopithecines were primarily herbivores. In comparison to modern humans, they were shorter with proportionately longer arms and shorter legs, and they had smaller brains. The average height of males was approximately 5 feet, 1 inch, with an average weight of approximately 100 pounds. Females were smaller, averaging approximately 4 feet 4 inches and 64 pounds. Their brain capacity, an important character used to differentiate hominids, averaged from 400 to 500 cubic centimeters (cc) (Figure 17.3).

Although anatomical features suggested that *A. afarensis* and other primitive australopithecines were bipedal, this issue was debated until two discoveries during the 1970s confirmed the upright, bipedal posture for these hominids. First, in 1974 a nearly complete specimen of *A. afarensis*, named Lucy, was discovered in rocks from 3.18 mya. Lucy was collected from Hadar, Ethiopia. The anatomical detail of this individual clearly indicated bipedalism, and any lingering doubt was eliminated when, in

1978, Mary Leakey discovered a remarkable set of hominid footprints at Laetoli, Tanzania. Footprints, clearly dated at 3.6 mya, are of complete bipedal aspect. Two individuals, one large and one small, left footprints along a beach. These footprints are either from an adult and a child or from a male and a female, and they are from *A. afarensis* or one of the other species of these earliest hominids. These footprints look identical, except somewhat smaller, to the footprints of modern humans.

From these most primitive hominids, two basic types of australopithecines evolved, one with a robust, strongly constructed jaw apparatus, and one with a more gracile, slightly built jaw. In the most conservative classification, the heavily-built forms are *Australopithecus robustus*, and the gracile forms are *Australopithecus africanus*. The oldest specimens of *A. africanus* are from 3.0 mya, and the oldest *A. robustus* are from 2.6 mya. These two species or species groups co-existed for at least 300,000 years.

THE GRACILE *AUSTRALOPITHECUS*

The gracile *Australopithecus*, *A. africanus*, had smaller males and larger females if compared with its ancestor, *A. afarensis*; and *A. africanus* is estimated to be slightly shorter and more slender than its counterpart *A. robustus*. Presumably, *A. africanus* should have been able to run more rapidly and was more agile than either of these other two species.

Australopithecus africanus was on average approximately 3 feet, 9 inches in height for females that weighed approximately 66 pounds. Whereas, males averaged 4 feet, 6 inches in height and 90 pounds in weight. Average cranial capacity in *A. africanus* was between 400 and 500 cc. The molar teeth of *A. africanus* were much smaller than those in *A. robustus*, undoubtedly indicating a much different diet.

Australopithecus africanus is first recorded from approximately 3.2 mya, and the youngest forms are from sediments of approximately 2.5 mya. It is this species of *Australopithecus* from which scientists believe that *Homo habilis* evolved, although as currently known these two species did not have overlapping ranges.

THE ROBUST *AUSTRALOPITHECUS*

The heavy-jawed *Australopithecus* with large molar teeth was first described as *A. robustus*. They are now known throughout east-central and southern Africa either as this species or as *A. boisei* or *A. aethiopicus*. Some scientists would even place *A. robustus* or all three of these robust species in a separate genus, *Paranthropus*. These australopithecines, referred to here as only *A. robustus*, were not substantially different in overall size from *A. africanus*. Males had an average height of 4 feet, 5 inches and an average weight of 90 pounds; females had an average height of 4 feet and a weight of 73 pounds. Their brain capacity averaged 500 to 600 cc.

The distinguishing aspects of *A. robustus* were especially large molar teeth and a heavily-built jaw that are interpreted to record a diet composed of quite coarse vegetable matter. The oldest known specimens of this robust australopithecine are from 2.6 mya, and the youngest specimens are from 1.4 mya. All are from Africa, and these forms left no known descendants.

HOMO HABILIS

As mentioned above, *H. habilis* is the oldest member of our genus, ***Homo***, and it evolved from and coexisted with *A. africanus* in eastern Africa. This and later species are placed in the genus *Homo* because of a brain capacity increased to greater than 600 cc. Although some youngest australopithecines are believed to have used very primitive tools, it was not until the appearance and dispersal of *H. habilis* that widespread usage of tools occurred. Numerous stone tools made of worked flint occur with *Homo* remains. The usage of tools undoubtedly made a substantial impact on the way of life of *H. habilis* and its descendants. Whereas australopithecines are considered herbivores, *Homo habilis* and later *Homo* species regularly incorporated meat into their diets and should be regarded as omnivores.

Again, the primitive *Homo* species is discussed here as *Homo habilis*. However, note that some scientists also recognize *H. rudolfensis* as a separate primitive species.

Homo habilis was a species with small individuals. Female individuals stood approximately 3 feet, 10 inches tall and weighed 70 pounds, whereas males stood at 4 feet, 4 inches and weighed 80 pounds. Although small for the genus, the brain capacity of *H. habilis* (averaging 600 to 750 cc) was larger than any australopithecines. Also, *H. habilis* had smaller teeth and a modified pelvis compared to *Australopithecus* individuals.

HOMO ERECTUS

Homo erectus was taller, leaner, and undoubtedly a more rapid runner than earlier hominids. Individuals had a larger face, a backward sloping forehead, and a protruding brow ridge. *Homo erectus* males were on average 5 feet, 10 inches tall and weighed 140 pounds; females averaged 5 feet, 3 inches and 120 pounds. The brain capacity average in this species was approximately 800 to 1,100 cc. The animals referred to here as *H. erectus* are subdivided by some to also include *H. ergaster*. Java man and Peking man are names given to some of the earliest discoveries of *H. erectus* in Indonesia and China, respectively.

Homo erectus was the first hominid to migrate out of Africa. This species first evolved approximately 1.8 mya in Africa from *H. habilis*; by 1.0 to 0.8 mya, *H. erectus* lived from Europe to Indonesia. Members of this species were good hunters and clearly migrated in groups. They utilized a diversified series of tools and were adaptable enough to survive north of Africa, even during colder glacial times. Both *H. sapiens* and the neandertals evolved from *H. erectus*.

HOMO SAPIENS

Primitive *Homo sapiens* first occurred approximately 800,000 years ago in Africa, and anatomically modern *H. sapiens* are present by approximately 125,000 years ago. *Homo sapiens*, considered anatomically modern, have more prominent jaws, barely visible brow ridges, and high foreheads (Figure 17.3). They have an average cranial capacity of 1,100 to 1,400 cc and have an average height of 5 feet, 9 inches (143 pounds) for males and 5 feet, 3 inches (120 pounds) for females. Brain enlargement in *H. sapiens* especially affected the forebrain or cerebral cortex that is concerned with observations, memory, and comparisons. The fossils discussed here as *H. sapiens* are subdivided by some, who would separate the primitive *H. sapiens* as *H. heidelbergensis* and *H. antecessor*. Also, as discussed below, some would include neandertals as members of *H. sapiens*. Early discoveries of *H. sapiens* in Europe were referred to as the Cro-magnon man.

Homo sapiens changed the face of Earth. Early members of our species shared Earth with one or two other hominid species, *H. erectus* and *H. neanderthalensis*. As discussed below, we have been the sole hominids on this planet for only approximately 30,000 years.

NEANDERTALS

In 1856 a fossil hominid was discovered in Germany that would become known as the **neandertal** man. This fossil caused a controversy during the nineteenth century and demanded a rethinking of our species' role among the hominids. We now know a tremendous amount about this fellow hominid, but this close relative still engenders controversy. Are neandertals a geographic variant of *Homo sapiens*, a subspecies (*H. sapiens neanderthalensis*), or the separate species *H. neanderthalensis*? Recent genetic information points to neandertals being a separate species, but a consensus on this issue is forthcoming.

We do know that neandertals were well adapted and restricted to cold-weather, northern climates during the Pleistocene and that they ranged from Britain to the Middle East. Neandertals did not live in Africa. These hominids were relatively short with short arms and legs and with stocky, robust bodies. Facially, they had a strong brow ridge, a sloping forehead, and a weak chin. On average, neandertal males stood 5 feet, 5 inches and weighed 185 pounds; females weighed 175 pounds and were 5 feet on average. Their brain capacity was slightly larger than *Homo sapiens* measured in actual size, but this capacity was slightly smaller in relation to their body mass.

This compact body was well adapted to conserving heat in the colder climates where neandertals lived. The oldest neandertals are 230,000 years old, and the youngest are approximately 30,000 years old. Neandertal tools were well crafted and diverse. Presumably, they had language and created artwork, and a few crude neandertal-carved art pieces are known. Because of the size of their prey, woolly rhinoceros and mammoths, neandertals must have hunted in groups, and their remains indicate that they traveled in small groups or family units. Different from other species except *H. sapiens*, neandertals cared for their sick and injured, as evidenced by neandertal remains with severe arthritis and healed injuries. The average lifespan of neandertals is estimated at approximately 30 years; few survived to be 40. The sapient aspect of neandertals is demonstrated by the fact that they buried their dead. These simple burials indicate a value on individual life and contemplation on the meaning of their existence.

Homo sapiens underwent what was perhaps the first technological revolution approximately 50,000 years ago when more elaborate and precise tools and projectile points became part of their tool kits. This preceded the migration of *H. sapiens* from Africa, so that by 40,000 years ago *H. sapiens* had spread throughout Europe and coexisted with neandertals. *Homo sapiens* and neandertals coinhabited Europe for approximately 10,000 years—more than 300 generations; in the Middle East they may have coexisted for as much as 40,000 years. What was the nature of their interaction? Some scientists argue that these two species or populations interacted peaceably. Perhaps they even interbred. In this view, neandertals gradually disappeared and were replaced by *H. sapiens*. Alternatively, their relationship has also been speculated to have been one of conflict; and in this view, *H. sapiens* survived through confrontation, probably as a result of better weapons, more agile bodies, and more sophisticated thinking. Whatever the means, for the first time since approximately 3.0 mya when *Australopithecus afarensis* and *A. africanus* (or *africanus-like*) first coexisted in Africa, *Homo sapiens* became the only hominid species on Earth.

AFRICAN ORIGINS

It is clear that hominids originated in Africa and evolved there in isolation for a considerable period of time. This evolution was not a simple, straight-line evolution, but it included a few to many side branches that ultimately became extinct without leaving descendant forms. Similarly, paleontological evidence demonstrates that the genus *Homo* evolved in Africa with the origination of *Homo habilis* (as used here) from *Australopithecus africanus*. Thus, Africa is commonly referred to as the "cradle of humanity." *Homo erectus* populations were the first hominids to migrate from Africa, approximately 1.0 to 0.8 mya. As mentioned above, these hunter-gatherers were quite adaptable and mobile. Relatively quickly, they migrated through Asia to Indonesia.

The origination of our species, *Homo sapiens*, is less certain. The fossil record indicates that *H. sapiens* first evolved in Africa, a primitive *H. sapiens* first appearing as early as 800,000 years ago, and anatomically modern *H. sapiens* at approximately 125,000 years ago. Development of a geographically widespread *H. sapiens* is a matter of debate with two hypotheses currently under discussion. The first hypothesis is commonly called the "Eve hypothesis," and it states that *H. sapiens* evolved exclusively in Africa and migrated out of Africa to populate Europe and Asia. Thus, *H. sapiens* underwent an origin and migration that paralleled the earlier one of *H. erectus*. This hypothesis is supported by genetic studies in which maternal DNA was analyzed from native populations throughout the world. Results suggested that all these native populations descended from a single woman, thus the "Eve hypothesis."

The alternative hypothesis is called the "multiregional hypothesis," which maintains that modern *H. sapiens* did not have an isolated origin in Africa. Instead, after first migrating from Africa, *H. erectus* populations continued to migrate across Europe, Asia, and Africa. This kept the populations in contact and interbreeding so that gene flow was maintained among populations. In this view, modern *H. sapiens* is thought to have evolved simultaneously throughout Africa, Europe, and Asia. Evidence for this hypothesis is morphological. The skulls of Chinese *H. erectus* share similarities with those of Chinese *H. sapiens* and not with other *H. sapiens* populations.

Resolution to this debate awaits further work. Additional fossil specimens of *H. erectus* and *H. sapiens* may help to better understand the morphologies of this transition throughout Europe, Asia, and Africa. Furthermore, perhaps details of tools and other artifacts can establish the movements of these peoples across Africa, Europe, and Asia prior to and during the emergence of *H. sapiens*.

THE HUMAN CONDITION

Why did humans evolve? What environmental factors favored the evolution of one group of primates toward obligate bipedalism, a larger brain, and the myriad of traits that separate us from other animals? Current discussion on this topic centers on whether hominids were specialists or generalists. The traditional viewpoint, most recently revised and argued by Steven Stanley, concludes that we are specialists, specialists for the African grasslands. When hominids first arose approximately 4.5 mya, Earth's climate was in a period of fluctuation, but generally it was becoming cooler and drier. Many heavily forested areas began to dissipate and be replaced by grassland habitats. This corresponds to a time of dramatic climatic change on Earth. The Isthmus of Panama emerged approximately 3.0 mya; and in addition to causing major changes in mammalian faunas in South America (see Chapter 16), this greatly affected patterns of oceanic circulation and brought about Pleistocene glaciations. Accompanying Pleistocene ice ages, the African climate became more arid, and grasslands and savannas replaced even more forests. In this view, *Homo* evolved during this interval of time. Whereas australopithecines were both arboreal and ground living, *Homo* evolved to live exclusively for this grassland habitat in the specialist hypothesis.

An alternative hypothesis proposed by Richard Potts views the climatic variability of the Neogene rather than any one specific habitat present during this time as the primary causal factor in hominid evolution. Hominids, both *Australopithecus* and *Homo*, evolved during a time of substantial climatic variability during the Neogene. Potts proposed that long-term success was accorded to hominids because of their adaptability. As adaptable generalists, hominids continued to evolve and succeed in a world of climatic variability. They migrated from Africa into numerous climatic zones. In this hypothesis, evolutionary trends should be viewed beyond a simple grasslands explanation, but be viewed as trends that were able to better adapt to uncertain environmental conditions. Eventually, man adapted to habitat conditions throughout the globe.

Whatever the causal process, we have inherited a complex mixture of physical traits from primate ancestors as well as newly acquired unique features that characterize us as a species. In some aspects the human skeleton is quite primitive and generalized. The presence of five digits on each limb, the number of bones per digit (two in thumb and big toe, and three in other digits), and the large number of teeth are all characteristics shared with primitive mammals. Most advanced herbivorous and carnivorous mammals have evolved in a reductionist way from this state, by decreasing the number of teeth and digits.

Some human features are related to an ancestral arboreal habit and an early insect diet, such as the opposable thumb and related manual dexterity, flat nails instead of claws, stereoscopic vision, color vision, and reduction in reliance on the sense of smell. An upright posture is interpreted by some to be related to tree dwelling. Thus, we carry with us abundant indications of our primate ancestry. The increasingly large forebrain of primates that controls the ability to remember, observe, and compare is surely the most remarkable human feature, which led to intelligence, language, art, culture, and social bonds that are unique to our species.

Many of these important human traits cannot be studied through the fossil record because they are not preservable in the ordinary sense of teeth or bones.

However, stone tools and other trace fossils record many behavioral traits of ancient hominids. The oldest tools predate *Homo sapiens* by nearly 2 million years. The youngest *Australopithecus* could probably control fire. Primitive neandertals certainly had some kind of language, practiced art, had burials, and surely had religious customs. Modern humans first controlled the growth of grasses for grains and domesticated sheep, goats, and other animals.

The unique aspects of humans relate largely to their elaborate social behavior. The underdevelopment of human young at birth is compensated for by long periods of postnatal parental care. Families, clans, and societies are more complex than the social structures of other social animals, such as apes, birds, or insects.

KEY TERMS

Australopithecus	*Homo*	primates
hominid	neandertal	prosimians

READINGS

Badgley, C. 1984. "Human evolution." In T.W. Broadhead, ed., *Mammals: Notes for a Short Course*. Studies in Geology 8, University of Tennessee Department of Geological Sciences. Pages 182–199. A summary of the evolution of humans and their near relatives.

Campbell, B.G. 1988. *Humankind Emerging*. Scott, Foresman. 522 pages. A popular book, based in part on the Time-Life series on physical anthropology, prehistoric man, and human evolution.

De Rousseau, C.J. 1990. *Primate Life History and Evolution*. Monographs on Primatology No. 14. Wiley-Liss. 366 pages. A compilation of articles on primates based on a conference held in 1987.

Johanson, D.C. 1989. *Lucy's Child*. Morrow. 318 pages. An account of recent developments in human paleontology in East Africa by one of the principal participants.

Lewin, R. 1987. *Bones of Contention: Controversies in the Search for Human Origins*. Simon and Schuster. 384 pages. A well-written survey of research on the origin of humans and human evolution including a discussion of the divergent theories of active researchers, written by a well-known science writer.

Martin, R.D. 1990. *Primate Origins and Evolution*. Princeton University Press. 804 pages. A very detailed and technical book on all the primates, with emphasis on living primates.

National Geographic. 1995–1997. A series of six articles on hominid evolution published in *National Geographic* that outlines ideas on their fossil record and evolution. Well illustrated, easy to read, and informative. Articles appear in the following issues: September 1995, January 1996, February 1997, May 1997, June 1997, September 1997.

Poirier, F.E. 1993. *Understanding Human Evolution*, 3rd ed. Prentice-Hall. 344 pages. A college-level textbook in anthropology.

Potts, R. 1997. *Humanity's Descent, The Consequences of Ecological Instability*. Avon. 336 pages.

Stanley, S.M. 1996. *Children of the Ice Age, How a Global Catastrophe Allowed Humans to Evolve*. Crown Publishing Group. 288 pages.

CHAPTER 18

Conclusion

We have now considered in a broad and introductory way the history of life on Earth through its 3.5-billion-year history. During this enormous span, a series of startling changes occurred in the nature and diversity of life; contrast a Precambrian vista, where only photosynthetic cyanobacteria and bacteria existed in the oceans, with the modern scene dominated by flowering plants, insects, and mammals. Strategies for making or catching food have also undergone conspicuous changes, as have the dominant groups that engaged in one form or another of food procurement.

It should now be clear that paleontology is almost a perfect blend of geology and biology. Paleontology is interdisciplinary. Physical, chemical, and biological aspects of Earth history must be taken into account if we are to arrive at a rational view of the history of life on Earth. If geological occurrences of fossils are given disproportionate emphasis, exciting biological concepts that could be applied to fossils may be overlooked. If fossils are considered strictly from a modern biological viewpoint, inaccurate judgements about past physical conditions or about relative ages of fossils may result.

There is much more to paleontology than simply describing and documenting new fossil species, although this is the basic data from which life's history is interpreted, and it is extremely important. However, using the best information possible, data synthesis is required in order to understand both the patterns in the history of life as well as the processes responsible for this history. The literature on fossils is so vast that no single individual can possibly master it all. Most paleontologists compromise by becoming a specialist in one or two groups of fossils. The important thing is not which group of fossils is studied or how large or complicated it is but, rather, how the study of fossils is viewed. The correct approach is that any group of fossils can shed light on important problems if the paleontologist is aware of such problems, asks the right questions of the fossil data, rigorously tests hypotheses, and concludes with viable solutions. Paleontology is an exciting and challenging field. Important new discoveries are made every year, and there are still many important questions without satisfactory answers. Paleontology is a vital component of several other fields of study, including the study of global climate change on Earth, the stratigraphy of the sedimentary rocks in the Earth's crust, and various aspects of energy exploration, especially for various petroleum products.

What will the future bring for paleontology? We cannot predict with absolute accuracy; however, current research always points the way that future research will proceed. First, the application of current biological concepts to the fossil record should continue to dominate paleontology. For example, ideas developed in the study of modern populations and communities of plants and animals have been successfully applied to fossils. Paleocommunity approaches should continue to be valuable in understanding both patterns and processes of global change. The fossil record is the record left by evolution, so this remains a rich source for understanding evolutionary processes. Equally important are new techniques. Various kinds of analytical equipment have important applications in paleontology, leading to the creation of new areas of study. The use of computers to handle large amounts of data or to compute theoretical solutions, isotopic methods to determine geochemical information, molecular techniques to probe the biochemistry of fossils, and CAT-scan X-radiography to examine fossil morphology are all examples of new technologies that are greatly advancing paleontology.

To some extent, paleontologists are limited by the vagaries of fossil localities, but this is also a source of unpredictable data. A new locality yielding unexpected fossils may shed considerable light on an important and otherwise intractable problem. Nearly all the Precambrian fossils were unknown to science earlier because we simply did not know how to look for them. So, paleontology, in part, depends on the availability of opportunities to collect fossils and on the imagination of the collector.

One of the most encouraging aspects of paleontology today concerns the outlook and training of future paleontologists. It was traditional for many years for paleobotanists and vertebrate paleontologists to be trained primarily as botanists and zoologists, respectively, and for invertebrate paleontologists to be trained mainly as geologists. This dichotomy is gradually disappearing, so that all paleontologists are receiving more thorough grounding in important aspects of both the geological and biological sciences. We are already seeing the results of this broader training, and it will be even more important during the future. The outlook for paleontology is stimulating and exciting. The next few decades will undoubtedly produce many new ideas, new theories, and new data that will greatly enhance our understanding of the life of the past.

Now that we have reviewed the history of life on Earth, does that history teach us any lessons about our present-day situation or what the future may hold? A famous quotation by George Santayana is that those who do not remember the past are doomed to repeat it. That statement may also be true for our understanding of past life.

The most obvious lesson surely has to do with extinction. Is our own species doomed to become extinct? As is commonly the case, there is not a single, simple answer to this question. We can say that extinction is the ultimate fate of every species that has ever appeared on this planet and that there is no evidence of any species surviving forever. Thus, *Homo sapiens* should eventually become extinct. However, we are the first species on Earth to understand this concept. Even though we only know part of the underlying factors, we may devise ways to prevent extinction of our own kind. We are also the first species to be able to control and consciously alter our environment in significant ways. At the present, we are altering our environment in ways that are seriously deleterious to our continued success as a species. Erosion of fertile topsoil, destruction of forests, fall of acid rain, increase in carbon dioxide emissions from the burning of fossil fuels, depletion of the ozone layer, and pollution of soil, air, and water all continue to pervasively alter Earth, and we are clearly putting ourselves in an ever increasingly precarious ecological situation. Life history has clearly shown that fragile ecosystems are susceptible to widespread extinctions.

The lesson to be learned seems clear. We have already altered the physical, chemical, and biological nature of the planet to the extent that it will never be the same as it was before the existence of modern humans. Instead of continuing to degrade our environment, we should actively improve the environment in order to maximize our long-term chances for survival. Hopefully, this will happen on a worldwide basis before deterioration becomes so severe that it cannot be reversed. Earth will clearly be here as a body of rock for many more millions of years. It remains to be seen whether forms of life as we know them today will persist into the distant future.

Glossary

abiotic synthesis (4)[1] process by which life evolved from nonliving material during the Precambrian.

absolute age (1) age of a rock in years, determined by radioactive dating. (See *relative age*.)

acritarchs (11) microscopic, organic-walled cysts thought to be formed by marine planktonic algae, perhaps related to dinoflagellates; range in age from late Precambrian to the present, but especially common during the Cambrian through the Devonian.

adaptive radiation (5) evolutionary pattern in which diversity of a group of organisms increases greatly and very rapidly, commonly resulting when a group breaks through into a new adaptive zone or otherwise fills vacated niche space.

aerobic (4) refers to an environment that has free oxygen.

Agnatha (12) the most primitive group (class) of fish, those without jaws.

ahermatypic (10) refers to a group of post-Paleozoic corals that do not contain symbiotic photosynthetic zooxanthellae algae and, thus, are not reef-building corals.

algae (2) general term for all water-dwelling, nonvascular multicellular plants (marine or nonmarine).

allopatric speciation (5) evolutionary process of new species formation where a new species diverges from its ancestral species due to geographic isolation of a small population; results in increase in number of species.

altered preservation (3) state of fossil preservation in which the fossil has been changed from its original composition and texture.

amino acids (4) simple organic molecules that form the basis for proteins in plant and animal tissues. Four amino acids are important constituents in DNA and RNA.

ammonite septum (12) type of septum of ammonoid cephalopods characterized by secondary wrinkles or sutures on every lobe.

ammonoid cephalopods (12) major extinct group of cephalopod molluscs characterized by an external shell and wavy partitions (septa) dividing the shell into chambers.

amniote egg (15) shelled egg of reptiles and birds that can be laid on land. The shell protects the embryo from drying out and allows respiration to occur. Tissues inside the shell nourish the embryo and take care of its wastes.

amphibian (13) class of chordates that is the oldest and most primitive of the tetrapods; typically has the initial life stages in water and the adult life stage on land; characterized by an egg that is essentially like a fish egg; includes living toads, frogs, and salamanders.

anaerobic (4) term applied to an atmosphere or environment that is reducing or lacking oxygen.

analogous (5) refers to parts of different organisms that have similar functions but not similar origin; for example, bird wings and insect wings are analogous.

anapsid (15) the most primitive group of reptiles, characterized by lack of an opening in the temple region of the skull; includes living turtles.

angiosperm (14) flowering plant.

Animalia (2) kingdom of life, organisms characterized as multicellular, heterotrophic, and sexually reproducing; includes organisms from sponges to man.

ankylosaurs (15) group of ornithischian dinosaurs with heavy bony plates on the upper surface of the body and a large bony knob at the end of the tail.

---

[1]Numbers in parentheses after each word or phrase refer to the chapter(s) numbers in which that word or phrase is important.

Annelida (2) phylum consisting of the segmented worms; includes earthworms, leeches, and marine polychaetes.

Anomalocaris (12) large (up to 50 cm) predator from the Cambrian, considered by most to be an arthropod. Best known from the Burgess Shale, where isolated parts of this organism were originally described as three separate animals.

aragonite (3) one mineral phase of $CaCO_3$; common composition of shells of most molluscs and scleractinian corals; not stable over geologic time.

araucarian (7, 14) primitive group of pines (conifers) now restricted to the southern hemisphere; common in the Triassic petrified forest of Arizona.

Archaea (2) domain of life that includes the prokaryotes distinguished from other prokaryotes (Bacteria) at a molecular level; also includes methanogenic and sulfur-dependent prokaryotes.

archaeocyaths (9) extinct phylum of sponge-like animals, confined to Cambrian rocks; built small patch reefs during the Cambrian.

Archaeopteryx (15) the oldest known fossil bird; found in Jurassic limestones in Germany.

Archaic mammalian fauna (16) see *Paleogene mammalian fauna*.

Archean (1) Interval of geologic time, an era of the Precambrian from approximately 4.0 to 2.5 bya; younger than the Priscoan and older than the Proterozoic.

Arthropoda (2, 9) phylum of animals that includes trilobites, insects, and crabs; characterized by jointed appendages.

articular bone (15) with the quadrate bone, the two bones that join the upper and lower jaws in reptiles; these become the two bones of the middle ear in mammals.

articulate brachiopods (10) class of brachiopods that includes those orders with the valves hinged with a tooth and socket articulation.

Artiodactyla (16) even-toed, herbivorous, hoofed mammals with an axis down the foot between the third and fourth digits; include many four-toed and two-toed advanced mammals, such as bison, camels, hippos, and pigs.

Australopithecus (17) early genus in the family Hominidae that includes man (Homo); the ancestral genus to Homo; first appeared approximately 4.4 to 4.2 mya.

autotrophs (2, 4) organisms capable of manufacturing their own food, such as cyanobacteria, some protists, and plants.

background extinction (5) more or less constant, expected rate of extinction; mass extinctions must have a rate significantly higher than background extinction to be recognized.

Bacteria (2) domain of life that includes prokaryotes distinguished from Archaea prokaryotes at a molecular level; includes most bacteria and cyanobacteria (blue-green algae).

banded iron formations (8) thick sequences of fine-bedded rocks with alternating layers of jasper (red colored silica-rich rock) and hematite (iron oxide) of Precambrian age.

barrier (7) any impediment to the migration of plants or animals, marine or terrestrial. For example, an ocean is a barrier to migration for land organisms and land is a barrier to marine organisms.

basin of deposition (3) area where sediments accumulate and eventually become sedimentary rocks; siliciclastics are transported into the basin of deposition, whereas non-siliciclastics are formed within the basin of deposition through either precipitation or biodeposition.

belemnites (7) extinct group of cephalopod molluscs that had an internal, cigar-shaped skeleton.

benthos (9) general name for all organisms, whether plant or animal, that live on or in the bottom sediment of an aquatic environment; applicable to either marine or fresh water organisms.

bentonite (3) rock formed through the diagenesis of volcanic ash from an ash fall.

bilateral symmetry (2) form of organism symmetry in which one and only one plane divides the organism into two equal halves that are mirror images of one another.

biofacies (3) assemblage of fossils that occur together and characterize a particular environmental setting.

biostratigraphic correlation (3) matching of the identity of rocks using their contained fossils; represents temporal identity.

biostratigraphy (3) science of determining the relative ages of rocks by the use of fossils.

bipedal (15) a term applied to an animal that uses only its hind legs for locomotion; contrasts with a four-legged (quadrupedal) organism.

Bitter Springs Formation (8) locality in central Australia that has yielded many Precambrian microfossils; the rocks at this locality are dated at 0.9 billion years in age.

bivalves (10) class of molluscs characterized by two hinged shells (valves); the valves are hinged along the dorsal margin. Also called pelecypods, they include clams, mussels, and oysters.

black chert (8) chert rock that contains abundant carbon and, hence, is black in color. Almost all Precambrian microfossils are from black cherts associated with stromatolites.

blastoids (7, 10) extinct group of stalked echinoderms present from the Ordovician through the Permian.

body fossil (1) direct evidence of ancient life as represented by a former part of that organism; also called morphologic fossil.

bony fish (12) general name for the advanced fish. Includes the ray-finned fish and lobe-finned fish; it excludes the primitive groups and sharks.

Brachiopoda (2, 10) phylum of invertebrate animals characterized by two shells (valves) that are situated on the top and bottom of the animal.

branch reduction (11) evolutionary trend in graptolite colonies in which the number of branches in the colony is gradually decreased, normally following a pattern of 8 to 4 to 2 to 1 branch.

Bryozoa (2, 10) phylum of exclusively colonial lophophore-bearing animals with a calcareous skeleton; colony form may be massive, cylindrical, branching, lacy, or ribbon-shaped. Mostly marine and very important during the Paleozoic. Individual animals that make up colonies are very small.

Burgess Shale (9) Middle Cambrian rock unit in British Columbia, Canada, that has yielded a soft-bodied fauna preserved as carbon films. This fauna contains many, many very unusual soft-bodied organisms that provide an important perspective on nonskeletonized marine life.

bya (1) billion years ago.

C^{12}/C^{13} ratio (8) measurement of relative composition of two isotopes of carbon that is lighter if the carbon is biogenic in origin; type of chemical fossil.

calcite (3) mineral composed of calcium carbonate ($CaCO3$); with aragonite the most common shell material of invertebrate animals. Brachiopod, bryozoan, echinoderm, rugose coral, and tabulate coral skeletons are composed of calcite.

Cambrian (1) oldest period of the Paleozoic Era; younger than the Precambrian and older than the Ordovician; from 543 to 510 mya.

Cambrian marine fauna (9) One of the three major faunas to populate the marine realm during the Phanerozoic; only lasted during the Cambrian; dominated by trilobites, inarticulate brachiopods, monoplacophorans, hyoliths, and primitive stalked echinoderms.

Carboniferous (1) period between the Devonian and the Permian in European classification. In North America the term Carboniferous is not used; it is replaced with the Mississippian and Pennsylvanian. The period is from 360 to 286 mya.

carbonization (3) mode of fossil preservation in which most of the organic material is destroyed, leaving behind a carbon-film impression of the organism.

carnassials (16) specialized premolar or molar teeth characterized by a high, bladelike shearing edge; present in advanced mammalian carnivores.

carnivore (16) exclusively meat-eating animal that pursues its prey.

cast (3) natural filling of a mold left behind in the surrounding rock after a fossil has been removed by solution.

cellular grade of organization (2) level of complexity of animals in which cells act independently of one another; includes Porifera.

cellulose (13) complex organic substance that comprises the cell wall in vascular plants.

Cenozoic (1) youngest era of geologic time, from 65 mya to the present. The term means "young life" or "recent life."

cephalopods (12) class of molluscs, including the living squids, octopi, and the pearly nautilus, as well as many fossil nautiloids, ammonoids, and belemnites.

ceratite septum (12) septum in ammonoid cephalopods with a suture pattern of intermediate complexity; every other lobe has secondary crinkles.

ceratopsians (15) group of ornithischian dinosaurs, including Triceratops, characterized by a frill and a parrotlike beak; all but the most primitive also have bony horns on the head.

chalk (11) type of limestone (sedimentary rock) composed largely of the skeletons of microscopic coccoliths; especially widespread during the Cretaceous Period.

chemical fossils (3, 8) chemical molecules, usually organic, that indicate former life and that are preserved in rocks. Precambrian occurrences of the degradation products of chlorophyll, pristane and phytane, are examples.

chemoautotrophs (2) organisms that synthesize their own food through reduction of various chemicals, such as sulfur.

chert (5) type of sedimentary rock formed by the chemical precipitation of silicon dioxide ($SiO2$); chemically the same as flint and quartz.

chitinophosphatic (3) term applied to the shell material of some invertebrates that is composed of alternate layers of chitin, an organic compound, and calcium phosphate ($CaPO_4$).

Chondrichthyes (12) cartilaginous fish, including sharks and rays, which lack a bony skeleton.

Chordata (2) phylum of animals that possess a dorsal nerve cord, notochord (a cartilaginous rod), and gill slits in the throat area during some stage of growth; includes tunicates, fish, amphibians, mammals, and others.

chromosomes (5) strands of genetic material composed of genes; present in the nucleus of eukaryotic organisms.

class (2) category in the classification of life that ranks below the phylum level and above the order level.

Cnidaria (2, 10) phylum of animals that includes corals, jellyfish, and hydroids; characterized by tissue grade of organization.

coal ball (14) concretion formed in coal, composed of calcium carbonate or pyrite, in which the remains of coal swamp plants are typically preserved in exquisite three-dimensional detail.

coccoliths (11) skeletons of tiny marine algae composed of calcium carbonate, one of the principal ingredients of chalk.

coelacanth (7, 12) group of marine lobe-finned fish represented today by a single living genus, *Latimeria*.

coelom (2) fluid-filled internal cavity that surrounds the vital organs in advanced groups of animals.

condylarth (16) primitive mammalian herbivores, especially common during the early part of the Cenozoic Era.

cone (14) cluster of modified leaves and their associated reproductive structures in vascular plants.

conglomerate (3) clastic sedimentary rock composed of rounded grains larger than those of sand; contains pebbles or cobbles.

conifer (14) group of mostly evergreen gymnosperms that were especially widespread and diverse during the Mesozoic Era.

conodonts (11) group of extinct animals that persisted from the Cambrian to the Triassic; known principally from microscopic, tooth-like parts of the body. They were small, marine, and nektonic. These were either some type of chordate or a separate phylum.

continental drift (6) movement of continents on Earth's surface; includes the idea that the continents were once one giant supercontinent that later broke and drifted apart.

convergent evolution (5) pattern of evolution in which two or more unrelated groups develop forms that become similar in appearance.

corals (10) group of Cnidaria, either colonial or solitary, calcified; Paleozoic groups were tabulates and rugosans, and the post-Paleozoic group is scleractinians; important in the formation of reefs today and many times during the past.

core (4, 6) central part of Earth, partly molten, very dense, and composed of metallic iron and nickel.

correlation (3) in stratigraphy, process by which units of rock are determined to be equivalent from area to area, either in terms of their physical character or their relative age.

corridor (7) broad path of migration and dispersal without significant barriers so that many kinds of animals can move from one area to another.

cosmopolitan (7) term used to describe a species or genus that has a very widespread geographic distribution.

creationism (5) idea that all observed records of ancient life on Earth can be explained by the story of creation in the Old Testament.

creodont (16) primitive carnivorous mammals during the Mesozoic and early Paleogene.

Cretaceous (1) period of geologic time that closes the Mesozoic Era, occurring between the Jurassic and the Paleogene Periods and lasting from 146 to 65 mya. The name is derived from the Latin word for *chalk* because chalk is a very common rock type from this period.

crinoids (10) class of echinoderms, usually with a stem or stalk on which the body sits with few to many feeding arms; the only living stalked echinoderms.

crust (4, 6) uppermost, rigid layer of Earth; includes the continents and the rocks that underlie the ocean.

crustaceans (12) group of arthropods, including crabs, lobsters, ostracods, and others; very important benthonic predators during the Mesozoic and Cenozoic.

cuticle (13) the outer layer on the above-ground parts of vascular plants; prevents desiccation; composed of a waxy biopolymer.

cyanobacteria (2) photosynthetic prokaryotes, formerly known as blue-green algae; among the most primitive organisms living on Earth and among the oldest fossils known.

cycads (14) group of gymnosperms, now mostly extinct, that had a short, barrel-shaped trunk and many long leaves issuing from the top; especially common during the Mesozoic Era.

cyst (11) a resting phase during the life cycle of a protist or plant; typically encased in a protective coating to weather severe conditions.

daughter element (1) in a radiogenic decay series, isotope that is the final, stable decay product from the radioactive parent isotope.

dentary (15) major bone in the lower jaw of vertebrates; the defining characteristic of mammals is having only this bone comprising the lower jaw.

deposit feeder (9) any animal that ingests sediment, either on or beneath the sea floor.

dermal denticles (12) bony elements in the skin of sharks; except for some fin supports, the only bone in these fish.

dermal layer (2) skin layer; one of the three basic cell layers of advanced animals—ectoderm, mesoderm, and endoderm.

deuterostomes (2) animals in which the first embryonic opening becomes the anus and the second embryonic opening becomes the mouth; examples are echinoderms and chordates.

Devonian (1) period of the Paleozoic Era occurring after the Silurian and before the Mississippian; named after Devonshire in southwestern England. The absolute age of this period is 408 to 362 mya.

diapsid (15) the ruling reptiles, including dinosaurs, snakes, lizards, and crocodiles; characterized by having two openings in the temple region of the skull.

diatoms (11) group of microscopic, freshwater and marine, planktonic protists with a skeleton composed of silica.

diatomite (11) rock type principally composed of diatoms.

dichotomous branching (13) branching pattern in which each branch is equal, resulting in a tuning fork or Y-shaped pattern; typical of many primitive plants.

dinosaurs (15) popular name for the two dominant groups of diapsid reptiles of the Mesozoic that include the predators and herbivores of the Mesozoic terrestrial ecosystem.

diploid (13) having twice the number of chromosomes that normally occur in a sex cell. All nonreproductive cells of most animals and land plants are diploid. (See *haploid*.)

diploporans (10) class of stalked echinoderms important during the lower Paleozoic; formally, one group of cystoids.

DNA (5) abbreviation for deoxyribonucleic acid, an organic molecule important for the storage and replication of genetic information within the nucleus of cells.

domain (2) largest category in the classification of life; ranks above kingdom.

Echinodermata (2) phylum of animals that includes starfish, echinoids, sand dollars, crinoids, and others.

echinoids (10) class of nonstalked echinoderms characterized by a spherical or flattened body with movable spines; includes sand dollars and sea urchins.

ectoderm (2) outer layer of tissue in an animal.

ectotherm (15) organism that relies on the external environment for internal body warmth; cold-blooded.

Ediacaran megabiota (8) late Proterozoic biota of soft-bodied organisms from south Australia and elsewhere. Debate on this biota is whether these fossils represent the oldest metazoans or an entirely different kind or organism.

endemic (7) native to and/or restricted to a specific geographic area or region; for example, kangaroos and many other marsupials are endemic to Australia.

endoderm (2) innermost layer of tissue in an animal.

endotherm (15) organism that maintains a nearly constant internal body temperature through metabolic heat; warm-blooded.

enterocoels (2) animals in which the internal cavity, the coelom, is formed from outpockets of the larval gut; includes the echinoderms and chordates.

Eocene (1) epoch of the Paleogene occurring after the Paleocene and before the Oligocene; lasting from 56 to 35 mya.

epifaunal (9, 10) refers to marine animals that live on or are attached to the sea bottom.

epoch (1) interval of geologic time, intermediate in duration between longer periods and shorter ages; in this book most commonly used to describe the subdivisions of the Cenozoic (e.g., Eocene Epoch).

era (1) one of the longest subdivisions of geologic time, consisting of one or more periods. Commonly used era names are Paleozoic, Mesozoic, and Cenozoic.

Eucarya (2) domain of life for all eukaryotic organisms; includes fungi, protists, plants, and animals.

eukaryotes (2) organisms having a membrane around the cell nucleus and highly organized chromosomes.

euryapsids (15) group of aquatic and semi-aquatic reptiles with a single opening high on the temple region of the skull.

eurypterids (12) lower and middle Paleozoic arthropods that were large and important predators in the ocean.

Eurydesma (6) large, thick-shelled marine clam restricted to Gondwana during the Permian and closely associated with glacial deposits.

event beds (3) sedimentary layers that were deposited by an identifiable, short-duration process such as a volcanic ash fall or a storm.

evolution (2) origin of new kinds of life from preexisting forms of life; the fundamental genetic and morphologic change in biological populations.

extinction (5) demise of any group of organisms.

facies (3) term used to distinguish a portion of a rock unit with a distinctive aspect (rock type or fossils) from another portion of the same general rock unit. Commonly synonymous with depositional environment.

facies fossil (3) fossil that occurs in a specific kind of rock and, by inference, in a specific kind of environment.

family (2) formal category in the classification of organisms that is more inclusive than a genus but less inclusive than an order (which may contain one or more families).

faunal province (7) large region characterized by a distinctive assemblage of animals.

ferns (14) group of spore-bearing vascular plants characterized by much-divided, large leaves.

Fig Tree Formation (8) Precambrian formation in Africa that has yielded some of the oldest fossils known; dated at 3.4 bya.

filter bridge (7) migration pathway through which only some animals can pass; some migrate successfully, whereas others cannot.

floral province (7) large region characterized by a distinctive assemblage of plants.

flower (14) reproductive structure of angiosperms, consisting of several series of specially modified leaves and their associated sex parts.

foraminifera (11) group of heterotrophic protists with pseudopodia. They commonly have a test (or shell), are mostly microscopic, and are important fossils.

formations (3) mappable bodies of rock with a definite upper and lower boundary; the basic divisions of stratigraphy that are given geographic names.

fossilization (3) process by which the hard parts or, much less frequently, the soft parts of an organism become buried in sediments and preserved as fossils.

fossil (1, 3) any direct indication preserved in rocks of the existence of former life.

Fungi (2) kingdom of life that includes organisms that are eukaryotic, reproduce with spores, completely lack cilia or flagella, and are heterotrophic through absorption of nutrients.

gametophyte (13, 14) haploid phase, produced by spores, in the life cycle of vascular plants; the phase that produces sex cells.

Gangamopteris (6) see Glossopteris.

gastropods (10) one of the classes of the phylum Mollusca, characterized by an unchambered, helically spiral shell; also called snails.

gene (5) unit of heredity consisting of approximately 1,500 nucleotide base pairs along a chromosome.

genetics (5) study of heredity among organisms.

genotype (5) genetic composition of an individual, as contrasted with the physical appearance (phenotype) of that organism.

genus (2) one or more closely related species that share common descent; pl., genera.

geologic mapping (3) technique by which the areal distribution of rock formations is depicted on a map.

geologic time scale (1) division of relative time in Earth history consisting of several long eras of time and smaller subdivisions.

gigantotherm (15) idea that explains that constant body temperature could have been maintained in dinosaurs as a by-product of the metabolism of very large organisms.

gill arches (12) tissues and bones that separate and support adjacent gills in fish.

ginkgo (7, 14) group of tree-sized gymnosperm plants with fan- shaped, parallel-veined leaves; especially common during the Mesozoic Era.

global change (1) subdiscipline in which Earth processes are considered on a global basis to attempt to explain climatic and other variations through Earth history.

Glossopteris (6) seed fern with large, elongate leaves restricted to Gondwana during the late Paleozoic differs from Gangamopteris in having a distinct midvein down the center of the leaf.

Gondwana (6, 7) southern supercontinent during the Paleozoic, consisting of South America, Africa, Australia, Antarctica, and peninsular India before these areas drifted apart; the southern half of Pangaea.

goniatite septum (12) simplest type of septum in ammonoid cephalopods, consisting of a wavy pattern without any secondary wrinkles.

graptolites (11) group of extinct colonial animals, some benthonic, some planktonic, placed in the phylum Hemichordata; important Paleozoic fossils.

Gunflint Chert (8) Precambrian formation along the northern shore of Lake Superior that has yielded black cherts and stromatolites containing microscopic fossils; from 2.1 bya.

gymnosperm (14) primitive seed plants in which the seeds are exposed; important land plants during the late Paleozoic and Mesozoic with such forms as conifers, cycads, ginkgos, and others.

hadrosaurs (15) family of ornithischian dinosaurs that were bipedal herbivores; includes the duck-billed dinosaurs.

half-life (1) amount of time necessary for a naturally occurring radioactive element to decay so that only one half of the parent atoms remain.

haploid (13) number of chromosomes in a sex cell, created by the process of meiosis that reduces the number of chromosomes to half that in a nonsex cell of the same species. (See *diploid*.)

hardgrounds (10) recently deposited sediments that have become cemented and exposed at the bottom of the ocean, common in limestones; become site for encrustation by many benthonic organisms.

hard parts (3) morphological features of organisms composed of mineralized materials so that their potential for preservation as a fossil is high; bones and shells.

Hemichordata (2) phylum of animals that have a dorsal nerve cord and gill slits during some growth stage but lack the notochord; common fossil hemichordates are graptolites.

herbivore (9) animal that feeds exclusively on plants.

hermatypic corals (10) corals that have symbiotic photosynthetic zooxanthellae algae in their soft tissues; responsible for building the framework of modern coral reefs.

heteromorph (12) means "different form"; applied specifically to ammonoid cephalopods that do not have the usual coiled shell form.

heterotrophs (2, 4) organisms that cannot manufacture their own food; applies to animals and some protists.

hierarchy (2) classification of life into categories from most inclusive (domain) to least inclusive (species).

Holocene (1) epoch of geologic time during the Cenozoic that includes the present time; geologic time younger than 10,000 years ago. Formerly known as Recent.

hominid (17) members of the primate family Hominidae that include humans and other bipedal primates.

Homo (17) genus that includes our own species, H. sapiens; "Homo" means "man."

homologous (5) structures or parts of organisms that have the same evolutionary origin but which may or may not have the same function.

hyomandibular arch (13) third gill arch of primitive fish that ultimately became the stapes earbone in tetrapod vertebrates.

ichthyosaurs (12) group of extinct, carnivorous marine reptiles that were dolphinlike or porpoiselike in aspect; flourished during the Mesozoic.

Ichthyostega (13) one of the oldest and most primitive tetrapods, or land vertebrates; an amphibian from Upper Devonian rocks of Greenland.

igneous rock (3) rock that was originally molten and that crystallized when it cooled; includes basalt, granite, and others.

inarticulate brachiopods (9, 10) term applied to a group of brachiopods having a chitinophosphatic or calcareous shell where the two valves are not hinged together.

incus (16) the anvil bone in the middle ear of mammals, derived from the reptilian quadrate bone.

infaunal (9, 10) refers to aquatic animals that live within the sediment rather than on the sea floor.

initial Earth (4) earliest phase of Earth development with a cooled solid Earth that lacked water and an atmosphere.

insectivores (16) group of insect-eating primitive placental mammals that are first found as fossils in Cretaceous rocks; includes moles and shrews.

irregular echinoids (10) major group of Mesozoic and Cenozoic echinoids (class Echinoidea, phylum Echinodermata) that had asymmetrical tests; mouth was central or subcentral on lower surface, anus was marginal or on the posterior side of the lower surface, spines were very short.

isolation (5) condition in which a group of plants or animals is separated for a considerable length of time from other groups; caused by geographic or ecologic factors.

isotope (1) one of two or more forms of an element having the same atomic number (same number of protons and electrons) but differing in atomic weight (having different numbers of neutrons).

jellyfish (11) member of the phylum Cnidaria; related to corals and sea anemones. Although they lack a skeleton, jellyfish are sometimes found as fossils.

Jurassic (1) period of geologic time during the Mesozoic Era occurring after the Triassic and before the Cretaceous. The absolute age is from 208 to 146 mya.

kingdom (2) formal category in the classification of organisms that is more inclusive than a phylum but less inclusive than a domain (which may contain one or more kingdoms).

K/T boundary (15) scientific shorthand for the boundary between the Cretaceous and the Paleogene. *K* is the abbreviation for Cretaceous, and *T* is the abbreviation for Tertiary, a former name applied to what is now most of the Paleogene and Neogene.

Laurasia (6) northern supercontinent of the late Paleozoic, consisting of North America, Europe, and Asia; the northern half of Pangaea.

lignin (13) complex organic compound in vascular plates that is incorporated into the cell walls and provides structural rigidity; also helps prevent drying out.

limestone (3) sedimentary rock composed primarily of calcite; commonly composed of fossil shells.

lithostratigraphic correlation (3) matching of identity of rocks using compositional and textural characteristics; represents depositional environments.

Litopterna (16) group of South American hoofed, herbivorous mammals that radiated during the Paleogene; many forms convergent on North American hoofed herbivorous mammals.

living fossil (7) name applied to the last surviving representatives of a once flourishing group of organisms.

lobe-finned fish (12) one of the main groups of bony fish; important as ancestors of land vertebrates, the amphibians.

lophophorates (2) group of invertebrates including two phyla common as fossils, the Brachiopoda and Bryozoa, that have a conspicuous feeding structure called the lophophore.

lophophore (10) loop-shaped ciliated feeding structure of brachiopods and bryozoans.

lycopod (14) group of spore-bearing plants characterized by a spiral arrangement of long slender leaves on the stem; conspicuous tree-sized plants in coal-swamp forests.

magnetic reversal (6) change in the relative position of Earth's north and south magnetic poles. Because reversal has happened many times during the past, it provides a means for dating the sea floor and proving sea-floor spreading.

malleus (16) a bone in the middle ear of mammals commonly called the hammer bone; derived from the articular bone of reptiles.

mammal-like reptiles (15) see *synapsids*.

mammoth (16) extinct elephant group that has complex molars; includes woolly forms.

mantle (4, 6) 1. middle layer of Earth between the core and crust; very thick and composed of dense silicate minerals. 2. tissue layer in molluscs that secretes the shell.

marsupials (16) pouched mammals, a group that first evolved during the Cretaceous; includes the opossum and kangaroo.

mass extinction (5) short periods during which excessively high rates of extinction prevail; extinction rate significantly higher than background extinction.

mass number (1) sum of the number of protons and neutrons in the nucleus of an atom. Different isotopes of the same element have different mass numbers.

mastodon (16) large-tusked extinct mammal related to elephants but typified by less specialized teeth than present in elephants and mammoths.

meiosis (14) process in which cell division takes place such that the resultant cells have half of the number of chromosomes as the parent cell; the process by which sex cells (haploid) are produced from nonsex cells (diploid).

mesoderm (2) middle layer of tissue in an animal; between endoderm and ectoderm.

Mesosaurus (6) small, fresh-water extinct reptile confined to South America and Africa; important in early debates on continental drift.

Mesozoic (1) era of geologic time from 245 to 65 mya; the age of dinosaurs and ammonoids. The word means "middle life."

metamorphic rock (3) either igneous or sedimentary rock that has been altered by heat and pressure; includes slate, marble, and quartzite.

metaphytes (8) macroscopic, multicellular plants; this general name applies to most land plants and macroscopic seaweeds.

metazoa (2) multicellular animals; includes all phyla of animals except sponges.

miacidids (16) primitive group of carnivorous mammals; evolved during the early Paleocene and eventually gave rise to canids (dogs) and felids (cats).

Miocene (1) oldest epoch of the Neogene, older than the Pliocene and younger than the Oligocene; absolute age is 23 to 5 mya.

Mississippian (1) period of the Paleozoic older than the Pennsylvanian and younger than the Devonian; absolute age is 362 to 323 mya. Named for outcrops along the upper part of the Mississippi River in Illinois, Iowa, and Missouri. Outside of North America equivalent to the Lower Carboniferous.

mitosis (14) process of cell division in which no change in chromosome number occurs, as during growth of an individual.

modern mammalian fauna (16) see *Neogene mammalian fauna*.

Modern marine fauna (9) one of three major faunas to populate the marine realm during the Phanerozoic, from the Triassic to the present; dominated by infaunal suspension feeders and deposit feeders; composed predominantly of bivalves, gastropods, bryozoans, crustaceans, irregular echinoids, sponges, foraminiferans, osteichthyes, chondrichthyes, marine reptiles, and marine mammals.

mold (3) impression left in the surrounding rock by the decay of organic material or dissolution of a shell.

Mollusca (2) phylum of animals that includes gastropods, bivalves, cephalopods, and others; typically unsegmented, mantle secretes a shell; marine and nonmarine.

monoplacophorans (9) class of primitive molluscs with a single, cap-shaped shell; first appeared and are important during the Cambrian, but are still living today in deep-ocean habitats.

monopodial branching (13) pattern of branching in land plants in which there is a single large central branch with smaller side branches issuing from it at intervals along the stem.

monotremes (16) egg-laying mammals for which there is virtually no fossil record, including living duck-billed platypus and the spiny echidna of Australia.

morphological fossil (8) see *body fossil*.

mosasaurs (12) large marine lizards of the Mesozoic.

mudstone (3) sedimentary rock composed predominantly of clay- sized particles.

multituberculate (16) one of the extinct groups of early mammals that inhabited the Mesozoic and early Cenozoic; larger than most contemporaneous mammals, from mouse- to woodchuck-sized and with specialized teeth like those of living rodents.

mummy (3) preservation in which soft tissues are preserved by a process of either freezing or dehydration.

mutation (5) any change in the structure or arrangement of genes or chromosomes in a cell.

mya (1) million years ago.

natural selection (5) process of evolution whereby certain members of a population, because of their genetic composition, are either reproductively successful or unsuccessful.

nautiloid cephalopods (12) group of cephalopods characterized by very simple sutures, represented today only by the chambered nautilus; important Paleozoic predators.

neandertal (17) name for a group of cold-weather hominids in Europe, Africa, and western Asia; sometimes regarded as a separate species, a subspecies of Homo sapiens, or simply a variety of our species.

nebular cloud (4) exceedingly large aggregation of gas and dust spinning slowly in space; e.g., the Milky Way.

nektonic (9, 11) term applied to aquatic animals that actively swim, as opposed to those that passively float (planktonic).

Neogene (1) period of the Cenozoic Era from 23 to 1.6 mya, younger than the Paleogene and older than the Quaternary.

Neogene mammalian fauna (16) late Cenozoic terrestrial; dominant herbivores were perissodactyls, artiodactyls, and proboscideans; dominant carnivores were cats and dogs.

neutron (1) one of the building blocks of atomic nuclei; lacks an electrical charge. Changes in the number of neutrons in atoms of an element result in different isotopes of that element.

non-siliciclastic (3) type of sedimentary rock composed of particles that were biogenically created or crystallized within the basin of deposition.

Notoungulata (16) group of primitive placental mammals common during the early Cenozoic; they were probably isolated in South America until their Pleistocene extinction.

Oligocene (1) one of the epochs of the Paleogene, younger than the Eocene and older than the Miocene; from 35 to 23 mya.

omnivore (9) organism that normally eats both animal and plant tissue.

order (2) one of the categories in the hierarchal classification of life consisting of groups larger than family and smaller than class.

Ordovician (1) period of the Paleozoic Era younger than the Silurian and older than the Cambrian; named after an early tribe in ancient Britain that lived in central Britain where these rocks were first described; from 510 to 439 mya.

organ grade of organization (2) level of complexity of animals in which tissues combine to form organs and organ systems; present in all animals except Porifera and Cnidaria.

original preservation (3) mode of fossilization in which the composition of the fossil is identical to the composition when the organism was alive.

Ornithischia (15) bird-hipped dinosaurs; all were herbivores; include the ornithopods, stegosaurs, ankylosaurs, and ceratopsians.

ornithopod (15) duck-billed dinosaurs and others; bipedal, herbivorous ornithischians; most common during the Cretaceous.

Osteichthyes (12) bony fish. Includes lobe-finned fish and ray-finned fish.

otic notch (13) indentation at the back of the skull, across which stretches the ear drum, located at the rear of the skull in amphibians but absent in reptiles and mammals.

oxygen isotope method (7) technique in which paleotemperatures can be determined by comparing the ratio of composition of O^{16} to O^{18}.

paleobiogeography (6, 7) study of the ancient distributions of plants and animals.

paleoecology (3) the science of interpreting ancient interactions between organisms and their environment and among different organisms.

paleogeography (7) study of the geographic distribution of ancient life.

Paleocene (1) oldest epoch of the Paleogene, younger than the Cretaceous Period and older than the Eocene; from 65 to 56 mya.

Paleogene (1) oldest period of the Cenozoic, younger than the Cretaceous and older than the Neogene; from 65 to 23 mya.

Paleogene mammalian fauna (16) early Cenozoic terrestrial fauna; dominant herbivores were condylarths, pantodonts, uintatheres, notoungulates, and litopterns; dominant carnivores were creodonts and miacidids.

paleogeographic map (3) map depicting the ancient distribution of land and sea and mountain ranges and other geographic features.

paleontology (1) study of ancient life on Earth.

Paleozoic (1) oldest era of the Phanerozoic, younger than the Precambrian and older than the Mesozoic; from 543 to 245 mya.

Paleozoic marine fauna (9) one of three major faunas to populate the marine realm during the Phanerozoic, from the Ordovician through the Permian; dominated by epifaunal suspension feeders; composed predominantly of brachiopods, bryozoans, corals, crinoids, cephalopods, starfish, graptolites, and ostracods.

Pangaea (6) supercontinent in which all the present continents were grouped into a single land mass.

pantodonts (16) primitive group of large herbivorous mammals that flourished during the early part of the Cenozoic Era and became extinct during the Oligocene.

pantotheres (16) one of the extinct groups of small Mesozoic mammals.

parallel evolution (5, 11) process by which two distinct and unrelated groups of organisms undergo a series of similar changes through time; examples of parallel evolution occur among the graptolites.

parasite (9) organism that gains its nutrition by taking it from a live host without killing it.

parent element (1) in a radiogenic decay series, the isotope that is radioactive and naturally decays to other elements through the loss of protons, neutrons, and electrons.

pedicle (11) horny and muscular stalk by which many brachiopods are attached to the sea floor.

pelagic (9) term applied to organisms that live in the water column, floating or swimming.

Pennsylvanian (1) period of the Paleozoic Era that occurs after the Mississippian and before the Permian Period; named after the coal-producing area of Pennsylvania. Basically equivalent to the Upper Carboniferous as used outside of North America; from 323 to 290 mya.

period (1) interval of geologic time represented by a system of rock; the principal subdivisions of an era that are further divided into epochs.

Perissodactyla (16) odd-toed hoofed mammals, generally with one or three functional toes; includes the horse, tapir, rhinoceros, and others.

Permian (1) final period of the Paleozoic Era, occurring after the Pennsylvanian and before the Triassic Period; named after the town of Perm in Russia; from 290 to 245 mya.

permineralization (3) process of fossilization in which pore spaces in the shell or other skeletal material is infiltrated by mineral matter making the hard part denser and heavier; also called petrifaction, as in petrified wood.

Phanerozoic (1) eon of geologic time including the Paleozoic, Mesozoic, and Cenozoic; from 543 mya to the present.

phenotype (5) outward physical aspect of an organism.

phloem (13, 14) vascular connective tissue in land plants.

photoautotrophs (2) organisms that metabolize their own food with the use of sunlight and complex molecules such as chlorophyll.

phyletic evolution (5) evolution that is a gradual change through time from one species population to another, phenotypically distinct species population. May or may not lead to an increase in the number of species.

phylum (2) major group of organisms; one or more phyla make up a kingdom; phyla are divided into classes; pl., phyla.

phytane (8) organic compound that, along with pristane, is a degradation product of chlorophyll; found in some Precambrian rocks, an indicator of early photosynthesis.

phytoplankton (11) floating photosynthetic organisms (principally protists and algae) that live in aquatic environments.

pistil (14) one of several female reproductive structures in the center of a flower within which seeds develop.

placental (16) term used for advanced mammals whose young develop within the mother's body. The young are born in an immature state and require nurturing.

placoderms (12) primitive jawed fish with heavy bony armor and more than two pairs of lateral fins.

planet (4) one of the larger bodies circling the sun, including Earth.

planktonic (9) term applied to any organism that floats in water.

Plantae (2) kingdom of life; organisms typically photosynthetic and multicellular; may be either aquatic or terrestrial.

plates (6) subdivisions of the outer rigid surface of Earth, the lithosphere, composed of the crust and uppermost mantle.

plate tectonics (6) mechanisms by which large parts of Earth's crust are formed, moved, and destroyed.

Pleistocene (1) one of two epochs of the Quaternary Period, younger than the Pliocene and older than the Holocene; from 1.6 to 0.01 mya; time of the latest ice ages.

Pleistocene extinctions (16) mass extinction approximately 10,000 years ago that affected large mammals in the northern hemisphere; debate exists as to whether this was due to climate change or the influences of early man.

plesiosaurs (12) group of marine aquatic reptiles from the Mesozoic Era typically with long necks, short heads, and large flippers.

Pliocene (1) epoch of the Neogene younger than the Miocene and older than the Pleistocene; from 5 to 1.6 mya.

pollen (14) fine, dustlike grains of seed plants that contain the male gametophyte and produce the sperm that fertilizes the plant egg.

Porifera (2) sponges; phylum of animals that have only a cellular grade of organization.

potassium-argon (1) end members of a radiogenic isotope series that are used to age date rocks in years. Radioactive potassium (K^{40}) is the parent isotope with which the decay series begins, and the stable daughter isotope of argon (Ar^{40}) is the end product.

Precambrian (1) rock, time, and Earth history prior to the beginning of the Paleozoic Era. From approximately 4.6 bya to 534 mya; includes the Priscoan, Archean, and Proterozoic Eras or the corresponding system of rocks.

predator (9) carnivorous animals that actively seek out their prey.

priapulid worms (12) phylum of predaceous worms; living today and known to be important predators at least during the Cambrian.

primary consumer (9) heterotrophs that feed directly on primary producers (plants or protists); herbivores.

primary producer (9) autotrophic; organisms that can manufacture complex organic molecules themselves through photosynthesis or otherwise.

primates (17) order of mammals that includes humans, apes, monkeys, and their relatives.

primitive Earth (4) phase of Earth development after the initial Earth, with more moderate temperatures, oceans, and an atmosphere composed of H_2O, CO_2, N, SO_2, CH_4, and NH_3.

principle of faunal and floral succession (3) stratigraphic concept proposed by William Smith stating that fossils occur in a predictable order in the stratigraphic record; due to evolution.

principle of lateral continuity (1, 3) stratigraphic concept proposed by Nicholas Steno stating that sedimentary strata continue for long distances geographically; concept true with the exception that distances can vary from being nearly continent-wide to the width of a stream channel.

principle of original horizontality (1, 3) stratigraphic concept proposed by Nicholas Steno stating that sedimentary strata were deposited in horizontal layers.

principle of superposition (1, 3) stratigraphic concept proposed by Nicholas Steno stating that in an undisturbed sequence of sedimentary strata, the oldest strata are at the bottom and the youngest strata are at the top.

Priscoan (1) oldest interval of the Precambrian, older than the Archean; from about 4.6 to 4.0 bya.

pristane (8) see *phytane*.

proboscideans (16) order of mammals that includes elephants, mammoths, and mastodons.

prokaryotes (2) simple organisms, generally microscopic, that lack a highly organized nucleus with a surrounding membrane; incudes the bacteria and cyanobacteria.

prosimians (17) less advanced and earliest appearing primates; includes the living lemurs and tarsiers.

Proterozoic (1, 5) interval of geologic time between the Archean and the Paleozoic eras; from about 2.5 bya to 543 mya.

Protista (2) kingdom of organisms; mostly one-celled organisms, some of which are autotrophs and some heterotrophs.

proton (1) positively charged particle in the nucleus of an atom.

protostomes (2) animals in which the first embryonic opening becomes the mouth and the second embryonic opening becomes the anus; examples are molluscs and arthropods.

punctuated equilibrium (5) mode of evolution in which new species arise rapidly from small subpopulations; between speciation events only random morphological change (stasis) occurs; stratigraphic pattern of species derived from allopatric speciation mode.

pyrite (3) mineral composed of iron and sulfur; may be a replacement mineral of fossil shell or bone.

pyritic conglomerates (8) conglomerates composed of detrital pyrite; occurs only in an anaerobic environment, otherwise the pyrite cobbles would oxidize.

quadrate bone (15) with the articular bone, the two bones that join the upper and lower jaws in reptiles; these become two bones of the middle ear in mammals.

quadrupedal (15) animal that walks on all four legs, as opposed to a bipedal animal that walks on only the hind two legs.

Quaternary (1) youngest period of geologic time, including the last 1.6 million years of Earth history; includes the Pleistocene and Holocene.

radial symmetry (2) form of symmetry in which more than one plane divides an organism into two equal halves.

radioactivity (1) property of certain isotopes that makes them naturally unstable so that they change into other elements by the discharge of particles from the nucleus.

radiolarians (11) group of heterotrophic protists that have a radially symmetrical silicon skeleton; microscopic and unicellular.

ray-finned fish (12) group of advanced bony fish with many fine parallel bones supporting the fins; includes the most common freshwater and saltwater fish today.

rays (12) one of the cartilaginous fish, the Chondrichthyes, which generally lack bones except in their teeth and dermal denticles; sharks also included in the Chondrichthyes.

Recent (1) see *Holocene.*

recrystallization (3) process of fossilization by which the original microstructure of fossil hard parts is destroyed by the growth of new crystals.

red beds (8) terrestrial deposits in an aerobic environment; coloration from oxidization of iron-bearing minerals.

reducing atmosphere (4) an atmosphere lacking oxygen (see anaerobic).

reefs (10) organically constructed structure raised above the sea floor with a rigid framework of skeletal material; in today's oceans constructed primarily by scleractinian corals.

regular echinoids (10) group of echinoids (class Echinoidea, phylum Echinodermata)

important throughout the Phanerozoic that had symmetrical tests; mouth was central on lower surface, anus was central the upper surface, spines were generally long.

relative age (1) geologic age of a rock or fossil defined in relation to the ages of other rocks and fossils rather than in years.

relative time scale (1) method for tracking history in which events are known in relationship to the order of other events, but dates are not known.

remnant magnetism (6) direction and polarity of Earth's ancient magnetic fields, preserved and recorded by small magnetic grains in rocks.

replacement (3) process of fossilization in which the original mineral material of a hard part is replaced by another kind of mineral.

rhipidistian (13) group of lobe-finned fish. The amphibians evolved from the rhipidistian crossopterygians during the Devonian.

rhizome (13) underground stem that serves to anchor some land plants; the original anchorage before true roots had evolved.

rhombiferans (10) class of stalked echinoderms important during the lower Paleozoic; formally one group of cystoids.

rhyniophytes (14) most primitive vascular plants; have terminal sporangia, dichotomous branching of stem, and lack true roots and true leaves.

RNA (5) abbreviation for ribonucleic acid, an organic molecule responsible for transferring information from DNA molecules to build proteins.

rubidium-strontium (1) one of the radioactive decay series that is used to date rocks. Radioactive rubidium (Rb^{87}) is the initial unstable parent isotope that eventually decays to the stable daughter isotope of strontium (Sr^{87}).

rudistids (7) group of large, thick-shelled bivalves that flourished in tropical waters during the Cretaceous Period, built reefs, and became extinct at the end- Cretaceous extinctions.

rugose corals (10) group of extinct Paleozoic corals, solitary or colonial, that had prominent septa; also known as horn corals.

sandstone (3) sedimentary rock composed primarily of sand- sized particles (between 1/16 and 2 mm diameter) regardless of their mineral composition; commonly composed of quartz particles.

Saurischia (15) lizard-hipped dinosaurs, including bipedal carnivores and quadrupedal herbivores; confined to the Mesozoic Era.

sauropods (15) giant, quadrupedal dinosaurs that evolved from a bipedal form; one group of Saurischia.

scavenger (9) animal that feeds primarily on dead animal bodies.

scleractinian corals (10) modern hard corals that began during the Triassic; include reef-building hermatypic as well as ahermatypic corals.

sea-floor spreading (6) process by which new sea-floor crust is formed along oceanic ridges and pushed progressively farther from the ridge as younger and younger crust is formed; thus, any given segment of the sea floor gradually spreads away from the ridge.

secondary consumers (9) those heterotrophs that feed on primary consumers; predators and carnivores.

sedimentary rock (3) any rock formed of sedimentary particles that are either preexisting (detrital grains) or that may be chemically precipitated, such as sandstone, shale, and limestone.

seed (14) reproductive product of advanced land plants that contains an embryo sporophyte and a food supply within a resistant seed coat.

septa (10) vertical partitions of skeleton in a rugose or scleractinian coral; partitions separating chambers in a cephalopod; sing., septum.

sessile (9) term applied to a fixed or immobile organism that is attached or cemented to the sea floor.

shale (3) fine-grained, laminated sedimentary rock composed mostly of clay-sized particles; commonly splits into thin beds or layers.

sharks (12) one of the cartilaginous fish, the Chondrichthyes, which generally lack bones except in their teeth and dermal denticles.

silica (3) silicon dioxide (SiO_2), a common mineral constituent of various organic hard parts among protists, algae, and sponges.

siliciclastic (3) type of sedimentary rock composed of particles that have been eroded from elsewhere and transported to a basin of deposition.

silicoflagellates (11) group of unicellular protists that have two flagella for locomotion and a siliceous skeleton.

siltstone (3) siliciclastic rock with silt-size clasts (between 1/256 and 1/16 mm in diameter).

Silurian (1) period of the Paleozoic Era occurring after the Ordovician and before the Devonian; named after an ancient British tribe that inhabited an area in central England where rocks of this age were first found; from 439 to 408 mya.

siphons (10) in bivalves, fused posterior portion of mantle that allows infaunal bivalves to maintain contact with water above the bottom.

size selection for food (10) among suspension feeders, one mode of niche subdivision by capturing and ingesting largely different size fractions of food particles.

skeleton (3) hard parts of any organism but used especially in describing protists and animals.

small shelly fauna (8) earliest skeletons known from the late Proterozoic and early Cambrian; elements of fauna include lightly skeletonized tubes, sclerites of armored worms, and others.

Smith, William (3) English engineer who, while designing canals in 1820, proposed the principle of faunal succession.

solar winds (4) energy in the form of subatomic particles released by the sun; swept away much of the inert gases when Earth was first forming.

species (2) one or more actually (or potentially) interbreeding natural populations that produce fertile offspring and are reproductively isolated from other such species.

sphenopsids (14) group of primitive vascular plants characterized by a jointed stem and whorls of leaves or reproductive structures at the intersections of the stem segments; common in Pennsylvanian coal swamps.

spherical symmetry (2) type of organism symmetry in which an infinite number of planes equally bisect the organism into equal parts.

sponge (10) phylum of animals that consists of simple assemblages of relatively undifferentiated cells that are without tissues and organs.

sporangia (13) small, organic sac within which spores are produced and held until released by a plant; a spore case.

spore (13) reproductive body produced by plants; the haploid (N) phase of the plant life cycle, produced by the sporophyte by meiosis and growing into the gametophyte by mitosis.

sporophyte (13, 14) conspicuous plant body of vascular plants; the diploid ($2N$) phase of the plant life cycle, produced by sex cells from the gametophyte that, in turn, produces spores.

spreading center (6) edge of one of the plates of Earth's crust where new crust is formed and spreads to either side; commonly represented by a high ridge or rise on the ocean floor.

squamosal (15) bone in the skull of mammals that articulates with the dentary bone that forms the lower jaw.

stalked echinoderm (10) suspension-feeding echinoderms elevated above the sea floor by a stem; includes crinoids, blastoids, rhombiferans, diploporans, and others.

stamen (14) one of the male, pollen-producing structures in the flower of angiosperms.

stapes (13) bone that is the single earbone of amphibians and reptiles; derived from the hyomandibular bone that was used for jaw support in fish.

stasis (5) in evolutionary terms, the idea that species populations persist for long periods of time without discernable genetic or morphologic change.

stegosaurs (15) suborder of ornithischian dinosaurs consisting of animals with two rows of large bony scutes along the back and long, bony spines on the end of the tail; confined to the Jurassic Period.

Steno, Nicholas (1) Dane who, in 1820, proposed the principles of lateral continuity, original horizontality, and superposition.

stomata (13) specialized cells concentrated on the underside of leaves that allow plants to breathe.

stratigraphy (3) study of layered sedimentary rocks, especially their formation and relationships.

stromatolite (8) finely laminated structures in rocks that consist of concentric laminations (cabbage-head structure) built by algae, although not part of their skeleton or hard parts; especially common during Precambrian and early Paleozoic rocks.

stromatoporoids (10) group of calcareous sponges especially important as reef builders; common during the Paleozoic.

subduction zone (6) edge of plates of Earth's crust where the crust is turned down into the mantle; commonly marked by a deep trench on the ocean floor.

suspension feeder (9) feeding mode among marine benthos in which particulate food is collected from the water column.

sweepstake route (7) dispersal pathway characterized by a formidable barrier that prevents most migrating organisms from crossing; migration to an island from a mainland (island hopping) is an example.

synapsids (15) mammal-like reptiles characterized by a single opening low on the temple region of the skull; common during the Permian and Triassic; during the Triassic they gave rise to mammals and became extinct.

tabulate corals (10) group of corals that were all colonial and generally lacked septa; conspicuous reef builders during the Paleozoic Era, but extinct since the Paleozoic.

temporal opening (15) one or more holes in the side or temple region of the skull, behind the eye, that are present in many reptiles; the number and arrangement are important in reptile classification.

Tethys Seaway (7) ancient ocean stretching from between Europe and Africa eastward across Asia; occupied the present site of the Alps, Himalayas, and Mediterranean Sea.

thecodonts (15) group of diapsid reptiles that gave rise to the dinosaurs, specifically the Saurischia; some were bipedal and lightly built.

therapsids (15) mammal-like reptiles; Permian to Triassic reptiles that evolved into mammals; include animals such as *Dimetrodon* and *Edaphosaurus*.

theropod (15) large, bipedal carnivorous dinosaur; a saurischian.

thin section (10) very thin slices through a rock or fossil that allow light to be transmitted; permits examination of the microstructural detail in fossils.

tiering (9, 10) vertical subdivision of space by organisms for either feeding or avoiding other organisms; allows for niche subdivision among suspension feeders and deposit feeders.

tillites (7) any sedimentary rocks deposited beneath a glacier.

tissue grade of organization (2) level of complexity of animals in which cells are organized into tissues; includes Cnidaria.

trace fossil (1, 3) any indirect evidence of the existence of former life, commonly indicative of animal activity. Common types include burrows, trackways, and fossil excrement.

tracheophytes (2) see *vascular plants*.

Triassic (1) initial period of time during the Mesozoic Era, occurring before the Jurassic and after the Permian; lasting from 245 to 208 mya.

triconodont (16) most primitive group of mammals during the Mesozoic Era; small with simple molar teeth; now extinct.

trilobites (9) group of extinct arthropods, characteristic of the Paleozoic Era, with a body divisible into a cephalon (head), thorax, and pygidium (tail).

uintatheres (16) rhinoceros-sized herbivorous mammals that were a conspicuous part of the North American Paleogene mammalian fauna.

ungulates (16) hoofed herbivorous mammals; included are both odd- and even-toed types.

uniformitarianism (3) principle that general laws of physics and chemistry applied during the past just as they do today, and that geologic processes we see today also shaped Earth during the past, although the rates and intensities of geologic processes have varied through time.

vagile (9) refers to organisms that move.

variation (5) departure from a normal or average condition. In biology the term is used to describe the variability in species populations.

vascular plants (2) fully terrestrial plants with vascular tissue, stomata, trilete spores, and cuticle; includes rhyniophytes, lycopods, sphenopsids, gymnosperms, angiosperms, and others. Also known as tracheophyles.

vascular tissue (2, 13) conductive and supportive tissues characteristic of land plants; includes both xylem and phloem.

volcanic outgassing (4) gaseous compounds that are emissions from volcanic eruptions, including H_2O CO_2, NH_3, CH_4, N, and SO_2.

Warrawoona Group (8) the oldest known locality with undisputed fossils; 3.5 bya from western Australia.

wood (14) secondary xylem that serves to strengthen and support the stems of large land plants, that is, trees and shrubs.

xylem (13, 14) one of the two important kinds of vascular tissue in vascular plants that provides for conduction of water and food and for support of the stem in air; secondary xylem is the woody tissue of trees and shrubs.

zonal fossil (3) fossil used to define time intervals for biostratigraphic correlation.

zooplankton (11) nonphotosynthetic organisms that float in the water; two major groups are the microscopic protists (radiolarians and foraminifera) and macroscopic organisms such as graptolites or jellyfish.

zooxanthellae algae (10) unicellular, photosynthetic, microscopic protists that reside in the tissues of hermatypic corals; important as a factor in coral reef growth.

zygote (13) first cell formed by fertilization in the development of any new plant or animal.

Index

[1]boldface page numbers refer to illustrations